Current Research in Systematic Musicology

Volume 7

AF803653

Series Editors

Rolf Bader, Musikwissenschaftliches Institut, Universität Hamburg, Hamburg, Germany
Marc Leman, University of Ghent, Ghent, Belgium
Rolf-Inge Godoy, Blindern, University of Oslo, Oslo, Norway

The series covers recent research, hot topics, and trends in Systematic Musicology. Following the highly interdisciplinary nature of the field, the publications connect different views upon musical topics and problems with the field's multiple methodology, theoretical background, and models. It fuses experimental findings, computational models, psychological and neurocognitive research, and ethnic and urban field work into an understanding of music and its features. It also supports a pro-active view on the field, suggesting hard- and software solutions, new musical instruments and instrument controls, content systems, or patents in the field of music. Its aim is to proceed in the over 100 years international and interdisciplinary tradition of Systematic Musicology by presenting current research and new ideas next to review papers and conceptual outlooks. It is open for thematic volumes, monographs, and conference proceedings.

The series therefore covers the core of Systematic Musicology,—Musical Acoustics, which covers the whole range of instrument building and improvement, Musical Signal Processing and Music Information Retrieval, models of acoustical systems, Sound and Studio Production, Room Acoustics, Soundscapes and Sound Design, Music Production software, and all aspects of music tone production. It also covers applications like the design of synthesizers, tone, rhythm, or timbre models based on sound, gaming, or streaming and distribution of music via global networks.

- Music Psychology, both in its psychoacoustic and neurocognitive as well as in its performance and action sense, which also includes musical gesture research, models and findings in music therapy, forensic music psychology as used in legal cases, neurocognitive modeling and experimental investigations of the auditory pathway, or synaesthetic and multimodal perception. It also covers ideas and basic concepts of perception and music psychology and global models of music and action.
- Music Ethnology in terms of Comparative Musicology, as the search for universals in music by comparing the music of ethnic groups and social structures, including endemic music all over the world, popular music as distributed via global media, art music of ethnic groups, or ethnographic findings in modern urban spaces. Furthermore, the series covers all neighbouring topics of Systematic Musicology.

More information about this series at http://www.springer.com/series/11684

Tim Ziemer

Psychoacoustic Music Sound Field Synthesis

Creating Spaciousness for Composition, Performance, Acoustics and Perception

Springer

Tim Ziemer
Institute of Systematic Musicology
University of Hamburg
Hamburg, Germany

ISSN 2196-6966 ISSN 2196-6974 (electronic)
Current Research in Systematic Musicology
ISBN 978-3-030-23035-7 ISBN 978-3-030-23033-3 (eBook)
https://doi.org/10.1007/978-3-030-23033-3

© Springer Nature Switzerland AG 2020
This work is subject to copyright. All rights are reserved by the Publisher, whether the whole or part of the material is concerned, specifically the rights of translation, reprinting, reuse of illustrations, recitation, broadcasting, reproduction on microfilms or in any other physical way, and transmission or information storage and retrieval, electronic adaptation, computer software, or by similar or dissimilar methodology now known or hereafter developed.
The use of general descriptive names, registered names, trademarks, service marks, etc. in this publication does not imply, even in the absence of a specific statement, that such names are exempt from the relevant protective laws and regulations and therefore free for general use.
The publisher, the authors and the editors are safe to assume that the advice and information in this book are believed to be true and accurate at the date of publication. Neither the publisher nor the authors or the editors give a warranty, expressed or implied, with respect to the material contained herein or for any errors or omissions that may have been made. The publisher remains neutral with regard to jurisdictional claims in published maps and institutional affiliations.

This Springer imprint is published by the registered company Springer Nature Switzerland AG
The registered company address is: Gewerbestrasse 11, 6330 Cham, Switzerland

Meiner verstorbenen Familie gewidmet:
Wolli, Mölli, die alten und die Zwillinge.
Wann sehen wir uns wieder?

Foreword

Music and space is one of the hot topics of contemporary research and applications. Wave Field Synthesis, 3D Audio in Gaming, ATMOS cinema audio standard, Concert Hall, Techno Club or Home Entertainment acoustics—designing a 3D audio space is one of the main fascinations and challenges today.

This book throws light on space and audio from many viewpoints. Modern methods of spatial acoustics, signal processing, concert hall design, auralization, or visualization are improving very fast at the moment. The book presents a comprehensive overview on all these aspects, with deep mathematical foundation of equations, often scarcely found in other textbooks today, like with the derivation of the Kirchhoff–Helmholtz integral and its connection to the Huygens principle. Such equations might be used daily, but their derivation and justification teach about their strength and restrictions. By approaching these methods anew from scratch, Tim Ziemer is able to give a deep understanding of the reasons for using these methods today.

The book also considers neuromusicology, psychoacoustics, semantics, and philosophy, giving fascinating insights into neural interactions and coding, semantic loads, synaesthetic relations, as well as historical and ethnical diversifications of the perception and treatment of music and space. Tim Ziemer describes the human ear to have developed from the hair cell line element of fish, by which they detect water flow around them and are therefore able to build swarms. So the ear "feels around" in space to detect where we are. Localization of a sound source on the other hand is strongly enhanced by a left/right brain interaction in the auditory pathway, another of the many tasks we are able to perform when it comes to spatial perception.

Historically, many examples of the use of space in music are known and discussed. Monteverdi placed different musical instrument sections in different rooms, therefore adding reverberations, like a modern mixing console adds delays and effects to different instrument groups. In the Stone Age in caves, certain places might be found to resonate best for different vowels. Ovid tells us about the metamorphosing of a nymph called Echo (from Greek echon, to behave like) who talked back everything one said. She turned into a stone, and so since then stones talk back.

Such echoes are also found in musical instrument, where impulses travel through and are reflected at different parts. So guitars, violins, or pianos are also echo chambers, and we hear them having an intrinsic space, an apparent source width (ASW). Indeed, we associate musical instrument sounds very often with spatial or tactile attributes like rough, flat, deep, open. A heavy metal guitar sound is known to be a plank, violins might sound hollow, pianos spacious and big, flutes small.

When comparing audio equipment, space is maybe the most important quality criteria today. Digital/analog converter (DAC) is considered good if they sound deep, are able to graduate musical instruments, and display them spatially. Otherwise, they might sound flat, small, and therefore dull. The same holds for loudspeaker and headphones. Like when using a 3D Virtual Reality headset in Gaming, we want to hear sound as differentiated as we do in our 3D real environment.

Tim Ziemer presents all these aspects, and much more. He is working in the field for many years and has developed fascinating applications on 3D audio. Maybe most remarkable is his idea of a psychoacoustic wave field synthesis. The use of rules of perception in audio signal processing has already been shown to increase sound quality tremendously. Most prominent might be MP3, where psychoacoustic rules have been added to an audio compression algorithm. The psychoacoustic wave field synthesis takes such advantages too. It uses the precedence effect of hearing, a very well-established and powerful effect, to improve the design of such a wave field.

This development was only possible, as Tim Ziemer has a deep understanding of all aspects of audio and space, acoustical, psychological, semantical, philosophical, or historical. His book is therefore also a bonanza for those who want to get inspired for developing new algorithms to improve spatial audio. Those who want to get to know about the whole picture find an excellent introduction with many references and sources.

The book is a wonderful example of systematic musicology as an interdisciplinary research field, combining musical acoustics and signal processing, music psychology and neurocognition, as well as music ethnology and related disciplines. To understand music, one needs to consider all these aspects and get the whole picture. Only then such inventions as presented by Tim Ziemer in this wonderful and inspiring book become possible.

Therefore, the book is a comprehensive overview on the research going on in spatial audio and music. I enjoyed reading it a lot and I hope it will inspire many of those working in the field or are interested in the very old and always new research topic of music and space.

Hamburg, Germany
March 2019

Rolf Bader

Preface

My work on spatial audio for music presentation started when I was a Magister's student at the University of Hamburg in 2008. My professor, Rolf Bader, said something like: "We have fifteen loudspeakers. Choose the wave field that you want and solve a linear equation system to create appropriate loudspeaker signals." What sounds so simple has really struggled me: *What is the wave field that I want? What is the sound impression that I desire? What listening experience do I want to offer? What needs to be done but has not been achieved yet? And how can I achieve it?* The task started a chain reaction. The quest for the ideal music listening experience has challenged philosophers, researchers, artists, and engineers for centuries. The topic has been approached from the viewpoint of aesthetics, music instrument building and synthesizer development, composition and performance, architectural acoustics, audio technology, psychophysics, ethnomusicology, music psychology and sociology, music theory, and many more. Interestingly, a lot of musical concepts, ideals, and open questions are related to spaciousness in music.

Many musicians are disappointed that even the best electric pianos do not manage to sound as grand as a grand piano. Pitch, loudness and dynamics, temporal and spectral fine structure, and even the haptics of electric pianos can come very close to the original and sound realistic. Some electric pianos even create a low-frequency vibration for the piano bench. This is supposed to make the playing conditions more realistic by adding information for the sense of touch. Some keyboards advertise with mystic *concert grand* buttons or *extended stereo width* functions whose involved signal processing remains a company secret. However, even inexperienced listeners can instantly tell whether they hear a real grand piano or a loudspeaker playback from an electric piano or a spatial audio system. Since temporal and spectral dynamics are reconstructed almost perfectly, the only aspect that is left is the spatial dynamics. Naturally, the room acoustics have a huge influence on the perceived spaciousness of the sound. But as a musicology student who had spent a lot of time in a free field room, I knew that a concert grand piano keeps sounding grand, even in the absence of notable room reflections. The large vibrating soundboard creates a complicated sound field. Here, interaural level and phase differences make it sound wide. Furthermore, the sound impression and the interaural differences slightly change when moving the head. Head motions are typical when listening to music or playing musical instruments. So I found my task: Capture and reconstruct the sound radiation characteristics of musical instruments to create a spatial and natural listening experience for listeners who can move while listening. Unfortunately, the state of the art in microphone array measurements and loudspeaker array technology only delivered moderately satisfying results for our constellation of comparably many microphones (128) and few loudspeakers (15, surrounding a listening area from three sides in an acoustically untreated room). So I adapted the conventional technology to my needs, facing the typical issues of sparsity, undersampling, and inverse problems. However, being a musicologist, I came up with solutions that consider the origin of the auditory system, the psychological organization of sound, ideas of music production, composition, and performance practice, and musical acoustics in terms of instrument acoustics, room acoustics, and especially psychoacoustics. The result could be called a *psychoacoustic sound field synthesis system for music presentation*. It does not aim at a perfect physical copy of a desired sound field. Instead, it delivers the cues necessary for the auditory system to localize the source, and experience its spatial extent and its natural coloration, which may be very different at different listening positions. The cues are delivered with the necessary precision, taking into account the temporal, spectral, and spatial accuracy of the auditory system.

General remarks and an overview about the structure and content of this book are given in Chap. 1. Some concepts of spaciousness in music are reviewed in Chap. 2, considering spaciousness in music psychology, composition and modern music production, music theory, and music information retrieval. The primary function of the auditory system is spatial hearing and a mental representation of the outside world. A treatise of the biology of the auditory system is presented in Chap. 3. The relationship between the physical outside world and its mental representation is discussed in Chap. 4. The sound radiation characteristics of musical instruments and microphone array methods to record them are gathered in Chap. 5. The radiated sound of musical instruments propagates through the listening room and reaches the listeners' ears directly, and after single or multiple reflections. The effect of room acoustics on the spatial listening experience is discussed in Chap. 6. Conventional stereophonic audio systems are reviewed against the background of presentation of spatial sound in Chap. 7. Wave field synthesis represents an alternative to stereophonic audio systems. It overcomes certain restrictions but comes along with new challenges for spatial music presentation. The approach is derived and discussed in Chap. 8. Finally, Chap. 9 introduces psychoacoustic sound field synthesis as a new paradigm in spatial audio technology development.

The presented psychoacoustic sound field synthesis approach for music is an exemplary case that illustrates the advantage of psychoacoustic considerations throughout the development of new spatial audio technology. With psychoacoustic sound field synthesis, it is possible to create a desired sound impression by means of a hybrid approach that includes physical and perceptual aspects of the sound field. Psychoacoustic control is a new paradigm that can serve for many more audio technologies. The approach is not restricted to music but can include applications with speech, sonification, and many more. The approach lays the foundation of a new generation of psychoacoustic audio technology and reconsideration of established audio systems.

Hamburg, Germany
May 2017

Tim Ziemer

Acknowledgements

Thanks to Rolf Bader and Albrecht Schneider who taught me all the basics and then gave me enough leeway to follow my own thoughts and find my own ways. Their courage to enter new fields and find unconventional solutions inspired me to consider the whole spectrum of systematic musicology and find approaches and solutions from the fields of biology, music theory, instrument acoustics, room acoustics and psychoacoustics, cognitive science and music psychology, electrical engineering, and computer science. I am grateful that they keep giving me advice and support me whenever they can.

I thank the interdisciplinary review committee of my dissertation. Besides Rolf Bader and Albrecht Schneider, Wolfgang Fohl, Georg Hajdu, Christiane Neuhaus, and Clemens Wöllner gave critical feedback from diverse viewpoints and valuable input for further research in the field of spatial audio for music, which culminates in this book.

A number of foundations and societies gave me financial support to present my research and work on this book. I thank the Claussen-Simon Foundation who supported me a lot during the finalization of this book and the German Academic Exchange Service who funded many of my conference travels.

My team at the University of Hamburg always had my back, especially Niko Plath, Florian Pfeifle, Michael Blaß, Christian Köhn, Orie Takada, Claudia Stirnat, Jost Fischer, Rolf Bader, Albrecht Schneider, Christiane Neuhaus, Marc Pendzich, Clemens Wöllner, Konstantina Orlandatou, Lenz Hartmann, Jesper Hohagen, and Henning Albrecht. It was inspiring for me to see how many disciplines are united in the field of systematic musicology. You gave me insights in your work, which leverages approaches from the field of biology, mathematics, digital signal processing, physics, ethnology, computer science, cognitive science, psychology, sociology, arts, culture, economics, politics, and humanities. You made me realize the power of interdisciplinarity.

Contents

Symbols and Abbreviations

Symbols

α	Angle between normal direction of wave front and secondary source
β	Absorption coefficient
$\gamma(t-\tau)$	Time window
$\Gamma(\omega, \varphi)$	Solution for the azimuth angle of the Helmholtz equation
$\Gamma_{\mathrm{Q}}(\omega, \varphi)$	Horizontal radiation characteristic of the source
$\Gamma_{\mathrm{Y}}(\omega, \varphi)$	Horizontal radiation characteristic of the secondary source
δ	Dirac delta
κ	Matrix condition number
ϕ	Phase
φ	Azimuth angle
φ_0	Viewing direction in the horizontal plane
φ_Q	Azimuth angle of the source in the head-related coordinate system
λ	Wavelength
Π	Function of radius
ϑ	Polar angle
Θ	Solution for the polar angle of the Helmholtz equation
ρ	Density
σ	Root-mean-square deviation
ω	Angular frequency ($\omega = 2\pi f$)
Ω	Width of the beamformer- or sound radiator lobes
$\Psi(\omega, \varphi, \vartheta)$	Spherical harmonics (solution for azimuth and polar angles of the Helmholtz equation)
a	Encoding factor
$\hat{A}$	Amplitude
$\hat{\mathbf{A}}$	Amplitude- or gain vector
$A(\omega)$	Complex amplitude
b	Decoding factor
B	Surround-/Rear-channel (“Back”)

BQI	Binaural quality index ($= 1 - IACC$)
BR	Bass ratio
c	Sound velocity
C	Center channel
C_{80}	Clarity factor (early to late sound ratio)
const	Constant
$d(\mathbf{Y})$	Windowing function
D	Manipulation factor turning source amplitude to transducer amplitude
dB	Decibels sound pressure level
$\mathrm{dB}_{\mathrm{SL}}$	Decibels spectrum level
$\mathrm{dB}_{\mathrm{SPL}}$	Decibels Sound Pressure Level
dir	Direct sound
e	Euler's number ($\approx 2.718281828\ldots$)
EDT	Early decay time
EEL	Earl ensemble level
eig	Eigenvalue
f	Frequency
$\mathbf{f}$	Steady differentiable vector function
F	Frontal channel ("Front")
FF	Far field
g	Special solution of the Green's function in time domain (impulse response)
$\tilde{g}$	General solution of the Green's function in time domain
G	Special solution of the Green's function in frequency domain (complex transfer function)
$\tilde{G}$	General solution of the Green's function in frequency domain
G_X	Sound strength
h	Index ("height")
H	Hallmaß (sound proportion)
$H_n^{(2)}(\omega, r)$	Spherical Hankel function of second kind and nth order
$\imath$	Imaginary unit $\sqrt{-1}$
$I_n(r)$	Spherical Bessel function of second kind and nth order
IACC	Interaural cross-correlation coefficient
IACF	Interaural cross-correlation function
ITD	Interaural time difference
ITDG	Initial time delay gap
$J_n(r)$	Spherical Bessel function of first kind and nth order
k	Wave number $\left(k = \frac{\omega}{c} = \frac{2\pi}{\lambda}\right)$
K	Propagation matrix
$K(\omega, \varphi)$	Propagation function
l	Index ("length")
L	Left channel ("Left")
$\mathbf{L}$	Position of the left loudspeaker
$L1$	Separation line between source area and source-free area

LFC	Lateral fraction coefficient
lg	Decadic logarithm
LG	Lateral strength
max	Maximum
min	Minimum
$\mathbf{M}_m(\mathbf{Q}, \varphi)$	Position of the microphones
NF	Near field
$p(t)$	Sound pressure in the time domain (sound signal)
$P(\omega)$	Sound pressure in the frequency domain (spectrum)
$P(t, \omega)$	Sound pressure in time–frequency domain
$P_n^m(cos\vartheta)$	Associated Legendre functions
pre	Predicted
Q	Primary source
$\mathbf{Q}$	Primary source position
Q'	Mirror source
$\mathbf{Q}'$	Mirror source position
$\mathbf{Q}_p$	Phantom source position
r	Radius
$\mathbf{r}$	Position vector in polar-/spherical coordinates
$\mathbf{r}'$	Mirror position of **r** in polar-/spherical coordinates
R	Right channel ("Right")
R	Position of the right loudspeaker
res	Resonance
RR	Reverberation ratio
RT	Reverberation time (decay time)
s	Loudspeaker basis
S	Surface of the source-free volume
$\tilde{S}$	Equivalent absorption area
$S1$	Separation plane between source volume and source-free volume (see Fig. 8.8b)
$S2$	Hemispherical separation surface between a source volume and a source-free volume (see Fig. 8.8b)
SL	Spectrum Level
SDI	Surface diffusivity index
SPL	Sound pressure level
st	Static
ST	Support
$\mathrm{ST}_{\mathrm{early}}$	Early support
$\mathrm{ST}_{\mathrm{late}}$	Late support
T	Transmission channel ("Transmission"/"Total"/"Track")
T_{fade}	Fading duration of the precedence fade
t_{e}	Echo threshold
t_s	Center time
TR	Treble ratio

U	Source volume (see Fig. 8.8)
$\mathbf{v}$	Sonic particle velocity in time domain
V	Source-free volume (see Fig. 8.8)
w	Index ("width")
$\mathbf{x}$	Position vector in Cartesian coordinates
$\mathbf{X}$	Listening position
X	Placeholder for a frequency or amplitude decay
$\mathbf{y}$	Loudspeaker matrix containing all loudspeaker locations
$\mathbf{Y}$	Secondary source position
$\mathbf{Y'}$	Mirrored secondary source position
Y_n	Spherical Bessel function of second kind and *n*th order (spherical Neumann function)
z	Critical band from the Bark scale
∇	Nabla operator
∇^2	Laplace operator

Abbreviations

AAC	Advanced Audio Coding (psychoacoustic compression algorithm)
AC-3	Adaptive Transform Coder No. 3 (psychoacoustic compression algorithm)
ADT	Artificial Double Tracking
ar.	Arithmetic
ASC	Audio Spectrum Centroid
ASW	Apparent Source Width
ATRAC	Adaptive Transform Acoustic Coding (audio compression algorithm)
ATSC	Advanced Television System Committee
BEM	Boundary Element Method
BWV	Bach-Werke-Verzeichnis (Bach works catalogue)
c#	C-sharp programming language
CD	Compact Disc (digital data medium)
CD-ROM	Compact Disc—Read-Only Memory (digital data medium)
CRC	Cyclic Redundancy Check (an error detection system)
CTC	Crosstalk Cancellation
DFT	Discrete Fourier Transform
DIN	Deutsches Institut für Normung (German Institute for Standardization)
DOA	Direction Of Arrival
DSP	Digital Signal Processing
DVB	Digital Video Broadcast
DVD	Digital Versatile Disc (digital data medium)
ER	Early Reflections

FDM	Finite Difference Method
FEM	Finite Element Method
FFT	Fast Fourier Transform
GPU	Graphics Processor Unit
GUI	Graphical User Interface
h/w/d	Height/width/depth
HDMI	High-Definition Multimedia Interface
Hi-Fi	High Fidelity (quality demand on audio playback systems)
HOA	Higher-Order Ambisonics
HRTF	Head-Related Transfer Function
ICLD	Interchannel Level Difference
ICTD	Interchannel Time Difference
ILD	Interaural Level Difference
ISO	International Organization for Standardization
ITD	Interaural Time Difference
JND	Just Noticeable Difference
K-H integral	Kirchhoff–Helmholtz integral
LA	Listening Area
LD	Laser Disc
LEV	Listener Envelopment
LFC	Lateral Fraction Coefficient
LFE	Low-Frequency Effects (subwoofer)
LR	Late Reflections
LSR	Least Squares Regression
MADI	Multichannel Audio Digital Interface
MDAP	Multiple Direction Amplitude Panning
MEM	Minimum Energy Method
MIR	Computational Music Information Retrieval
MLP	Meridian Lossless Packing (lossless audio coding format)
MP3	MPEG II audio layer 3 (psychoacoustic audio compression algorithm)
NAH	Near-field Acoustical Holography
NFC-HOA	Near-field Compensated Higher-Order Ambisonics
NWDR	NordWestDeutscher Rundfunk (German broadcasting company regulated by public law)
ORTF	Office de Radiodiffusion Télévision Française
PA	Public Address loudspeakers
PC	Personal Computer
PCM	Pulse Code Modulation
RC	Radiation Characteristics
RMS	Root-Mean-Square (Comparative value for the power of amplifiers)
SACD	Super Audio Compact Disc
SDDS	Sony Dynamic Digital Sound

SPL	Sound Pressure Level
TV	Television
VBAP	Vector Base Amplitude Panning
VCA	Voltage-Controlled Amplifier
VOG	Voice Of God loudspeaker
WFS	Wave Field Synthesis

List of Figures

List of Tables

Chapter 1
Introduction

1.1 General Remarks

The present book describes a new approach to spatial audio: *psychoacoustic sound field synthesis*. The technical implementation of this approach enables a natural, spatial, music listening experience. Several listeners are free to move around and enjoy the instrumental music in a listening area. The implementation of the sound radiation characteristics of musical instruments makes the experienced sound wide, natural and vivid. Due to the combination of a physical sound field synthesis core with psychoacoustic considerations the instrumental sound is experienced as natural in terms of source location and width, pitch and tuning, loudness and dynamics, timing, and timbre. Computational efforts to achieve this are low and compatibility to many established spatial audio systems is given, like stereo, 5.1 and wave field synthesis. Implementing the radiation characteristics in spatial audio is a necessary step towards an immersive music listening experience. Considering psychoacoustics already during the process of developing the spatial audio system is a new paradigm that can be transferred to all types of audio technology, like instrument building and synthesizer design, music mixing practice, and audio compression.

Music is inherently spatial. The relationship between music and space is a topic that has fascinated already the early music philosophers. Music evolved over centuries in terms of concepts, composition, instrument building, performance practice and technology. A lot of this evolution is of spatial nature. The relationship between music and space is not only reflected in musical thinking and ideals in composition, mixing and mastering. It is also an inherent part of performance practice, concerts and music playback. The relationship between hearing and space can be traced back to the evolutionary origin of the ear and is evident at several stages of auditory processing. The often complicated sound radiation characteristics of musical instruments affect the perception of source extent and naturalness. For example, monophonic high fidelity recording and loudspeaker presentation of a grand piano lack its original spaciousness. Even though timing, pitch, dynamics and timbre may sound perfectly original, a loudspeaker does not sound as *grand* as a grand piano

© Springer Nature Switzerland AG 2020

T. Ziemer, *Psychoacoustic Music Sound Field Synthesis*, Current Research in Systematic Musicology 7, https://doi.org/10.1007/978-3-030-23033-3_1

and even inexperienced listeners can tell them apart by hearing. The main reason for this is that the spatial sound radiation characteristics of the source are not considered in most spatial audio systems. Implementing natural room acoustics can bring back some of the original source width and vividness. But as it is uncertain which frequency region and other sound features affects spaciousness perception the most, conventional stereophonic sound setups have only limited capabilities of manipulating spatial audio and presenting realistic auditory scenes. To overcome this drawback advancements in audio recording and playback over the last century were almost exclusively spatially motivated. Stereo virtually displaced monophonic audio for music presentation due to its superior spaciousness: Phantom sources can be placed at different angles, the perceived source extent can be manipulated and stereophonic reproduction of room acoustics is much more immersive and enveloping. The success of conventional audio systems has been attributed to the reliable cues that they deliver for a robust and intersubjective perception of source location and width, loudness, dynamics, and timbre.

Advancements in spatial audio culminate in wave field synthesis. After the conceptualization in the 1980s, implementations and studies in the 1990s, research and development have reached a climax in the early 2000s which peaked in market-ready wave field synthesis systems. Wave field synthesis systems recreate the sound field of virtual monopole sources with a high accuracy within a large listening area. Sources can be placed statically or even move around through almost the entire horizontal plane. Wave field synthesis systems have reached a point at which higher physical accuracy is only a matter of a higher number of loudspeakers and a better acoustical treatment of the listening room. Extensions to two-dimensional loudspeaker arrays to synthesize three-dimensional sound fields are straight-forward and mostly hindered by computational efforts and the impracticability of installing planar or hemispherical loudspeaker arrays around a listening area.

Even though researchers still presents refinements of methods and extensions of existing audio systems, the progress is stagnating. And the need to install one loudspeaker every couple of centimeters is impractical for many event locations in the entertainment industry. Therefore, solutions with a lower number of loudspeakers are needed, which brings back psychoacoustic considerations to enable a natural auditory scene and make sure that physical synthesis errors are inaudible. Psychoacoustic considerations can give researchers food for thought and get the development of spatial audio systems back in motion.

This book introduces the novel concept of psychoacoustic sound field synthesis for a natural, spatial listening experience with music. The radiation characteristics of musical instruments are represented by 128 directional radiation factors for each critical frequency band. The concept is implemented in an audio system which creates a natural, spatial sound impression and a precise source localization for listeners within an extended listening area. The computational efforts to achieve this are comparably small. The method is scalable and can be applied on various loudspeaker setups, even with irregular loudspeaker spacing. Simulations and listening tests provide a proof of concept. Physical accuracy of the synthesized sound field is postponed after perceptual precision.

The book at hand considers the relationship between music and space from the viewpoint of multiple disciplines to argue for psychoacoustic sound field synthesis as a new approach to spatial audio technology. Due to the interdisciplinarity, a wide audience can benefit from this book, from novice music enthusiasts over audio engineers to researchers in the field of spatial audio, be it from the field of biology, audiology or psychophysics, physics, electrical engineering, communication technology or computer science, music theory, music therapy or musicology.

1.2 Intersection of Space and Music

Music and space are linked in many ways. In this book every chapter is dedicated to the relationship between music and spaciousness from another point of view:

Spatial Concepts of Music:

Ideas of dimensionality, geometry, size, distribution and locality can be found through many languages and cultures, when composing, imagining, describing, analyzing or interpreting music.[1] Models of music perception, as developed by the musicologists Albert Wellek and Gerhard Albersheim, contain ideas of *musical space* and a *perception space* ("Empfindungsraum").[2] Space also plays an important role in music production, composition and performance practice. *Symmetric figures*can be found from Johann Sebastian Bach's baroque fugues to Pierre Boulez's 20th century serial music, spatially distributed orchestras from 16th century Venetian polychorality to modern loudspeaker ensembles. *Pseudostereo* techniques are common practice in recording studios.[3] The existence of space in music perception, composition and performance is reflected in concepts of music theory—like the *circle of fifth* or the *melodic map*—and of *music information retrieval* as in matrix operations, multidimensional scaling and source separation.[4] The spatial concepts in the creation, analysis and psychology of music are the topics of Chap. 2.

Biology of the Auditory System:

The strong interrelationship between music and space can be traced back to the evolutionary origin of the *ear* which seems to lie in the *lateral line system* of fish, known to exist in the earliest vertebrates.[5] Its original function was spatial orientation rather than communication or cultural expression.[6] Orientation by means of source identification and localization is only possible due to the spatial distribution of sensory organs like the lateral line or the ear. Even the frequency of sound waves is spatially

[1] See e.g. Kurth (1990) and Griebsch (2000).

[2] See e.g. Albersheim (1939) and Schneider (1989).

[3] See e.g. Stoianova (1989); Mazolla (1990); Kaiser (2012), and Ziemer (2017a).

[4] See e.g. Lerdahl (2001) and Brandenburg et al. (2009).

[5] See e.g. Coombs et al. (1992), p. 267 and Gans (1992), p. 7.

[6] See Fay (1992), p. 229.

encoded in the inner ear.[7] Chap. 3 gives an insight into the evolution of the auditory system from fish to humans.

Psychoacoustics:

The spatial organization of frequencies in the inner ear is one reason of various *psychoacoustic* effects such as *masking*.[8] Although psychological mechanisms, which allow for spatial orientation by acoustical signals, are not fully understood, our capability of localizing sound sources has been extensively investigated and is well known today.[9] The way people represent the auditory outside world in a mental map for matters of orientation is called *auditory scene analysis* and is especially known from a phenomenological point of view.[10] Psychoacoustic knowledge has already been integrated in audio systems and audio compression formats and has potential to improve existing applications[11] or it can act as the basis for new audio technology. It is therefore extensively discussed in Chap. 4.

Spatial Sound of Musical Instruments:

The human auditory system is not only capable of localizing sources. It is also able to distinguish original musical instruments from loudspeaker playback by perceiving differences in their spatial *sound radiation characteristics*.[12] The sound radiation of musical instruments can be very complex due to the interplay of vibrating, radiating and reflecting parts of the instrumental body and enclosed air. Many methods have been developed to measure these radiation characteristics for a better understanding of instrumental sound, such as far field recordings, beamforming and near field recordings.[13] The spatial sound of musical instruments is treated in Chap. 5.

Spatial Acoustics:

In natural listening situations the direct sound of musical instruments is enriched by reflections, diffractions and scattering from room surfaces, the *room acoustics*. This led to considerations about dimensions, geometries and other architectural features of rooms for musical performance.[14] It has been found that the perceived musical sound quality of a room is especially dependent on spatial attributes.[15] Chapter 6 gives an overview about objective and subjective aspects of room acoustics.

---

[7] See e.g. Zwicker and Fastl (1999), p. 29.

[8] See e.g. Fastl (1977) and Gelfand (1990).

[9] See Blauert (1997).

[10] See Bregman (1990) and Theile (1980).

[11] See e.g. Blauert (2008) and Fastl (2010).

[12] See e.g. Warusfel et al. (1997), p. 1.

[13] See e.g. Maynard et al. (1985); Kim (2007).

[14] See e.g. Ahnert and Tennhardt (2008), Fuchs (2013), Knudsen (1988) and Blauert and Xiang (2009).

[15] See e.g. Beranek (1996, 2004) and Kuhl (1978).

Conventional Stereophonic Sound:

An important task of *high fidelity* audio systems is to play back these spatial features as natural as possible.[16] The technological progress of conventional audio systems from *mono* over *stereo* to *surround sound* and *immersive audio* is based almost entirely on advanced facilities to add spatial features to sound (re-)production. Typically, these audio systems aim at creating a psychologically equivalent sound field at the listening position, rather than a physical replication. This approach has certain advantages and disadvantages,[17] discussed extensively in Chap. 7.

Wave Field Synthesis:

Quick advancements in computer technology and digital signal processing made it possible to go another way, aiming at physically recreating a natural sound field. *Sound field synthesis* disclosed new possibilities concerning spatial sound reproduction.[18] Applications typically create virtual sources at arbitrary positions that can be localized accurately by multiple listeners within a large listening area. Approaches exist to recreate not only the desired source location but also the desired sound radiation characteristics.[19] *Wave field synthesis* is a widespread sound field synthesis approach. Its theory and applications are presented in Chap. 8 with a focus on the reconstruction of the complicated spatial sound radiation of musical instruments.

Psychoacoustic Sound Field Synthesis:

Throughout this book the sound radiation characteristics of musical instruments is investigated from the perspective of psychoacoustics, instrument acoustics and room acoustics, stereophonic audio systems and sound field synthesis. On this basis, a theoretic framework is developed to measure, store and recreate the sound radiation characteristics of musical instruments by means of *psychoacoustic sound field synthesis*. The main novelty is the extensive implementation of psychoacoustics throughout the complete procedure, including the *precedence fade* for a distinct localization. Furthermore, the *radiation method* is introduced, a method to implement the radiation characteristic of the loudspeakers to make calculations more robust and improve the precision of the reconstructed wave field. The evaluation of the implemented approach acts as a proof of concept. Furthermore, it validates hypotheses from the field of psychoacoustics and auditory scene analysis in a musical context. The approach is validated by means of simulations, physical measurements and listening tests. Implementation of psychoacoustic sound field synthesis can serve as a tool for research in the field of instrument acoustics, digital signal processing, psychoacoustics and music psychology, and for musical applications. The principle can be applied with conventional stereophonic audio systems and existing sound

---

[16]See Verheijen (1997), p. 9, Pulkki (2008), p. 747, Schanz (1966), pp. 8–18, Berkhout et al. (1993), p. 2764 and Faller (2009).

[17]See e.g. Blauert (2008) and Ziemer (2017a).

[18]See e.g. Berkhout (1988), Verheijen (1997), Ahrens (2012), Ziemer (2016, 2018).

[19]See e.g. Avizienis et al. (2006), Baalman (2008), Corteel (2007), Ziemer (2017a), Ziemer and Bader (2017).

field synthesis setups. Chapter 9 describes this psychoacoustic sound field synthesis approach for music and gives an outlook to future developments in the field of psychoacoustic sound field synthesis.[20]

References

Ahnert W, Tennhardt H-P (2008). Raumakustik. In: Weinzierl S (ed) Handbuch der audiotechnik, Chap. 5. Springer, Berlin, pp 181–266. https://doi.org/10.1007/978-3-540-34301-1_5

Ahrens J (2012) Analytic methods of sound field synthesis. Springer, Berlin. https://doi.org/10.1007/978-3-642-25743-8

Albersheim G (1939) Zur Psychologie der Ton- und Klangeigenschaften (unter Berücksichtigung der 'Zweikomponententheorie' und der Vokalsystematik). Heitz & Co

Avizienis R, Freed A, Kassakian P, Wessel D (2006) A compact 120 independent element spherical loudspeaker array with programable radiation patterns. In: Audio engineering society convention 120, May 2006. http://www.aes.org/e-lib/browse.cfm?elib=13587

Baalman M (2008) On Wave Field Synthesis and electro-acoustic music, with a particular focus on the reproduction of arbitrarily shaped sound sources. VDM, Saarbrücken

Beranek LL (1996) Acoustics. American Institute of Physics, Woodbury (New York). Reprint from 1954 edition

Beranek LL (2004) Concert halls and opera houses: music, acoustics, and architecture, 2nd edn. Springer, New York. https://doi.org/10.1007/978-0-387-21636-2

Berkhout AJ (1988) A holographic approach to acoustic control. J Audio Eng Soc 36(12):977–995. http://www.aes.org/e-lib/browse.cfm?elib=5117

Berkhout AJ, de Vries D, Vogel P (1993) Acoustic control by wave field synthesis. J Acoust Soc Am 93(5):2764–2778. https://doi.org/10.1121/1.405852

Blauert J (1997) Spatial Hearing, Revised edn. MIT Press, Cambridge, MA, The pychophysics of human sound source localization

Blauert J (2008) 3-d-Lautsprecher-Wiedergabemethoden. In: Fortschritte der Akustik—DAGA '08, Mar 2008, Dresden, pp 25–26

Blauert J, Xiang N (2009) Acoustics for engineers. Troy lectures, 2nd edn. Springer, Berlin. https://doi.org/10.1007/978-3-642-03393-3

Brandenburg K, Dittmar C, Gruhne M, Abeßer J, Lukashevich H, Dunker P, Gärtner D, Wolter K, Grossmann H (2009) Music search and recommendation. In: Furht B (ed) Handbook of multimedia for digital entertainment and arts, Chap. 16. Springer, Dordrecht, pp 349–384. https://doi.org/10.1007/978-0-387-89024-1_16

Bregman AS (1990) Auditory scene analysis. MIT Press, Cambridge, MA

Coombs S, Janssen J, Montgomery J (1992) Functional and evolutionary implications of peripheral diversity in lateral line systems. In: Webster DB, Fay RR, Popper AN (eds) The evolutionary biology of hearing, Chap. 15. Springer, New York, pp 267–294. https://doi.org/10.1007/978-1-4612-2784-7_19

Corteel E (2007) Synthesis of directional sources using wave field synthesis, possibilities, and limitations. EURASIP J Adv Signal Process. 2007:Article ID 90509. https://doi.org/10.1155/2007/90509

Faller C (2009). Spatial audio coding and MPEG surround. In: Luo F-L (ed) Mobile multimedia broadcasting standards. Technology and practice, Chap. 22. Springer, New York, pp 629–654. https://doi.org/10.1007/978-0-387-78263-8_22

[20]Details on the approach and each single step can be found in the literature, like Ziemer (2009, 2011a,b,c,d, 2014, 2015a,b, 2016, 2017a,b,c, 2018), Ziemer and Bader (2015a,b,c,d, 2017).

Fastl H (1977)Temporal masking effects: II. critical band noise masker. Acustica 36:317–331. https://www.ingentaconnect.com/contentone/dav/aaua/1977/00000036/00000005/art00003
Fastl H (2010) Praktische Anwendungen der Psychoakustik. In: Fortschritte der Akustik—DAGA 2010, Berlin, pp 5–10
Fay RR (1992) Structure and function in sound discrimination among vertebrates. In: Webster DB, Fay RR, Popper AN (eds) The evolutionary biology of hearing, Chap. 14. Springer, New York, pp 229–263. https://doi.org/10.1007/978-1-4612-2784-7_18
Fuchs H (2013) Applied acoustics. Concepts, absorbers, and silencers for acoustical comfort and noise control. Alternative solutions-innovative tools-practical examples. Springer, Heidelberg. https://doi.org/10.1007/978-3-642-29367-2
Gans C (1992) An overview of the evolutionary biology of hearing. In: The evolutionary biology of hearing, Chap. 1. Springer, New York, pp 3–13. https://doi.org/10.1007/978-1-4612-2784-7_1
Gelfand SA (1990) Hearing. An introduction to psychological and physiological acoustics, 2nd edn. Informa, New York and Basel
Griebsch I (2000) Raum-Zeit-Aspekte beim Zustandekommen vermittelnder Dimensionen. In: Böhme T, Mehner K (eds) Zeit und Raum in Musik und Bildender Kunst. Böhlau, Cologne, pp 139–150
Kaiser C (2012) 1001 mixing tipps. mitp, Heidelberg
Kim Y-H (2007) Acoustic holography. In: Rossing TD (ed) Springer handbook of acoustics, Chap. 26, pp 1077–1099. Springer, New York. https://doi.org/10.1007/978-0-387-30425-0_26
Knudsen VO (1988) Raumakustik. In: Winkler K (ed) Die Physik der Musikinstrumente. Spektrum der Wissenschaft, Heidelberg, pp 136–149
Kuhl W (1978) Räumlichkeit als Komponente des Raumeindrucks. Acustica 40:167–181. https://www.ingentaconnect.com/contentone/dav/aaua/1978/00000040/00000003/art00006
Kurth E (1990) Musikpsychologie. G. Olms, Hildesheim, 2. nachdruck der ausgabe berlin 1931 edition. https://doi.org/10.2307/932010
Lerdahl F (2001) Tonal pitch space. Oxford University Press, Oxford. https://doi.org/10.1093/acprof:oso/9780195178296.001.0001
Maynard JD, Williams EG, Lee Y (1985) Nearfield acoustic holography: I. Theory of generalized holography and the development of NAH. J Acoust Soc Am 78(4):1395–1413. https://doi.org/10.1121/1.392911
Mazolla G (1990) Geometrie der Töne. Elemente der mathematischen Musiktheorie, Birkhäuser, Basel
Pulkki V (2008) Multichannel sound reproduction. In: Havelock D, Kuwano S, Vorländer M (eds) Handbook of signal processing in acoustics, Chap. 38. Springer, New York, pp 747–760. https://doi.org/10.1007/978-0-387-30441-0_38
Schanz GW (1966) Stereo-Taschenbuch. Stereo-Technik für den Praktiker, Philips, Eindhoven
Schneider A (1989) On concepts of 'tonal space' and the dimensions of sound. In: Sprintge R, Droh R (eds) MusicMedicine. International society for music in medicine IV, international musicmedicine symposium October 25–29, 1989, California
Stoianova I (1989) Textur/Klangfarbe und Raum. Zum Problem der Formbildung in der Musik des 20. Jahrhunderts. In: Morawska-Büngeler M (ed) Musik und Raum. Mainz, Vier Kongressbeiträge und ein Seminarbericht, pp 40–59
Theile G (1980) Über die Lokalisation im überlagerten Schallfeld. PhD thesis, University of Technology Berlin, Berlin
Verheijen E (1997) Sound reproduction by wave field synthesis. PhD thesis, Delft University of Technology, Delft
Warusfel O, Derogis P, Caussé R (1997) Radiation synthesis with digitally controlled loudspeakers. In: Audio Engineering Society Convention 103, Sep 1997
Ziemer T (2009) Wave field synthesis by an octupole speaker system. In Naveda L (ed) Proceedings of the second international conference of students of systematic musicology (SysMus09), Nov 2009, pp 89–93. http://biblio.ugent.be/publication/823807/file/6824513.pdf#page=90

Ziemer T (2011a) Wave field synthesis. Theory and application. (magister thesis), University of Hamburg
Ziemer T (2011b) A psychoacoustic approach to wave field synthesis. In: Audio engineering society conference: 42nd international conference: semantic audio, Ilmenau, Jul 2011, pp 191–197. http://www.aes.org/e-lib/browse.cfm?elib=15942
Ziemer T (2011c) Psychoacoustic effects in wave field synthesis applications. In: Schneider A, von Ruschkowski A (eds) Systematic musicology. Empirical and theoretical studies. Peter Lang, Frankfurt am Main, pp 153–162. https://doi.org/10.3726/978-3-653-01290-3
Ziemer T (2011d) A psychoacoustic approach to wave field synthesis. J Audio Eng Soc 59(5):356. https://www.aes.org./conferences/42/abstracts.cfm#TimZiemer
Ziemer T (2014) Sound radiation characteristics of a shakuhachi with different playing techniques. In Proceedings of the international symposium on musical acoustics (ISMA-14), Le Mans, pp 549–555. http://www.conforg.fr/isma2014/cdrom/data/articles/000121.pdf
Ziemer T (2015a) Exploring physical parameters explaining the apparent source width of direct sound of musical instruments. In Jahrestagung der Deutschen Gesellschaft für Musikpsychologie, Oldenburg, Sep 2015, pp 40–41. http://www.researchgate.net/publication/304496623_Exploring_Physical_Parameters_Explaining_the_Apparent_Source_Width_of_Direct_Sound_of_Musical_Instruments
Ziemer T (2015b) Spatial sound impression and precise localization by psychoacoustic sound field synthesis. In: Deutsche Gesellschaft für Akustik e.V., Mores R (eds) Seminar des Fachausschusses Musikalische Akustik (FAMA): "Musikalische Akustik zwischen Empirie und Theorie", Hamburg, pp 17–22. Deutsche Gesellsch. f. Akustik. https://www.dega-akustik.de/fachausschuesse/ma/dokumente/tagungsband-seminar-fama-2015/
Ziemer T (2016) Implementation of the radiation characteristics of musical instruments in wave field synthesis application. PhD thesis, University of Hamburg, Hamburg, July 2016. http://ediss.sub.uni-hamburg.de/volltexte/2016/7939/
Ziemer T (2017a) Source width in music production. Methods in stereo, ambisonics, and wave field synthesis. In: Schneider A (ed) Studies in musical acoustics and psychoacoustics, vol 4. Current research in systematic musicoogy, Chap. 10. Springer, Cham, pp 299–340. https://doi.org/10.1007/978-3-319-47292-8_10
Tim Ziemer (2017b) Perceptually motivated sound field synthesis for music presentation. J. Acoust. Soc. Am. 141(5):3997. https://doi.org/10.1121/1.4989162
Ziemer T (2017c) Perceptual sound field synthesis concept for music presentation. In: Proceedings of meetings on acoustics, page paper number 015016, Boston, MA. https://doi.org/10.1121/2.0000661
Ziemer T (2018) Wave field synthesis. In: Bader R (ed) Springer handbook of systematic musicology, Chap. 18, Berlin, Heidelberg, pp 175–193. https://doi.org/10.1007/978-3-662-55004-5_18
Ziemer T, Bader R (2015a) Complex point source model to calculate the sound field radiated from musical instruments. In: Proceedings of meetings on acoustics, Oct 2015, vol 25. https://doi.org/10.1121/2.0000122
Ziemer T, Bader R (2015b) Implementing the radiation characteristics of musical instruments in a psychoacoustic sound field synthesis system. J Audio Eng Soc 63(12):1094. http://www.aes.org/journal/online/JAES_V63/12/
Ziemer T, Bader R (2015c) Implementing the radiation characteristics of musical instruments in a psychoacoustic sound field synthesis system. In: Audio engineering society convention 139, page paper number 9466, New York. http://www.aes.org/e-lib/browse.cfm?elib=18022
Ziemer T, Bader R (2015d) Complex point source model to calculate the sound field radiated from musical instruments. J Acoust Soc Am 138(3):1936. https://doi.org/10.1121/1.4934107
Ziemer T, Bader R (2017) Psychoacoustic sound field synthesis for musical instrument radiation characteristics 65(6):482–496. https://doi.org/10.17743/jaes.2017.0014
Zwicker E, Fastl H (1999) Psychoacoustics. Facts and models, 2nd edn. Springer, Berlin. https://doi.org/10.1007/978-3-662-09562-1

Chapter 2
Spatial Concepts of Music

Concepts of space play a major role in music. Music theories are models for an analysis of musical compositions as premise of semantic interpretation and comparison. For western music, they are mainly based on scores. Music Information Retrieval (MIR) pursues the same goal, based on operations on digital audio signals and is often heading to automatic music search to make music easily accessible for consumers using music data mining, expert systems and automatic music recommendation. Both, conventional music theory and MIR, have a tradition of spatial thinking, representation, and reasoning, as will be illustrated in the following sections. Ideas of room and space permeate music thoroughly, from the music theory and music information retrieval to the creative processes of composition and performance to music perception.

2.1 Space in Music Psychology

Space strongly affects the perception, imagination, and expression of music. The psychologist Révész (1937) as well as musicologist Kurth (1990) extensively dealt with this phenomenon and put forward several hypotheses, which were revised and discussed, e.g. by Schneider (1989) and by Griebsch (2000). Based on their works, the relationship between space and music will be discussed in the following.[1]

The musicologists Ernst Kurth and Hans Mersmann state that the musical evolvement in time can be understood as static and kinematic energy which is perceived by the listener as movement in an imaginary space, neither visible nor palpable.[2] Kurth as well as the music psychologist Albert Wellek use the term "musical space";

[1]Particularly based on Révész (1937), Kurth (1990), Schneider (1989), and Griebsch (2000).

[2]See Schneider (1989), p. 113 and Kurth (1990), p. 119.

© Springer Nature Switzerland AG 2020

T. Ziemer, *Psychoacoustic Music Sound Field Synthesis*, Current Research in Systematic Musicology 7, https://doi.org/10.1007/978-3-030-23033-3_2

Wellek added "hearing space" and "tonal space", to describe the human perception of music.[3] The perception of space and material in music does not occur by chance but is necessary to make music imaginable, despite its absence of physical form, its "Wesenlosigkeit" (unsubstantiality/bodylessness).[4] According to Kurth, that is why spatial terms are used in most languages to describe music. For the philosopher and mathematician Thom (1983), spatial form is essential for understanding as such, not only restricted to music.[5] It is a feeling, rather than an imagination, and below the level of clear sensation. Likewise, the philosopher, psychologist and musicologist Carl Stumpf considers the spatial character in musical perception immanent but only a matter of pseudo-locality.[6] To him, as well as to Thom, it seems to be in the nature of people to imagine, explain and express music in spatial manners. Blauert (1974) observed in his investigations in the field of psychoacoustics that subjects are mostly able to describe sound stimuli in terms related to space and spaciousness.[7] According to Griebsch (2000), four dichotomies exist in all cultures, three of which are of spatial nature[8]:

- up/down
- center/border
- inside/outside
- active/passive.

Wellek thinks a primal synesthesia, "Ursynästhesie", is the origin of this association—Griebsch (2000) rephrases it as primal analogy, "Ur-Analogie"—which can be understood by every human being.[9] Christian (von Ehrenfels, 1890) similarly suggests an intermodal analogy arguing that a melody transposed along the dimension of pitch remains its identity just as an object transposed along a dimension of physical space.[10] This connection of space and dimensionality with qualities of music is also a basic idea of some spatial concepts in music theory as will be discussed in Sect. 2.4. Qualities in music are based on *Gestalt* principles which will be discussed in detail in Sect. 4.5.

Not only in Western notation of music, the horizontal dimension is associated with time and the vertical dimension with pitch. According to Wellek, intensity and timbre can be considered as far-near dimension, and is applied in this way in compositions from Jan Pieterszoon Sweelinck's baroque "Echo-Fantasy" (around 1617) to György Ligeti's avant-garde piece "Lontano" (1967).[11] These dimensions are illustrated in Fig. 2.1. Karlheinz Stockhausen uses a concept of sound layers, "Klangschichten",

[3] See Schneider (1989), p. 121.

[4] See Kurth (1990), p. 116.

[5] Cf. Godøy (1997), p. 90, translating Thom (1983), p. 6.

[6] See Révész (1937), p. 150.

[7] See Blauert (1974), p. 75.

[8] See Griebsch (2000), pp. 143f.

[9] See Griebsch (2000), pp. 144ff.

[10] See von Ehrenfels (1890).

[11] See Schneider (1989), p. 114.

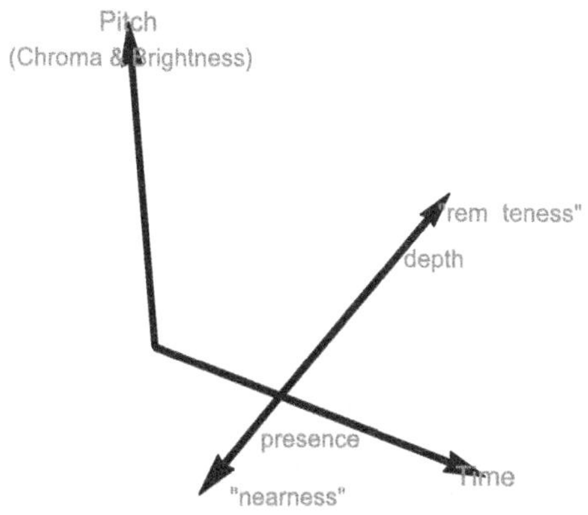

Fig. 2.1 Three dimensions in music, according to Wellek. After Schneider (1989), p. 115

to move into fore- and background by the use of loudness in his composition "Kontakte".[12] The Romantic composer and music theorist Arnold Schönberg followed the idea that melodies could be transposed, rotated and reflected in the same way as objects in physical space.[13] Kurth considers three dimensions sufficient for matters of analogy only, but emphasizes the role of an inner-geometry which is affected by intervals, chords and melodic continuous form. Ethnomusicologist Erich Moritz von Hornbostel understood a small-large dimension as a result of volume and density, but considers neither of them, nor vividness, size, hardness or diffuseness orthogonal dimensions.[14]

Zbikowski (2002) explains the usage of terms of physical space for music by means of "conceptual metaphors".[15] People tend to use cross domain mapping to explain entities and relations of the target domain (music) by terms of a source domain (e.g. physical space). Griebsch (2000) additionally highlights the importance of "tertium comparationis", an impairing dimension with common characteristics of analogy pairs between the two domains. The less dimensions there are and the more analogy pairs exist, the better the representation works. The same thing is done e.g. between physical space and emotions ("I'm feeling up"), creativity ("think outside the box") consciousness ("He fell asleep"), or health ("She's in top shape").[16]

As Révész stated, there seems to be no phenomenological similarity between tone movement and directions in a visual and tactile room. The generalized statement that high frequencies are located at a higher vertical position by the auditory system[17] has not much to do with everyday-life experience and conflicts with findings from listening tests as conducted by Blauert (1974) and discussed in detail in Sect. 4.4. On a piano low to high tones are played from left to right, a cello demands a lower

[12] See Motte-Haber (2000), p. 35.

[13] See Deutsch (1985), p. 131.

[14] See e.g. Schneider (1989), p. 108 and Révész (1937), pp. 164f.

[15] Zbikowski (2002), especially pp. 63–73.

[16] See e.g. Zbikowski (2002), p. 65 for some of the above-mentioned and further examples.

[17] See e.g. Lewald (2006), p. 190.

fingering for a higher pitch on the same string, a trombone-player may stretch and contract his or her arm to play different pitches. Obviously, these mechanisms to change pitch are not responsible for terms like "high" and "low" notes. Therefore, the displacement of the larynx for singing variable pitches is sometimes considered as origin of these terms for musical pitch, especially in view of the reasonable hypothesis that music is originated in vocal singing. Révész (1937) experienced in a self-experiment with plucked ears at an orchestral rehearsal that he felt vibrations of low frequencies in lower parts of his chest and localized higher frequencies in or even above the head. This convinced him of the vibration hypothesis which states that the localization of the felt vibrations—not the heard—within or around the body are the origin of the termination for "high" and "low" tones. On the other hand, although these terms are widely used, e.g. in English, German, Chinese, Indian and Hebrew, other cultures use different terms. In Bali and Java, musicians name pitches "small" and "large", the Suyá—a Native American folk—say "young" and "old", the standard Greek terms are "oxis" and "barys", which mean sharp/pointy and heavy.[18] Analogies and conceptual metaphors can explain why different metaphors are used in different cultures, though comprehensible throughout most of them. According to the biologist Lewald (2006), such metaphors initially stem from a lack of proper terms but establish and thus become proper and understandable terms themselves.[19] This also explains the use of non-spatial visual terms, like brightness, brilliance or sound color, and tactile vocables, such as roughness or sharpness. Since it explains all sorts of spatial and non-spatial metaphors, it seems to contradict the immanence of space in music. Still, an explanation providing the idea of immanence is given by the musician and musicologist Albersheim (1939). He differentiates the objective, empirical-real space in which our physical life takes place and a subjective perception room, the "Empfindungsraum".[20] Blauert (1997) agrees, differentiating between the sound event in the physical space and the auditory event in the "perceptual space".[21] For Albersheim (1939), prominence of unextended spots and their relations to each other and to an extension continuum provide space. Pitches and intervals take place in a perception space just as positions and distances in the objective room. For him, with this interpretation of a room, space is immanent in music and consequentially spatial terms are used to describe aspects of it.

In fact, psychoacoustic investigation revealed that the perception qualities such as timbre seem to take place in a perceptional space, often referred to as "timbre space".[22] Musicologist Bader (2013) summarizes the different approaches and the results from over twenty studies concerning timbre discrimination as well as identification and auditory discrimination of musical instruments.[23] In listening tests, subjective judgments about timbre similarity could be explained to more than 70% by

[18]See Zbikowski (2002), pp. 63 and 72f.

[19]See Lewald (2006), p. 190.

[20]See Albersheim (1939), p. 60, 71 and pp. 59ff.

[21]See e.g. Blauert (1997), p. 373. More details can be found in Sect. 4.4.1.

[22]See e.g. Donnadieu (2007), Troiviainen (1997), p. 338 and Bader (2013), p. 351.

[23]See Bader (2013), pp. 329–379.

multidimensional scaling with three dimensions all of which were highly correlated to physical parameters.[24] These were e.g. derived from similarity or dissimilarity judgments of sound pairs or triplets from synthetic sounds, natural instruments and artificially created "hybrid instruments", whose physical features lie in between real instrumental sounds.[25] Most of the researchers found similar dimensions crucial for the characterization of timbre. On the spectral side, there are brightness—often quantified by the audio spectrum centroid (ASC)—bandwidth and the balance of spectral components, which play a major role in timbre perception. On the temporal side, features of the initial transients—like duration of attack, onset-synchrony and fluctuations of partials—influence the perceived timbre character of musical sounds. Also, spectro-temporal features—like presence of high partials in the transients, spectral flux or the amplitude envelope—seem to play a role in timbre perception and when it comes to identifying and discriminating musical instruments. The same counts for specific characteristics such as the vibrato of violins. Brightness was found the most prominent feature in the evaluation of timbre, explaining most of the data. The psychologist Garner (1974) generally explains this phenomenon as follows: "[...] one dimension is more discriminable than the other, and [...] the more discriminable dimension will be used as the basis of classification."[26] Hence, he does not only describe the dominance of one dimension over the other but also addresses another important aspect: Although continuous dimensions seem to exist, classification of timbre can be observed. Donnadieu (2007) found that in listening tests where subjects were advised to freely group "similar" sounds, the resulting groups were mainly based on similarity in sound creation mechanisms and resonators.[27] A similar categorical grouping was also found by Lakatos (2000) in an investigation of percussive timbres.[28]

Timbre is considered, and sometimes even defined, as aspect of sound quality which is independent of pitch, loudness and duration.[29] Although timbres can be allocated within a low-dimensional perception space, it is considered an emergent quality—which is more than just the position that results from its magnitudes on several dimensions—and can rather be considered a "holistic image of the entire musical object".[30] It is possible that the dimensions which span the timbre space are not fully orthogonal nor orthonormal.[31] This may partly explain the slight differences in the denotation of dimensions or physical measures to quantify them.

In summary—if immanent or not, perception room or conceptual metaphor—space seems to be very suitable to imagine and express experienced aspects and phenomena of sound and music. Spatial terms are used in most languages to describe

[24]See Bader (2013), p. 331.

[25]See Bader (2013), p. 335.

[26]See Garner (1974), p. 113.

[27]Cf. Donnadieu (2007), pp. 300f, referring to her doctoral thesis, Donnadieu (1997).

[28]See Lakatos (2000).

[29]See Ando (2010), pp. 92 and 120.

[30]See Godøy (1997), p. 89. Emergent qualities are discussed in more detail in Sect. 4.5.

[31]See e.g. Bader (2013), pp. 359ff.

music and are even understandable between different cultures. The perception of space does not only arrive when listening to music. It also strongly affects the creative process of music composition and performance as will be discussed in the following section.

2.2 Space in Composition and Performance Practice

Symmetry as compositional technique subsisted from Johann Sebastian Bach's (1685–1750) fugues to Pierre Boulez's (1925–2016) serial music.[32] Also in performance practice, spatial considerations can be found through most epochs: Already in baroque performance practice, space-related ideas were implemented, like the distribution of choirs in Venetian polychorality.[33] In 1837 Hector Berlioz's "Requiem" was composed for choirs and brass ensembles, distributed in all four cardinal directions in the Dôme des Invalides in Paris. The famous Romantic composer and conductor Gustav Mahler thought of spatially distributed instruments to perform "Symphonie fantastique", composed by the Romantic French composer Hector Berlioz.[34] Wallace Clement Sabine, a pioneer in the field of room acoustics,[35] argued that room acoustics strongly influences the work of composers and that architecture shapes music. For example romanesque churches with their long reverberation time led to the technique of creating harmonies between successive notes.[36] Compositional parameters like duration of notes and tempo of sequences are affected by architectural acoustics, namely the strength and duration of the reverberation. Acoustician Jürgen Meyer (1986) found evidence that the room acoustics of original concert halls have a great influence on compositions. He prove it on compositions of the Classical composer Joseph Haydn (1732–1809) and concert halls that still exist.[37] Furthermore, he emphasizes the inter-relation between the organ and the room acoustics of churches concerning room resonances, early reflections and reverberation.[38, 39]

Room and space are central aspects in the musical avant-garde, observable in dodecaphony, serialism, electronic music and other innovative composition techniques. A liberation of hierarchy and authority cleared the way for a new compositional inner-

[32]See e.g. Mazolla (1990), pp. 84ff.

[33]See Stoianova (1989), p. 36.

[34]See Motte-Haber (2000), p. 35.

[35]More details on his work in the field of room- and concert hall acoustics, see Chap. 6.

[36]See Forsyth (1985), p. 3.

[37]See Meyer (1986).

[38]See Meyer (2003). Of course, the room not only influences compositions but performance as well. The contemporary conductor and composer Bräm (1986) describes the influence of room acoustics on tempo not from a compositional but from a conductor's point of view. Depending on the room acoustical conditions he had to adjust the tempo of Wolfgang Amadeus Mozart's "Jupiter Sinfonie" in a range of twenty percent and more to accomplish a desirable result. Room acoustics are discussed in detail in Chap. 6.

[39]See Bräm (1986), p. 9.

room of music just as multifunctional performance-halls allowed for new ways of musical perception.[40] Especially in electroacoustic music, space is often explicitly considered in imagination, mentality and intention. Karlheinz Stockhausen foresaw space-music as a rising trend, so did Bernd Alois Zimmermann, Luigi Nono, Boulez and Edgar Varése.[41] On the one hand this thinking is manifested in ideals and visions of composition and performance practice: Varèse considered spatial music as sounding movements in a room, having four dimensions: The horizontal temporal dimension, the vertical, spectral dimension, the dynamics as depth and a fourth dimension he describes as sound projection similar to a spotlight.[42] To Zimmermann, an essential requirement for a modern theater is an omni-mobile architecture where stages spherically surround the audience area. Both stage and audience need to be able to move, turn towards and away from each other, interchange and even interfuse.[43] On the one hand it is implemented compositions and performance practice: Stockhausen used a rotary table to create rotary sounds including a Doppler effect and the effect of facing and turning away from a listener. In 1958 Varèse and the Greek composer Yannis Xenakis created pieces for "Poème électronique", a collaborative total artwork presented at the world fair in Brussels in collaboration with the architect Le Corbusier, illustrated in Fig. 2.2. It was a Gesamtkunstwerk of architecture, pictures, light, and sound and contains a "room polyphony" with 350–425 loudspeakers.[44] The performance of "Réponse" (1981), a composition for a chamber orchestra surrounded by six solo instruments and loudspeakers, by Boulez utilizes microphones to record the solo instruments, a real-time DSP processor to manipulate the sounds, and loudspeakers to play back the results. As can be seen in Fig. 2.3, the relation of spatial center and surrounding becomes a compositional element as well as sound movements through the room.[45]

According to the avant-garde composer Erik Satie, music is a part of the room, like furniture.[46] The composer Bill Fontana creates art he calls "sound sculptures", painter Julius (Heidelberger Kunstverein) composed "Musik für einen gelben Raum – presto" (Music for a yellow room) which is intended to "paint" the room yellow by music.[47] The Swiss composer Walter Fähndrich designs music for spaces "giving the impression that the music is generated by the room itself"[48] and even created an exemplary sound catalog. This is an approach to shape or remodel our natural and technical acoustical environment, the so-called *soundscape*. The German composer

[40] See Nauck (1997), pp. 19–20.

[41] See Motte-Haber (2000), p. 31.

[42] See Stoianova (1989), p. 41.

[43] See Kirchmeyer and Schmidt (1970), p. 279.

[44] See Kirchmeyer and Schmidt (1970), p. 20, Barthelmes (1986), p. 85 and Motte-Haber (2000), p. 35.

[45] See Boulez and Gerzso (1988) for more details. The composition makes use of amplitude based panning, which is explained in detail in Chap. 7.

[46] See Barthelmes (1986), p. 85.

[47] See Barthelmes (1986), pp. 77, 81 and 86.

[48] See Fähndrich et al. (2010).

Fig. 2.2 Poème Électronique by architect Le Corbusier and composers Varèse and Xenakis. Photo by Hagens (2018), provided under Creative Commons License

Hans Otte composes meditative soundscapes, as a counterpart to landscape paintings.[49] The influence of pleasant sound to complement or mask encumbering noise has become a topic in psychoacoustical, psychological and sociological research.[50] In the time period from 1950 to 1979 the Swedish composer Allan Patterson created 16 room-symphonies using functionalism of a musical space to create micro- and macro forms.[51] For her "Hörbild" (audio picture) sound installation, composer and sound designer Sabine Schäfer created a loudspeaker ensemble, as shown in Fig. 2.4, which became a sort of instrument for her.[52] In 1998 Edwin von der Heide created an *acoustic window* leading to the auditory impression of a distal industrial landscape. He realized that by recording such a landscape with a rectangular array consisting of 40 microphones and playing the recordings back through a loudspeaker array with the exact same arrangement.[53] Compositions for modern audio systems, such as wave field synthesis installations, often make full use of the new possibilities that these systems offer in terms of source distribution and movement. All these sorts of

[49] See Barthelmes (1986), p. 85.

[50] See e.g. Genuit (2003) and Bockhoff (2007). Masking will be discussed extensively in Sect. 4.3.

[51] See Stoianova (1989), p. 40.

[52] See Schäfer (2000), pp. 251f. You can listen to pieces composed for the loudspeaker instrument at Bern University of the Arts.

[53] See Weinzierl (2008), p. 37. This technique is known as the "acoustic curtain", which is a basic concept of wave field synthesis as will be discussed in detail in Chap. 8, especially in Sect. 8.1.

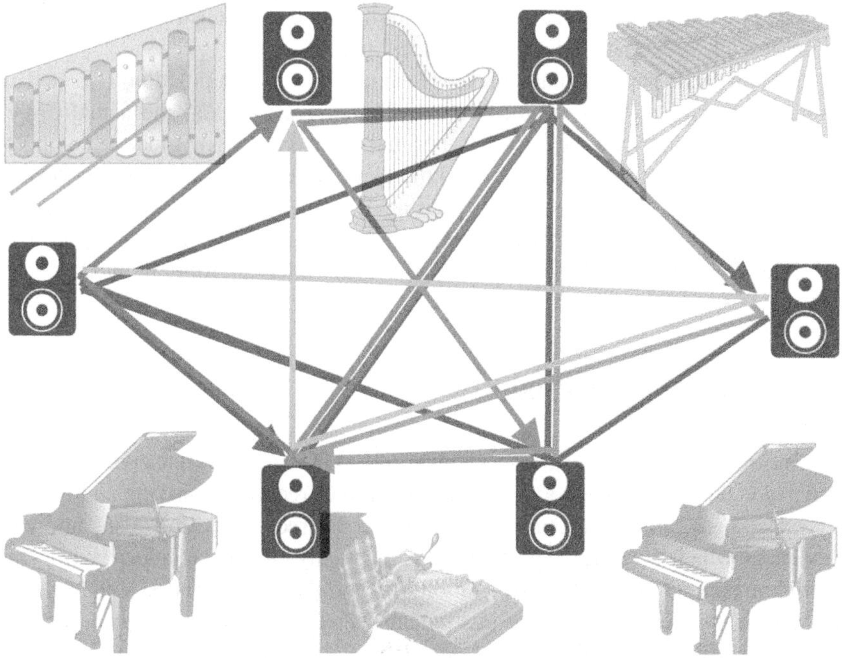

Fig. 2.3 Setup for "Réponse" by Boulez. An ensemble is placed in the center, surrounded by audience, solo instruments and loudspeakers. The loudspeakers use amplitude panning to let manipulated solo passages "wander" through the room (indicated by the arrows). After Boulez and Gerzso (1988), pp. 178f, which is a translation of Boulez and Gerzso (1988)

Fig. 2.4 "Hörbild"; a loudspeaker ensemble created by Sabine Schäfer in 1995 as music performing sound sculpture. She continued to use the installation as a musical instrument for her compositions. Photo by Felix Groß with friendly permission by Sabine Schäfer

composition and performance can be summarized under the term *sound art*, which comprises all sorts of spatial acoustic conceptions.[54]

Through most periods space played a central role in composition, reflected in scores, arrangement of ensembles or intended expression. Yet, the actual objective room, as described in Sect. 2.1, became more central in modern compositions, especially against the background of advanced technology, like electroacoustic performance and digital signal processing. Together with new forms of musical performance—such as sound sculptures, soundscapes and happenings—it gave rise to new compositional methods and ideas to actively integrate room in music.

2.3 Space in Music Production

One aim of recording is to conserve music, so that it can be transported, copied, distributed and played back easily. The main aim of recording, mixing and mastering music is to capture and tune the sound in a desired way to make an "auditory statement" which is neither physically nor perceptually identical to live music.[55] This procedure comprises e.g. the optimization of spectral balance, loudness and dynamics and the spatial distribution of instruments as well as the intensity and diffusity of room acoustics. Typically, this is done by an audio engineer after the completion of composition and arrangement. This procedure is often considered as crafts and the audio engineer tries to manipulate the sound as to underline the aesthetic idea of the work without impairing the composition. In some music pieces the sound impression plays an even more important role than the composition, orchestration or arrangement. Here, the sound design is a crucial part of the creative process. Music producers or Tonmeisters create, tune and manipulate sounds, often with little attention on lyrics, melody or harmony. Many developments in audio technology only target at shaping sound to create an "audioscape".[56]

A major part of the sound tuning has to do with spaciousness. According to the professional audio engineer Bobby Owsinski (2014) "...a mixer's main task is to make things sound bigger and wider."[57] To achieve this, many recording and mixing techniques have been established. These are presented successively in the following subsections, based on a number of textbooks and scientific articles.[58] This section concentrates on spatial audio recording and mixing for stereo. The history

[54] See e.g. Baalman (2008), Chap. 4 for an overview about wave field synthesis compositions and Weinzierl (2008), p. 37, for more information on sound art.

[55] Direct quote from Rogers (2004), p. 31 and indirect quote from Maempel (2008), pp. 232, 238, p. 240 and 245.

[56] See Zagorski-Thomas (2014), p. 124.

[57] See Owsinski (2014), p. 49.

[58] Namely Owsinski (2014), Levinit (2004a), Maempel (2008), Hamidovic (2012), Fouad (2004), Pulkki (2004), Levitin (2004b), Rogers (2004), Kaiser (2012a, b, 2013), Otondo (2008), Faller (2005), Mores (2018), Ziemer (2017) and Cabrera (2011).

and principles of stereophonic audio systems are discussed against the background of spaciousness in Chap. 7.[59]

2.3.1 *Space in Recording Techniques*

Omnidirectional microphones transduce forces acting on a membrane of some square-millimeters or -centimeters into voltage. Thus, they create a signal proportional to the sound pressure in the sensitive frequency region of the microphone which often lies between almost 0 and 20 to 50 kHz. Microphones continuously record the sound pressure which includes most aspects of instrumental sound, like pitch and melody, the dynamics and timbre. However, they record sound at the very microphone position only. As sound pressure is a scalar, the recording does not contain much information about the origin or propagation direction of a wavefront.

One hint about the location of the sound source relative to the microphone is the low-frequency boost of near sources. Due to acoustic short circuits low frequencies tend to stay in the near field of sources, if their body is small compared to the wavelength. Only little low frequency energy is propagated to the far field. Consequently, a microphone can only record the low-frequency content if it is proximate to the source.[60] Microphones with a cardioid-shaped sensitivity exaggerate this effect. Here, low frequencies cause another acoustic short-circuit at the microphone membrane. The effect is stronger for remote sources, i.e., cardioid microphones record less low frequency content the further the source lies away from the microphone. The short-circuit is compensated by a signal processing chain that boosts low frequencies. So a low-frequency boost is a clue for a proximate source. At the same time very high frequencies attenuate stronger with increasing distance than lower frequencies. This effect is both measurable and audible. Due to the short wavelength heat can transfer from zones of high pressure to zones of low pressure. This way some energy diffuses as heat instead of traveling as a sound wave. Another distance cue is given by the ratio of direct to reverberant sound. Near the source the direct sound is dominant and masks a lot of the room's response.[61] Here, room reflections are only audible between successive notes. The extreme opposite can be observed with remote sources. Early reflections can be even louder than the direct sound and the reverberation tail can mask parts of the direct sound and smear transients, like note onsets, modulations and offsets. These monophonic audio parameters are comparable with monaural auditory cues. They do carry some information about the spatial attributes of the sound scene and they are widely used by music producers

[59] A broader overview about spaciousness in music recording and mixing for stereo, surround, ambisonics and wave field synthesis can be found in Ziemer (2017).

[60] More information on the sound radiation characteristics of musical instruments is given in Chap. 5, especially in Sect. 5.2.

[61] An overview of masking and room acoustics is given in Chaps. 4 and 6.

and audio engineers.[62] But when it comes to horizontal source angle localization and the perception of source extent and listener envelopment, binaural cues are of major importance.[63]

Recording techniques try to capture binaural cues for spatial hearing in terms of localization, source extent and acoustical properties of the performance room. In the near field of the instrument, direct sound dominates the recording and the room reflections are comparably soft. Here, stereo recording techniques mainly aim at capturing a sample of the natural sound radiation characteristics of the source.[64] Due to interferences of waves emanating from different parts of an instrument's body and enclosed air the sound is slightly different in each direction. These differences create incoherent signals at the listener's ears which affects the perception of source extent. Further away from the instrument, the difference between microphone recordings can indicate the incidence angle of the wave fronts. The ratio of reverberation to direct sound indicates distance whereas differences in arrival time or amplitude indicates the source angle. Direct sound and early reflections especially affect the perception of source extent whereas late reflections have an influence on the perception of listener envelopment.

Several stereo recording techniques are illustrated in Fig. 2.5, recording a drum. They are based on microphone pairs. The directionality of each microphone is depicted by the shape of its head. The three types are omnidirectional, bidirectional (figure-of-eight), and cardioid. The point right on the drum membrane represents a piezo electric contact microphone. The left channel is dyed blue, the right channel is dyed red. Recording techniques where the microphone positions almost coincide are called *coincident techniques*. The A-B recording techniques works with two omnidirectional microphones spaced between 20 and 80 cm. It is a so-called *spaced technique*. In the near field the microphones may record slightly different spectra, depending on the radiation characteristics of the sound source. In the far field, inter channel time differences (ICTD) occur, when the instrument is deflected to one side. The X-Y microphone pair consists of two cardioid microphones. Their positions should almost coincide, facing 45° to the left and 45° to the right. Sometimes, super- or hypercardioids are applied for X-Y recordings. Close to an instrument, the spectra may again be slightly different. In the far field, laterally deflected sound sources create inter channel level differences (ICLD) denoting the source position. As the microphone positions almost coincide, the recordings are in phase. When summing them up, they barely create destructive interference. Thus, they are mono compatible. The so-called ORTF recording technique is a combination of A-B and X-Y. It is named after the Office de Radiodiffusion Télévision Française. Two cardioid microphones have a distance of 17 cm and an opening angle of 110°, creating slightly incoherent

[62] See e.g. Fouad (2004), p. 150, Rogers (2004), pp. 32f. Details on mixing in music production are given in the subsequent section.

[63] The psychoacoustics of spatial hearing are discussed in Chap. 4, especially in Sect. 4.4, as well as in Chap. 6, particularly Sect. 6.2.

[64] The sound radiation characteristics of musical instruments as well as recording techniques applied in research to capture them are discussed in detail in Chap. 5.

signals when placed in the nearfield. In the far field ICLD and ICTD occur when the source is deflected to either side in the horizontal plane. The Blumlein recording technique is related closely to the X-Y method. Two figure-of-eight microphones have an angle of 90° to each other. Considering pure direct sound, ORTF creates larger ICLDs than X-Y for deflected sources. However, it also records waves arriving from behind, so a larger portion of rear ambient sound is recorded. The recording technique has been patented in the early 1930s by Blumlein (1933).[65] For mid-side stereo recordings (MS) one omnidirectional or cardioid microphone is combined with a collocated figure-of-eight microphone. The neutral axis of the bidirectional microphone coincides with the direction of highest sensitivity of the cardioid. This way, frontal sound is only recorded by the first microphone, whereas the second microphone mainly captures laterally arriving wave fronts. The first recording contains the monaural portion of sound and is routed to both stereo channels. The second recording contains the difference between left and right and thus resembles the "interaural" differences as heard by listeners. It is added to the left and subtracted from the right channel. Here, the gain of the bidirectional recording can be manipulated to make the sound wider or narrower. In all recording techniques, the degree of ICLDs and ICTDs depends on the position and radiation characteristics of the source as well as on the amount and properties of the recording room reflections. More details on these stereo microphoning techniques can be found in the literature.[66]

According to the cognitive scientist and record producer Levitin (2004b), recording technology in the 1970s reached a point where high-fidelity recordings of bands was no challenge anymore, and so the challenge became to create something larger than life. He calls it "a sort of auditory impressionism"[67] where the sound can be more important than the notes. In the recording studio, several additional recording techniques have been tried out to enlarge sound sources. In the illustration, a light piezo transducer is sticked to the membrane with clay while another microphone is placed inside the drum shell. This approach is illustrated in Fig. 2.5. Likewise, guitars are often picked up with one microphone near the neck and another microphone near the sound hole. The first recording has a brighter sound and contains more fingering noise. The second microphone recording sounds warmer because it is dominated by the Helmholtz resonance around 100 Hz. According to Levinit (2004a) this gives the listener the feeling of being confronted with a huge instrument or having the head inside the guitar.[68] Melody and harmony instruments are often recorded twice. This procedure is called *overdubbing* and occurred in the 1960s. Due to slight differences in tune, timing, articulation and micromodulations, the recordings are incoherent. The degree of coherence changes dynamically, creating a natural variance. When each recording is played by an individual loudspeaker, the instrument sound broader than coherent loudspeaker signals. The broadening effect increases when adding a delay of some milliseconds between them or when high-passing one of the record-

[65] See Blumlein (1933).

[66] See e.g. Kaiser (2012b), pp. 33–43, Friedrich (2008), Chap. 13, Ziemer (2017) and Mores (2018).

[67] See Levitin (2004b), p. 14.

[68] See Levinit (2004a), p. 157.

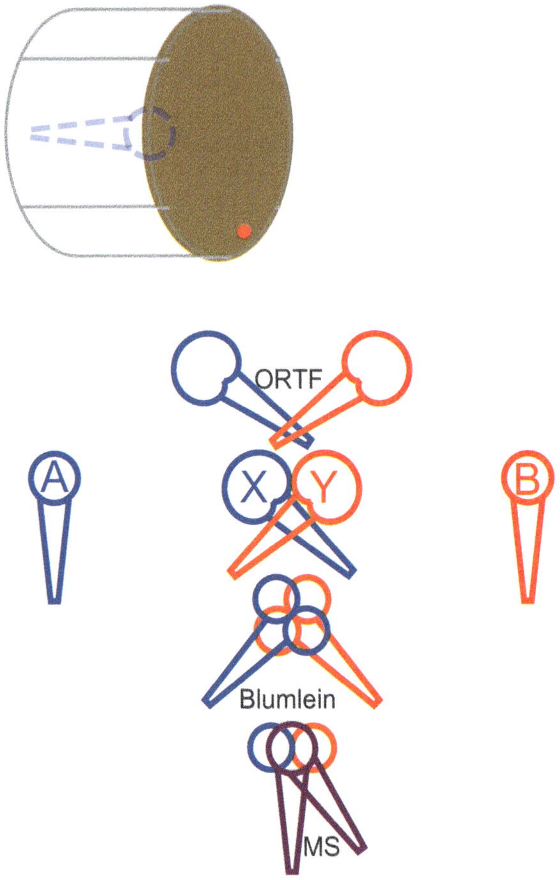

Fig. 2.5 Stereo recording techniques capturing different portions of the radiated drum sound. After Ziemer (2017), p. 309

ings.[69] This techniques is usually not applied for the rhythm section as it may smear transients and the timing may become blurred. Instruments of a drum kit are often picked up individually by one ore more microphones in the near field. In recording studio practice this is called *close-miking* and captures only little reverberation and crosstalk. If the desired drum sounds cannot be achieved by audio effects, like equalizers, filters and compressors, the recordings can be used to trigger a sampler with more favored sounds and the samples can be mixed more or less to the original recording.[70] Additional overhead microphones, usually an A-B pair with a large distance, records the whole drum set including the room reverberation. This is supposed to make the sound more homogeneous and tie the individual recordings together again.

---

[69]Especially guitars and vocals are treated that way. See e.g. Maempel (2008), p. 236, Hamidovic (2012), pp. 52, 57 ans 67 and Kaiser (2012a), p. 113 and pp. 116–127 for details on overdubbing.

[70]This hybrid approach became popular in the 1980s, see e.g. Levinit (2004a), p. 148 and 150, Hamidovic (2012), p. 27 and Kaiser (2012a), pp. 89f.

The natural room reverberation gives them more depth, inverted phase makes it sound larger.[71]

If an instrument had been recorded with one microphone only, it can be broadened by playing the recording back over a loudspeaker in a reverberant room and picking it up by a stereo-recording technique. Here, the radiation pattern of the loudspeaker and the incoherent room reflections yield slightly decorrelated recordings.[72]

In all recording techniques the distance between the source and the microphones affects the ratio of direct-to-reverberant sound. Many audio engineers prefer to record instruments in rather dry, i.e. almost anechoic, rooms. Artificial reverberation can easily be added later in the music production chain. This is common practice since the 1960s.[73] The opposite procedure, i.e., dereverberation or blind source separation, is a much more difficult task.[74] Other audio engineers prefer to record in a reverberant environment if they have a room with appropriate acoustic properties at hand.

2.3.2 Space in Mixing Techniques

Numerous mixing techniques increase the perceived spaciousness of musical instruments. They are either applied additionally to the stereo recording techniques or they are applied when the recording is already completed or the music is electronic rather than acoustic. Often, these techniques create a stereophonic signal from a mono source. Thus, the techniques can be summarized under the term *pseudo stereophony*. Like the recording techniques, they aim at making stereo channel signals less coherent. Many authors reveal their bag of tricks for "Making Instruments Sound Huge".[75] Spatial terms belong to the most-used terms in assessments of music mixes and it has been found that width is a major contributor to preference of music mixes and audio quality ratings.[76] But a certain degree of spaciousness is not only desired from a creative and aesthetic point of view. It has also been found that perceived audio quality and perceived spatial impression exhibit high correlation in headphone music listening, indicating that spaciousness is an important contributor to perceived audio quality.[77]

A simple but effective method to broaden up a source is to divide a spectrum and pan each half to an individual channel, e.g. by means of one low- and one high-pass filter.[78] The result can be seen in Fig. 2.6. The two signals are similar enough to

[71] According to Mozart (2015), p. 82 and Hamidovic (2012), pp. 20f.

[72] This is proposed e.g. in Faller (2005).

[73] See e.g. Stevenson (1968) discussing flexible room acoustical properties of a tv studio.

[74] Dereverberation and blind source separation by means of microphone array techniques is treated e.g. in Bader (2014).

[75] This is actually the name of a section in Levinit (2004a) on pp. 157ff.

[76] See Man and Reiss (2017), Wilson and Fazenda (2015, 2016).

[77] See Gutierrez-Parera and López (2016).

[78] Proposed e.g. in Faller (2005) and Hamidovic (2012).

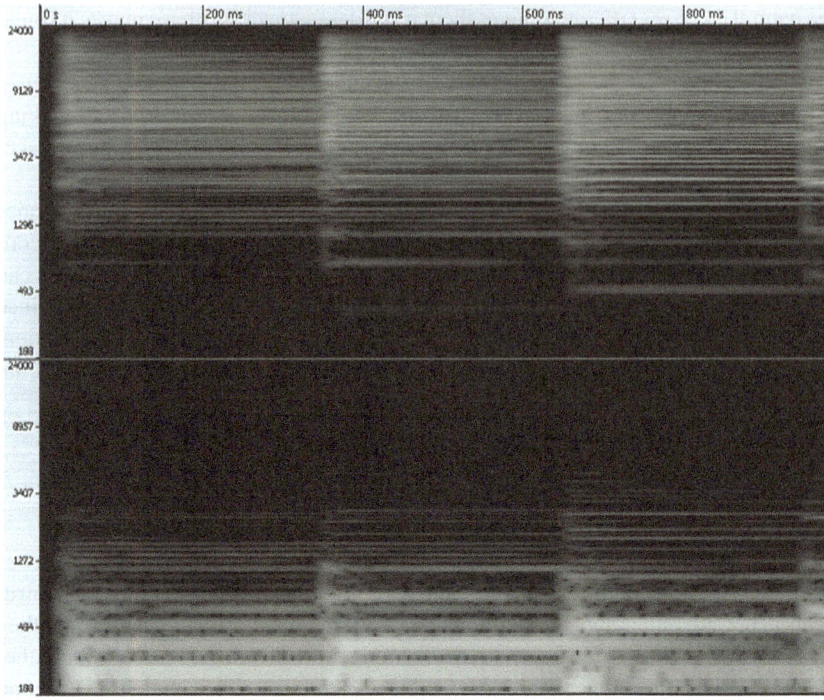

Fig. 2.6 Pseudostereo by high-passing the left (top) and low-passing the right channel (bottom)

be heard as one but so incoherent that they sound spacious.[79] Levinit (2004a) and his former student Rogers (2004) argue that simply applying a compressor already makes instruments sound larger. The auditory system applies a similar mechanism if sources are very loud. So a compressor shall evoke the illusion of a very near, large and loud source. The practicist Kaiser (2012a) differentiates between degrees of dynamic compression. He uses hard compression for a hard, small sound impression and soft compression for a broader sound.[80]

Levinit (2004a) proposes to record electric guitars via line cable as well as via microphones at the top cabinet, the bottom cabinet and with room microphones. To increase the spatial spread of an overdub recording, one channel can be pitch-shifted by one octave. This way spectral fusion is still strong but in addition to the slight temporal incoherence the spectral coherence is reduced. To increase the explosive quality and loudness of percussive instruments, a gated reverb can be used. It creates a sudden attenuation shortly after the onset of the reverberation.[81]

[79]This psychological effect can be explained by auditory scene analysis principles, discussed in Sect. 4.5.

[80]See Levinit (2004a), p. 158 and Rogers (2004), p. 35 versus Kaiser (2012a), p. 32.

[81]All these mixing procedures are proposed in Levinit (2004a).

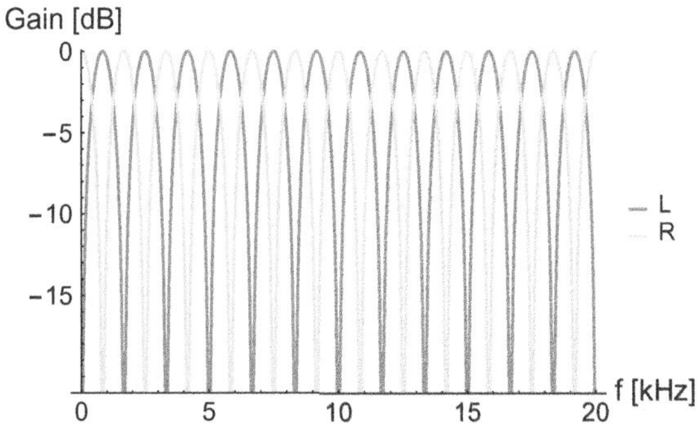

Fig. 2.7 Pseudostereo by applying complementary comb filters on the left and the right channel. From Ziemer (2017), p. 312

Another commonly applied pseudo-stereo technique is to either apply complementary comb filters on both channels or to use an all-pass filter network which randomizes the phase of frequency components in one channel.[82] Both methods create two decorrelated versions from one original mono file. The first method is illustrated in Fig. 2.7. The lobes of the left channel filter coincide with the notches of the right channel filter. The consequence of this processing is that frequencies are panned to different angles in the stereo panorama in terms of amplitude based panning. This panning pattern repeats over frequency. The second method is illustrated in Fig. 2.8. The amplitude spectra of both channels remain identical but the phase of each frequency the right channel is randomized. This randomization may have a drastic effect on the time signal which becomes audible especially during transients. If many frequencies start in phase, this may steepen attacks and become audible as a click. Frequencies starting out of phase smear and may sound weak or crescendo-like. For this method, the filter with the best transient behavior and the desired spaciousness may be found by trial-and-error. In the illustrated example may not be the best phase-randomization choice because the note onsets are definitely smeared. Another static decorrelation between channels can be achieved by using pitch changers, slight delays or individual compressors whereas individual chorus effects create more dynamic decorrelations.[83]

Adding phase inverted chorus effects individually on both channels has also been proposed to create a dynamic spatial sound.[84] This way the original monophonic sound is modified in opposite ways, so the original monophonic signal is enriched dynamically by synchronized but incoherent spectral changes. This approach comes close to artificial double tracking (ADT). In analogue recording studios a track is recorded with a tape recorder. The tape is then modulated by wow and flutter effects resulting in time variant time-, phase- and frequency differences compared to the original track. Again, these two tracks are routed to individual loudspeakers. The effective sound is a time-variant comb filter effect.

[82] Both methods are proposed e.g. in Cabrera (2011) and Faller (2005).

[83] See e.g. Hamidovic (2012), pp. 57 and 67 or Owsinski (2014), p. 50.

[84] See Cabrera (2011).

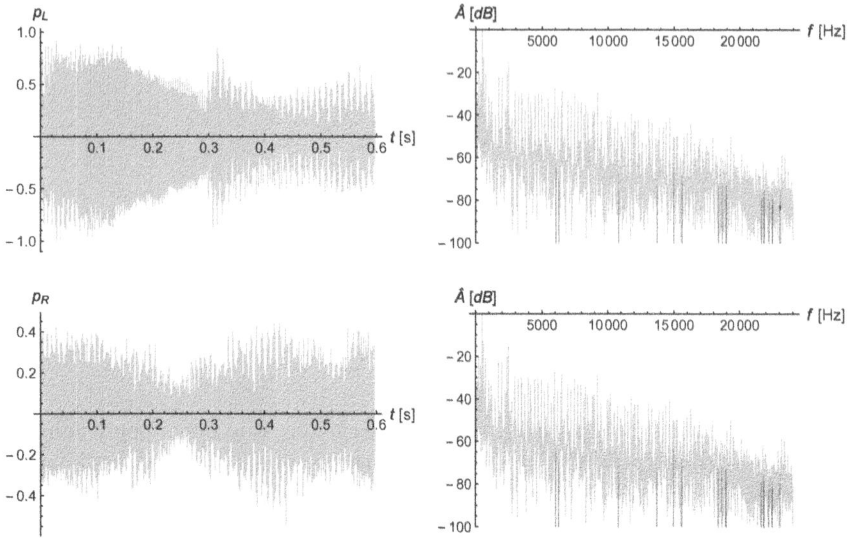

Fig. 2.8 Pseudostereo by phase randomization. The original recording is routed to the left channel (top). The phases of all frequency components of the original recording are randomized and routed to the right channel (bottom). The amplitude spectra (right) remain identical but the time series (left) changed, e.g. the steep attack at 0.3 s is lost

A number of spatial mixing tips is given by Kaiser (2012a). In contrast to the other authors he also denotes the necessity of making instruments sound narrow if desired. Sometimes, drums are panned narrowly to make them sound more realistic compared to distributing them all over the stereo panorama. The tom toms are an exception, typically being panned from left to right. He also likes to play with the spaciousness of drums, e.g., increasing the reverb before the chorus to create a spatially washy sound. Suddenly making the drums dryer and spatially precise at the beginning of the chorus gives it a powerful presence. Changing the width of instruments, the spatial distribution of sources and the perceived size of rooms is an important dramatic tool. For him, predelays, rhythmical delay effects as well as the frequency region above 15 kHz affects the perceived depth of instruments. The beating that result from chorus effects make sounds broader. Another trick of his is to pan the reverberation to a different location than the direct sound to make the source slightly wider. Other recording studio effects, like split-harmonizer with an inter-channel difference of ±9% and a delay of 20 ms increases the width and depth of lead-vocals. He pans stereo recordings of grand pianos hard to the left and the right, upright piano only slightly to the left and right and pans electric pianos slightly to one side to give them an appropriate width. Of course, the monaural distance cues stated above can be created by equalization. Near sources have more bass frequencies, distant sources have attenuated treble frequencies. Kaiser (2012a) likes to use autopan and dynamic delays to give the sources some motion and make the sound more vivid. Complementary equalizers improve the source separation. He also believes to control the vertical dimension by equalization and reverb.[85] The mixing engineer Mozart (2015) agrees that audio engineers can create a three-dimensional mix by the use of different reverbs and delays and groups.[86] Often the low frequency region of instrumental sounds stays mono whereas pseudostereo-effects, like stereo chorus, are applied on higher frequencies.[87] One reason to keep the left and the right channel in phase is that phase shifts may create wide regions where the superimposed waves create destructive interferences in the listening room. As a consequence, the sound is hollow. Furthermore, when summing up the loudspeaker signals—as often done in mono mixdowns—the signals may also cancel out each other. Compatibility to mono is not the only restriction. In electronic dance music for example, it is common to use only subtle and static panning because loudspeakers in discotheques and other party locations may have a rather wide distribution of loudspeakers which comes not even close to the standardized stereo triangle.[88]

Even though pseudostereo effects aim at increasing the auditory source width, they tend to affect other sound characteristics as well, such as loudness, roughness, sound color and the sharpness of transients. These psychological sound impressions are not orthogonal but somewhat related. Many audio engineers try to balance a mix in terms of three dimensions as illustrated in Fig. 2.9. The stereo panning largely defines the horizontal source angle. Volume and audio effects influence the perceived distance. Perceived width is naturally inherent in distance perception. A near instrument is imagined as being larger than an instrument that is far away. A piano with little reverberation sounds near. However, to sound large as well, the direct sound needs to be decorrelated. This is what the natural radiation characteristic of a piano does: it radiated slightly incoherent version in all directions. Together, effects applied on direct sound and the characteristics and intensity of the reverberation influence the perceived near-far dimension which contributes to the perception of size. Instruments with prominent high frequency content seem to lie on top of the mix: nothing stands in their way, as if they were placed above the rest of the instruments. Although this multidimensional idea is partly metaphoric, it also envisions the audio parameters mainly used to tune the dimensionality of a music mix. The similarity to the three dimensions in music according to Albert Wellek, Fig. 2.1, is obvious.

2.4 Space in Music Theory

Spatial thinking plays a major role in the perception of music as well as in the creation process. This discovery is reflected in manifold music theory approaches. They describe, analyze and interpret music in terms of spatial parameters, relations and organization. In "A generative Theory of Tonal Music" by the musicologist Fred

---

[85] See Kaiser (2012a) for these and other mixing techniques.

[86] See Mozart (2015), pp. 175 and p. 178.

[87] See Hamidovic (2012), p. 49.

[88] Cf. Owsinski (2014), p. 51. Details on stereo are given in this chapter.

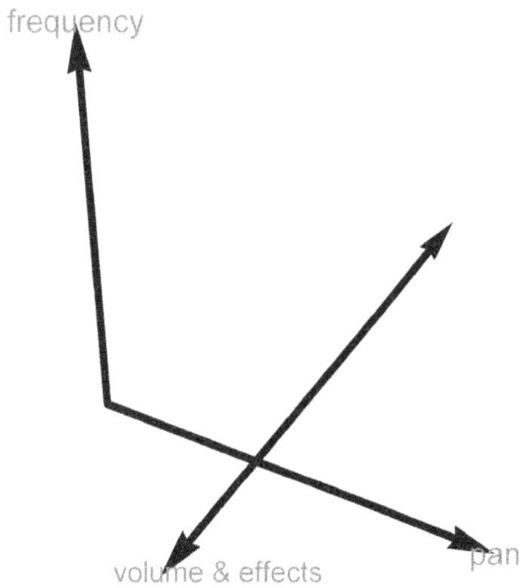

Fig. 2.9 Three dimensions in music mixes and the audio parameters to control them. After Edstrom (2011), p. 186

Lerdahl and the philosopher and linguist Ray Jackendoff states that musical input is psychoacoustically organized in a "musical surface", which has a "metric grid" or "metrical structure".[89] Expressing tonal hierarchies by geometric models to correlate them with intuitive musical distance led to the invention of many spatial models, such as the *circle of fifth*.[90] Lerdahl (2001) emphasizes the expression of tonal hierarchy by geometric models and spatial distance in both, music theory and music psychology.[91] There are two-dimensional pitch space models, like Leonhard Euler's *Tonnetz* or Gottfried Weber's approach, which use a similar structure as can be seen in Fig. 2.10. The Tonnetz and related models are also referred to as "two-dimensional lattice".[92] In fact, Weber's model combines Johann David Heinichen's *regional circle* and David Kellner's *double circle of fifths*, which can be seen in Fig. 2.11. Later, Moritz Wilhelm Drobisch suggested an extension of the circle of fifth by octave-representation on a vertical axis, leading to a higher dimensional helix structure. This idea was widened by the contemporary cognitive scientist Roger Shepard, who combined semitones and fifths cycles, yielding a double-helix-structure called *melodic map*, illustrated in Fig. 2.12. Here, the plane in the center divides the tones in to one group belonging to that scale and one not belonging to that scale. The melodic map illustrates an important observation that Roger Shepard made concerning pitch perception. It has a cyclic *chroma*-dimension which repeats every octave, as well as a rectilinear height dimension. For example the notes C1 and C2 have the same chroma but different

[89] See Lerdahl and Jackendoff (1983) and Lerdahl (2001), p. 3 and p. 8.

[90] See Lerdahl (2001), p. 42.

[91] Lerdahl (2001), p. 42.

[92] See e.g. Martins (2011), p. 126.

A♯	E♯	B♯	F×	C×	G×
F♯	C♯	G♯	D♯	A♯	E♯
D	A	E	B	F♯	C♯
B♭	F	C	G	D	A
G	D♭	A♭	E♭	B♭	F

d♯	F♯	f♯	A	a	C	c
g♯	B	b	D	d	F	f
c♯	E	e	G	g	B	b
f♯	A	a	C	d	E♭	e♭
b	D	d	F	f	A♭	a♭
e	G	g	B♭	b♭	D♭	d♭

Fig. 2.10 Two-dimensional models of tonal hierarchy. Left: Euler's "Tonnetz" (1739); a primitive representation of tonal hierarchy, representing degree of tonal relationship by proximity. Right: A more advanced model by Weber (1821–24), considering also parallel keys. After Lerdahl (2001), p. 43 and 44

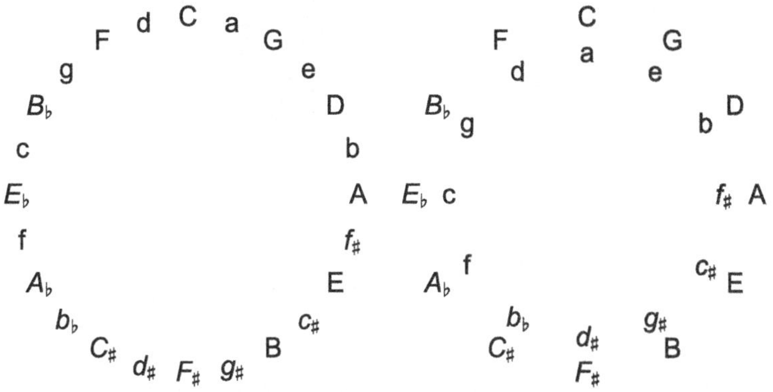

Fig. 2.11 Circular models of tonal hierarchy. Left: "Regional circle" by Heinichen (1728), right: "double circle of fifths" by Kellner (1737), adjusting distances between parallel keys. After Lerdahl (2001), p. 43

height.[93] Shepard even goes some steps further with his pitch model. He wraps the double helix around a torus and then adds a new height dimension so that the double helix is wrapped around a helical cylinder. This yields a five-dimensional map for pitch relations. These concepts can be found in Shepard (1982). Psychologist Carol Krumhansl and associates introduced a model based on cognitive proximity of "pitch classes, chords, and regions in a relation to an introduced tonic".[94] This model is derived from non-metric multidimensional scaling of similarity judgment of tones to an introduced tonic.[95] The resulting model is also illustrated in Fig. 2.12. All models are explained in detail and illustrated in Lerdahl (2001), pp. 42ff.

Ideas of nineteenth- and twentieth-century harmonic theorists that extend the idea of a Tonnetz to describe a framework of tonal relations, transpositions, voice leading

[93] See Shepard (1964), Burns (1981), Ziemer et al. (2018), Leman (1995), pp. 23ff and Sect. 4.5.4.1 for a detailed description of pitch perception and its components height and chroma.

[94] See e.g. Lerdahl (2001), p. 45 and Krumhansl et al. (1982).

[95] See e.g. Deutsch (1985), p. 138.

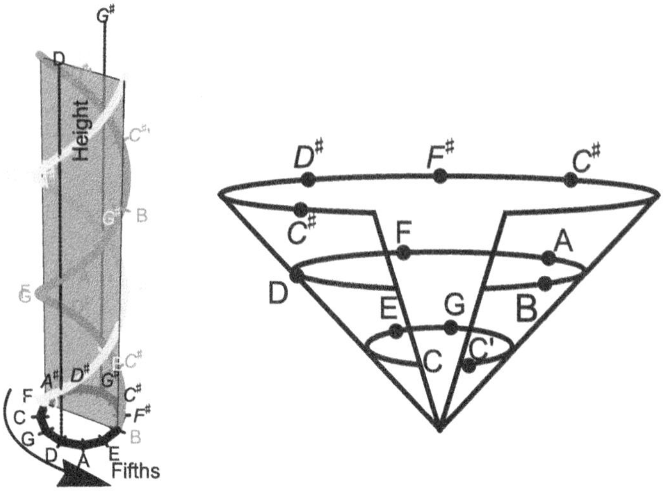

Fig. 2.12 Left: Shepard's "melodic map" (1982), extending Drobisch's helix representation (1855) to a double helix to include semitone relationships. Right: model of cognitive proximity by Krumhansl (1983), p. 40. After Lerdahl (2001), p. 44 and 46, Shepard (1982), p. 362 and Krumhansl et al. (1982)

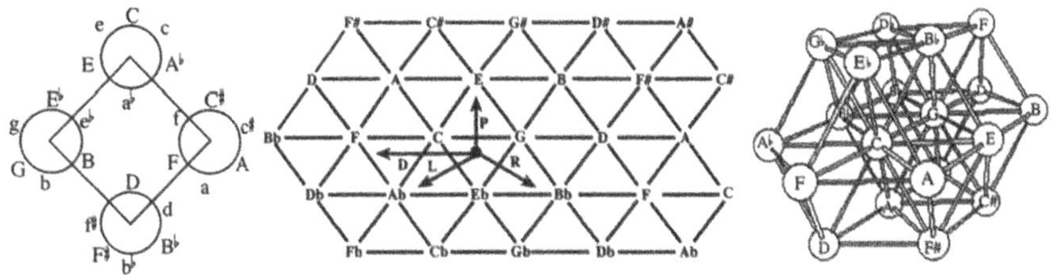

Fig. 2.13 Left: Richard Cohn's hyper-hexatonic space, center: Brian Hayer's table of tonal relations or Tonnetz, Right: A region within a three-dimensional Tonnetz with different intervals (4, 7 and 10 semitones) per step along each axis. From Cohn (1998), p. 172 and p. 175, and from Gollin (1998), p. 198, with friendly permissions by Richard Cohn and by Edward Gollin

etc., are summarized under the term "neo-Riemann theory".[96] Figure 2.13 illustrates three multi-dimensional concepts as part of the neo-Riemann theory.

In current music analyses of contemporary compositions there is still a notable number of spatial approaches, like the "voice leading space", "triadic space", "interactive trichord space", "transformational space" and many more.[97] A logical continu-

[96]See e.g. Cohn (1998) and Nolan (2003), named after the musicologist Hugo Riemann, not the mathematician Bernhard Riemann.

[97]See Cohn (2003), Cook (2009), analyzing works of the contemporary British Composer Gavin Bryars, Lind (2009), analyzing a piano work by the Canadian 20th century composer Clermont Pépin, and Roeder (2009), analyzing a string quartet of the contemporary English composer Thomas Adès.

ation of this geometric thinking in music analysis are the myriads of multidimensional representations of musical parameters in computer based music information retrieval approaches which will be outlined in the subsequent section.

2.5 Space in Music Information Retrieval

Music Information Retrieval (MIR) comprises computational analyses and interpretation of music. Tasks in this field can be beat tracking, tempo estimation, melody recognition, automatic music transcription, genre classification, source separation, lyrics recognition, computational auditory scene analysis and many more. Often such tasks are performed blind, i.e., without prior knowledge about any parameters, like artist, instrumentation, sheet music, recording setup. The analysis may be content-based. This means that audio files or scores are analyzed. Other approaches do not look at the musical piece itself. They may be based on user-generated tags, access patterns or purchase behavior.

Content-based analyses start with feature extraction of musical pieces. Audio files are often Pulse Code Modulated (PCM-) files, i.e. vectors of discrete digits, each representing a relative sound pressure at one point in time. They are represented as two-dimensional spaces as illustrated in Fig. 2.14. Not many standard works about the field of MIR exist, so an overview about works in that field is given, based on some specific studies and overview articles and book chapters.[98]

Low-level features can be directly extracted from audio files. For example plotting the temporal course against its derivative yields a phase space diagram from which the entropy can be interpreted. The noisier the signal, the more irregular the phase space plot. Periodic trajectories denote periodic oscillations, as demonstrated in Fig. 2.15.

This and other operations are usually not applied to a whole musical piece but to shorter overlapping time windows $\gamma\left(t-\tau\right)$.

$$P\left(t,\omega\right)=\mathrm{DFT}p\left(t\right)\gamma\left(t-\tau\right) \tag{2.1}$$

Here, $P\left(t,\omega\right)$ is the spectrum in time-frequency domain, transformed via discrete Fourier transforms DFT from the time windows of the signal in time domain $p\left(t\right)$.[99] This transformation yields a spectrogram. An illustration of a spectrogram is given in Fig. 2.16. After the transformation to time-frequency domain another low-level feature can be extracted: The audio spectrum centroid (ASC) of any time frame. It is

[98]Mainly Brandenburg et al. (2009), Wang and Plumbley (2005), Park et al. (2011), Cobos et al. (2011), Gärtner (2011) and Lee et al. (2011).

[99]The Fourier transform is explained in more detail in Sect. 5.1.3.

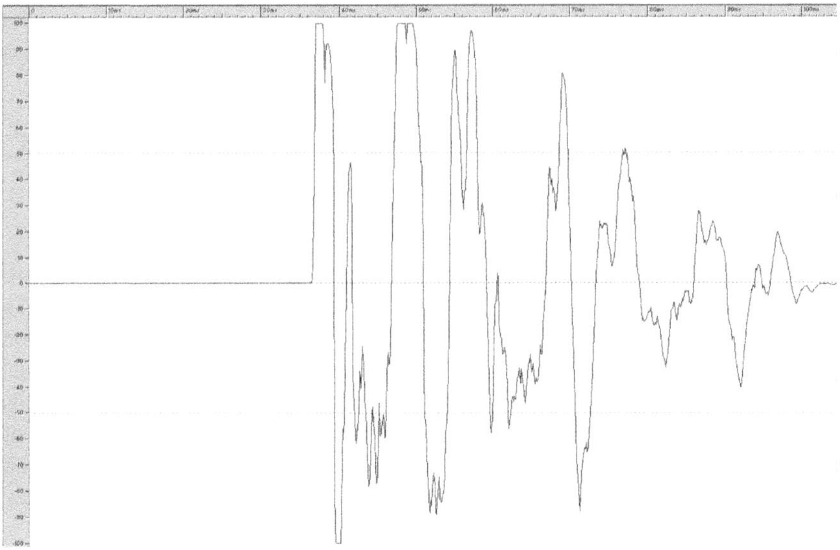

Fig. 2.14 Typical two-dimensional representation of a PCM-file. The horizontal dimension represents the time, the vertical dimension the relative sound pressure

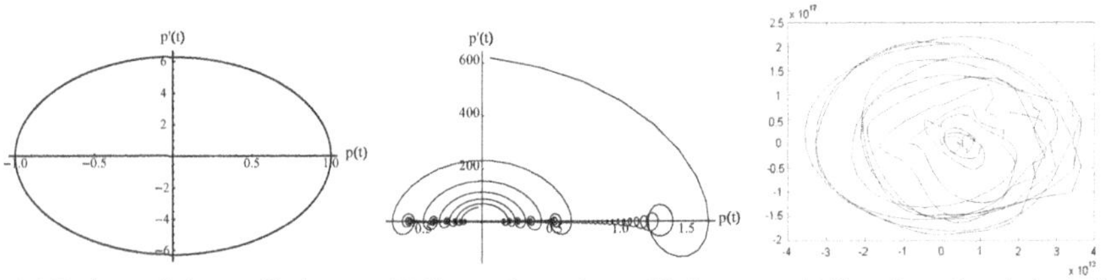

(a) *Undamped sine oscillation* (b) *Damped complex oscillation* (c) *Transient of a tubular bell sound*

Fig. 2.15 Phase space plots of a undamped sine (left), damped complex sound (center) and the first 20 ms of a tubular bell sound (right)

the center of gravity and can be calculated for the whole spectrum and for narrower frequency regions to describe the spectrum of an audio signal:

$$\text{ASC} = \frac{\int_{-\tau}^{\tau} t \; p(t) \, \mathrm{d}t}{\int_{-\tau}^{\tau} p(t) \, \mathrm{d}t} \tag{2.2}$$

The spectral centroid is assumed to be closely related to auditory brightness perception, which is an important part of timbre.[100] Other low-level features are the

[100] See, e.g., Ziemer et al. (2016), Donnadieu (2007), Bader (2013), pp. 352f, Troiviainen (1997), and Brandenburg et al. (2009), p. 359.

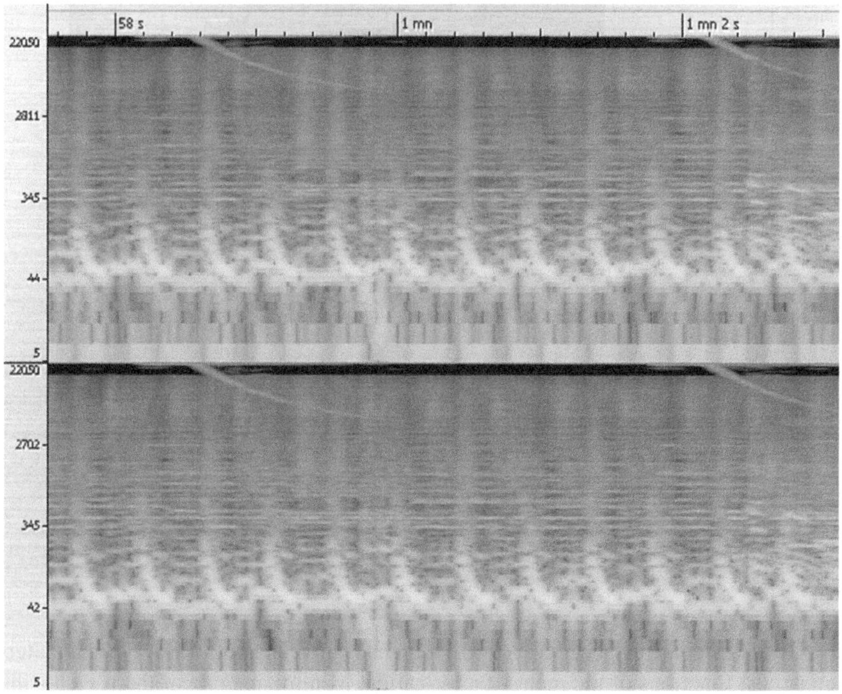

Fig. 2.16 Spectrogram of a dance music track excerpt. The abscissa is the time dimension, the ordinate is a logarithmic frequency scale and the pressure amplitude is coded by brightness from −96 dB (black) to 0 dB (white) relative to the highest possible amplitude of 2^{16} in a PCM file with a sample depth of 16 bits. The repetitive pattern comes from the 4-on-the-floor-beat and the resonance filter in the high frequency region looks like a falling star

spectral rolloff, *flatness* and *flux*, and the *zero crossing rate*.[101] These features do not have a semantic meaning for listeners and they are at best fairly related to auditory perception. Still, they are often used as a basis for genre recognition and music recommendation systems.[102]

Some approaches for melody and chord recognition start by approximating the spectrogram by non-negative matrix factorization

$$\mathbf{P} \approx \mathbf{WH}. \tag{2.3}$$

Here, **P** is the matrix containing the discrete values of the signal in time-frequency domain $P(t, \omega)$. **W** and **H** are non-negative matrices which are chosen to approximate P with minimum reconstruction error.

---

[101]These features are described on more detail in Tzanetakis and Cook (2002) and in Guaus (2009), pp. 72ff.

[102]See e.g. Tzanetakis and Cook (2002), Baniya et al. (2014), Yaslan and Cataltepe (2006), Bogdanov et al. (2010), Guaus (2009), Ziemer et al. (2016).

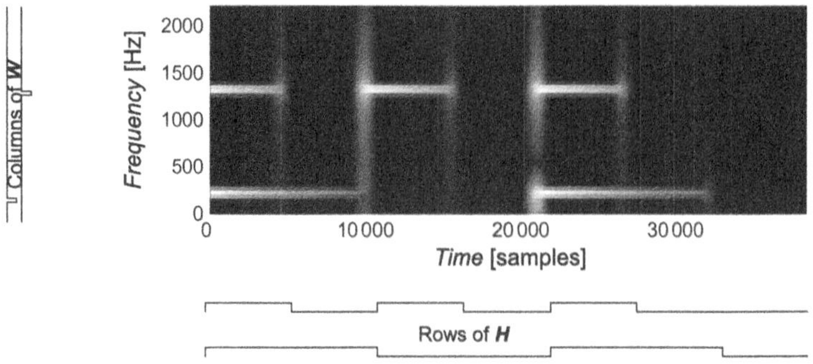

Fig. 2.17 Non-negative matrix factorization of an artificial signal, separating two frequencies. After Wang and Plumbley (2005), p. 2

$$\begin{aligned} C_1 &= \min ||\mathbf{P} - \mathbf{WH}||_2 \\ C_2 &= \min \sum_{m,n} \left(\mathbf{P}_{m,n} - (\mathbf{WH})_{m,n}\right)^2 \end{aligned} \tag{2.4}$$

C_1 is the Euclidean distance, C_2 is the divergence between **P** and **WH**. Both are alternative cost functions of the matrix factorization. Minimizing one of them yields the optimized factorization, i.e. the minimum reconstruction error. Figure 2.17 illustrates such an optimized matrix factorization. It can be used to separate two frequencies from a spectrogram. This can serve as a first step towards pitch recognition. If several pitches are found at the same time this can help to recognize chords. Finding several successive pitches is a first step towards melody recognition. Tracking multiple pitches over time can serve for advanced tasks, like key and mode recognition, source separation, and recognition of melodic lines in polyphonic pieces. Note, however, that the output of the non-negative matrix factorization is not sufficient for such tasks. They require additional audio analyses or meta data to gain knowledge about scale or the overtone series of inharmonic instruments. Without such information music in the Indian shruti scale could mistakenly be transferred to Western diatonic scale and loose a lot of the original information. The inharmonic overtone series of bells could be mistaken for multiple pitches and yield chords and melodies that deviate a lot from the original scores.

Therefore, it may be beneficial for several tasks if they were musically informed. Some mid-level features already include musical knowledge. Yet, no semantic meaning can be derived from the extracted feature. An example for mid-level features are chroma based histograms like the chord histogram in Fig. 2.18. It leverages low-level feature extraction, like pitch extraction and chord recognition. These are combined with musical knowledge, in this example the diatonic scale. Together, this informed feature extraction can serve for key and mode recognition. From the example one could assume that the piece is in C major scale. However, it could also be a piece in

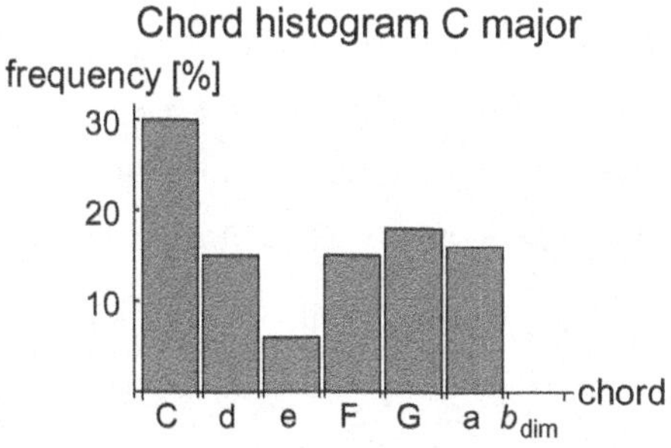

Fig. 2.18 Chord histograms of a musical piece in C major scale

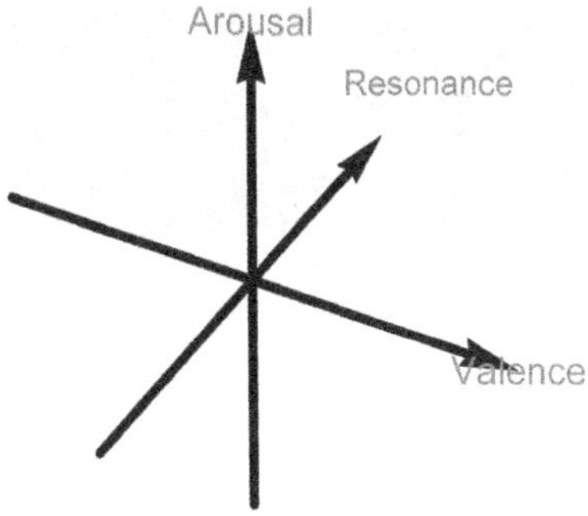

Fig. 2.19 Psychological mood space, a model to arrange emotions in two-dimensional space

the ionian mode or an enharmonic equivalent, like B ♯ major. Approaches exist to leverage mid-level features for tasks such as genre recognition.[103]

High-level features have a direct semantic meaning based on musical knowledge. As they are abstract rather than directly measurable, MIR approaches tend to try to derive them from lower-level features. Examples for high-level features include genre, musical structure and instrumentation. High-level tasks include musical audio stream separation, polyphonic music transcription, reverberation time estimation, tempo estimation and source separation.[104]

Many attempts have been made to describe the mood of a musical piece in the two dimensions *valence* and *arousal* from the psychological model of emotions which is illustrated in Fig. 2.19.[105] Features, such as tempo, syncopation, energy density or entropy—as derived e.g. from spectrograms or a phase space—can be considered to indicate arousal. Mode and timbre, as derived, e.g., from the chord histogram and the ASC, may indicate valence. Sometimes, a third dimension called *resonance* is added and related to low level features.[106]

[103] See e.g., Rosner and Kostek (2018).

[104] See Wang and Plumbley (2005), Park et al. (2011), Cobos et al. (2011), Gärtner (2011) and Lee et al. (2011).

[105] Cf. Russell (1980), Myers (2008), pp. 570ff, Frenzel et al. (2009), Nagel et al. (2007), and Deng and Leung (2012) for more details on the model and applications in music and emotion analysis.

[106] See e.g. Deng and Leung (2012).

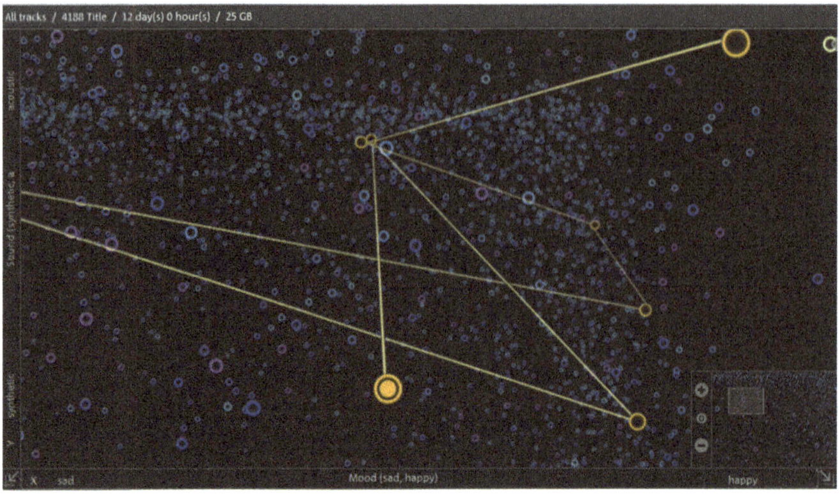

Fig. 2.20 Representation of similarity of musical pieces in a three-dimensional semantic space with the dimensions happy-sad, acoustic-synthetic, calm-aggressive integrated in the music player and -recommender mufin. From Magix AG (2012), with the permission of Magix Software GmbH

Such a representation is called "joint semantic embedding space"[107] and uses a low-dimensional embedding space for annotating, retrieving and suggesting music with software via semantic interpretations. It was adopted in the *outa space* from the commercial *mufin music player* and in the *3D music universe* in *MP3 deluxe*, both by MAGIX Software GmbH. Figure 2.20 is a screenshot of the mufin software. There are attempts for music indexing and browsing, playlist generation and music recommendation based on retrieved mood or genre similarity.[108] Here, the common problem is that low level features are barely related to human sound perception nor to auditory scene analysis principles or our concepts of musical organization.

Genre is a good example of a high-level feature. It has a direct meaning that can be understood by a listener. Unfortunately, there are no ultimate definitions of genres that everybody would agree on. Genre definitions may depend on cultural and educational background, music scenes and subscenes, location and personal experience. Furthermore, genres could be considered as typologies rather than as classes. Typologies allow for overlaps, whereas classes are exclusive. One song could only belong to one genre class but to several genre type. This lack of a ground truth is the crux of the matter. Many researchers in the field of MIR have defined their own ground truth and tuned their algorithms to replicate it. For example they may assign one out of 10 genre labels to a data set consisting of 1,000 songs. Then they try to

[107] See Weston et al. (2011).

[108] See e.g. Rauber et al. (2002), Gartner et al. (2007), Logan (2002) and many more for music indexing, exploration and browsing and for playlist generation, and Deng and Leung (2012), Shao et al. (2009), Bogdanov et al. (2010), Logan (2004) etc. for content-based music recommendation. An overview can be found in Òscar Celma (2010).

replicate these assigned genres by means of audio analysis and machine learning. For example they use 500 songs as training data and validate their approach on the other 500 songs. However, it is doubtful that the resulting method will succeed with any other data than the used data set. Such a method is referred to as a "horse" that does not solve the problem of genre recognition but simply creates a desired output from a given input.[109] Such a horse finds statistically significant relationships withing the given data set, i.e., between extracted feature magnitudes and the genre label. But if they are not of causal nature, these relationships will probably not be found in other data sets. They are irrelevant to the given task. It could be demonstrated that many MIR algorithms failed to work on their very own training data it the data was transformed in practically inaudible ways.

A solution could be do concentrate on more meaningful features. A number of psychoacoustic models exists, imitating the traveling wave in the cochlea, the critical bandwidth and the resulting neural excitation pattern. This way, sound characteristics, like *loudness*, *roughness*, *sharpness* and *tonalness* can be retrieved at least for artificial test signals.[110] To date only a few studies leverage auditory models to explain inter-subjective music judgments from psychoacoustic audio features.[111]

Due to the discrete character of digital data and the common use of vector representation of music in PCM files, it is reasonable to make a spatial representation of the data. Beyond that, many goals of MIR, like blind source separation and computational auditory scene analysis, are to retrieve spatial information from audio files. Other parameters to retrieve physical and semantic information are not of spatial character per se but are suitable for a spatial representation due to their multi dimensional nature and different degrees of relationships, e.g., using spatial representations to reveal relationships between sounds, musical pieces, genres etc. Music and space are not only closely related from a conceptual point of view as extensively discussed in this chapter. The origin of the auditory system as well its physiology already reveal a strong link between sound and space as will be discussed in the next chapter.

References

Albersheim G (1939) Zur Psychologie der Ton- und Klangeigenschaften (unter Berücksichtigung der 'Zweikomponententheorie' und der Vokalsystematik). Heitz & Co., Leipzig et al

Ando Y (2010) Auditory and visual sensation. Springer, New York. https://doi.org/10.1007/b13253

Aures W (1985a) Der sensorische Wohlklang als Funktion psychoakustischer Empfindungsgrößen (the sensory euphony as a function of auditory sensations). Acta Acust United Acust 58(5):282–290. https://www.ingentaconnect.com/content/dav/aaua/1985/00000058/00000005/art00006

[109] See Sturm (2014).

[110] See e.g. Zwicker (1958), Aures (1985b), Daniel and Weber (1997), Leman (2000), von Bismarck (1974), Aures (1985a), Aures (1985c). An extensive overview over psychoacoustic models can be found in Zwicker and Fastl (1999).

[111] See e.g. Leman et al. (2005), Rauber et al. (2002), Ziemer et al. (2016) and Panagakis et al. (2009). Some further studies are summarized and discussed in Richard et al. (2013).

Aures W (1985b) Ein Berechnungsverfahren der Rauhigkeit (a procedure for calculating auditory roughness). Acta Acust United Acust 58(5):268–281. https://www.ingentaconnect.com/content/dav/aaua/1985/00000058/00000005/art00005

Aures W (1985c) Berechnungsverfahren für den sensorischen Wohlklang beliebiger Schallsignale (a model for calculating the sensory euphony of various sounds). Acustica 59(2):130–141. https://www.ingentaconnect.com/content/dav/aaua/1985/00000059/00000002/art00008

Baalman M (2008) On Wave Field Synthesis and electro-acoustic music, with a particular focus on the reproduction of arbitrarily shaped sound sources. VDM, Saarbrücken

Bader R (2013) Nonlinearities and synchronization in musical acoustics and music psychology. Springer, Berlin. https://doi.org/10.1007/978-3-642-36098-5

Bader R (2014) Microphone array. In: Rossing TD (ed) Springer handbook of acoustics. Springer, Berlin, pp 1179–1207. https://doi.org/10.1007/978-1-4939-0755-7_29

Baniya BJ, Ghimire D, Lee J (2014) Automatic music genre classification using timbral texture and rhythmic content features. ICACT Trans Adv Commun Technol 3(3):434–443

Barthelmes Barbara (1986) Musik und Raum–ein Konzept der Avantgarde. In: Bräm Thüring (ed) Musik und Raum. Eine Sammlung von Beiträgen aus historischer und künstlerischer Sicht zur Bedeutung des Begriffes <Raum> als Klangträger für die Musik. GS-Verlag, Basel, pp 75–89

Blauert J (1974) Räumliches Hören. Hirzel, Stuttgart

Blauert J (1997) Spatial hearing. The pychophysics of human sound source localization, revised edn. MIT Press, Cambridge

Blumlein AD (1933) Improvements in and relating to sound-transmission, sound-recording and sound-reproducing systems. https://worldwide.espacenet.com/publicationDetails/biblio?II=10&ND=3&adjacent=true&locale=en_EP&FT=D&date=19330614&CC=GB&NR=394325A&KC=A

Bockhoff M (2007) Soundscapes in der abendländischen Malerei. In: Fortschritte der Akustik—DAGA'07. Stuttgart, pp 857–858

Bogdanov D, Haro M, Fuhrmann F, Gómez E, Herrera P (2010) Content-based music recommendation based on user preference example. In: WOMRAD 2010 workshop on music recommendation and discovery, colocated with ACM RecSys

Boulez P, Gerzso A (1988) Computer als Orchesterinstrumente. In: Winkler K (ed) Die Physik der Musikinstrumente. Spektrum der Wissenschaft, Heidelberg, pp 178–184

Brandenburg K, Dittmar C, Gruhne M, Abeßer J, Lukashevich H, Dunker P, Gärtner D, Wolter K, Grossmann H (2009) Music search and recommendation. In: Furht B (ed) Handbook of multimedia for digital entertainment and arts, chapter 16. Springer, New York, pp 349–384. https://doi.org/10.1007/978-0-387-89024-1_16

Bräm T (1986) Der Raum als Klangträger. Gedanken zur Entstehung und zum Inhalt dieses Buches. In: Bräm T (ed) Musik und Raum. Eine Sammlung von Beiträgen aus historischer und künstlerischer Sicht zur Bedeutung des Begriffes <Raum> als Klangträger für die Musik. GS-Verlag, Basel, pp 6–14

Burns EM (1981) Circularity in relative pitch judgements for inharmonic complex tones: the shepard demonstration revisited, again. Percept Psychophys 30(5):467–472. https://doi.org/10.3758/bf03204843

Cabrera A (2011) Pseudo-stereo techniques. C sound implementations. Csound J 14. http://csoundjournal.com/

Celma Ò (2010) Music recommendation and discovery. Springer, Berlin. https://doi.org/10.1007/978-3-642-13287-2

Cobos M, Vera-Candeas P, Carabias-Orti JJ, Ruiz-Reyes N, López JJ (2011) Blind estimation of reverberation time from monophonic instrument recording based on non-negative matrix factorization. In: Audio engineering society conference: 42nd international conference: semantic audio, pp 69–78, Jul 2011

Cohn R (1998) Introduction to neo-riemannian theory. A survey and a historical perspective. J Music Theory, 42(2):167–180. https://doi.org/10.2307/843871

Cohn R (2003) A tetrahedral graph of tetrachordal voice-leading space. Music Theory Online, 9(4). http://www.mtosmt.org/issues/mto.03.9.4/mto.03.9.4.cohn.pdf
Cook SA (2009) Moving through triadic space. an examination of bryars's seemingly haphazard chord progressions. Music Theory Online 14(1). http://www.mtosmt.org/issues/mto.09.15.1/mto.09.15.1.cook.html
Daniel P, Weber R (1997) Psychoacoustical roughness: implementation of an optimized model. Acta Acust United Acust 83(1):113–123. https://www.ingentaconnect.com/contentone/dav/aaua/1997/00000083/00000001/art00020
Deng JJ, Leung C (2012) Emotion-based music recommendation using audio features and user playlist. In: 2012 6th international conference on new trends in information science and service science and data mining (ISSDM), pp 796–801, Oct 2012
Deutsch D (1985) Verarbeitung und Repräsentation von Tonkombinationen. In: Bruhn H, Oerter R, Rösing H (eds) Musikpsychologie. Ein Handbuch in Schlüsselbegriffen. Urban & Schwarzenberg, Munich, pp 133–140
Donnadieu S (1997) Représentation mental du timbres des sons complexes et effects de contexte. PhD thesis, Université Paris V, Unpublished
Donnadieu S (2007) Mental representation of the timbre of complex sounds. In: Beauchamp JW (ed) Analysis, synthesis, and perception, chapter 8. Springer, New York, pp 271–319. https://doi.org/10.1007/978-0-387-32576-7_8
Edstrom B (2011) Recording on a budget. How to make great audio recordings without breaking the bank. Oxford University Press, Oxford, New York (NY)
Faller C (2005) Pseudostereophony revisited. In: Audio engineering society convention 118. Barcelona, p 5
Fähndrich W, Meyer T, Lichtenhahn E (2010) Music for spaces. http://www.musicforspaces.ch/en/F2.html. Accessed 14 Mar 2013
Forsyth M (1985) Buildings for music. The architect, the musician, and the listener from the seventeenth century to the prenent day. MIT Press, Cambridge. https://doi.org/10.2307/3105495
Fouad H (2004) Spatialization with stereo loudspeakers: understanding balance, panning, and distance attenuation. In: Greenbaum K, Barzel R (eds) Audio Anecdotes, vol II. A K Peters, Natick, pp 143–158
Frenzel AC, Götz T, Pekrun R (2009) Emotionen. In: Wild E, Möller J (eds) Pädagogische Psychologie. Springer, Berlin, pp 205–231. https://doi.org/10.1007/978-3-540-88573-3_9
Friedrich HJ (2008) Tontechnik für Mediengestalter. Töne hören—Technik verstehen—Medien gestalten. Springer, Berlin
Garner RW (1974) The processing of information and structure. Lawrence Erlbaum, New York
Gartner D, Kraft F, Schaaf T (2007) An adaptive distance measure for similarity based playlist generation. In: 2007 IEEE international conference on acoustics, speech and signal processing, vol 1, April 2007, pp I–229–I–232. https://doi.org/10.1109/ICASSP.2007.366658
Genuit K (2003) SoundScape—Eine Gefahr für Missverständnisse! In: Fortschritte der Akustik—DAGA'03. Aachen, pp 378–379
Godøy R-I (1997) Knowledge in music theory by shapes of musical objects and sound-producing actions. In: Leman M (ed) Music, gestalt, and computing. Springer, Berlin, pp 89–102. https://doi.org/10.1007/bfb0034109
Gollin E (1998) Some aspects of three-dimensional 'tonnetze'. J Music Theory 42(2):195–206. https://doi.org/10.2307/843873
Griebsch I (2000) Raum-Zeit-Aspekte beim Zustandekommen vermittelnder Dimensionen. In: Böhme T, Mehner K (eds) Zeit und Raum in Musik und Bildender Kunst. Böhlau, Cologne, pp 139–150
Gärtner D (2011) Tempo estimation from urban music using non-negative matrix factorization. In: Audio engineering society conference: 42nd international conference: semantic audio, Jul 2011, pp 208–215
Guaus E (2009) Audio content processing for automatic music genre classification: descriptors, databases, and classifiers. PhD thesis

Gutierrez-Parera P, López JJ (2016) Influence of the quality of consumer headphones in the perception of spatial audio 6:4. https://doi.org/10.3390/app6040117
Hamidovic E (2012) The systematic mixing guide. Systematic Productions, Melbourne
Hagens W (2018) Expo 1958 Philips pavilion. https://en.wikipedia.org/wiki/Philips_Pavilion#/media/File:Expo58_building_Philips.jpg
Kaiser C (2012a) 1001 mixing tipps. MITP, Heidelberg
Kaiser C (2012b) 1001 recording tipps. MITP, Heidelberg
Kaiser C (2013) 1001 mastering tipps. MITP, Heidelberg
Kirchmeyer H, Schmidt HW (1970) Aufbruch der jungen Musik. Von Webern bis Stockhausen, Gerig, Cologne
Krumhansl CL (1983) Perceptual structures for tonal music. Music Perception Interdiscip J 1(1):28–62
Krumhansl CL, Bharucha JJ, Kessler EJ (1982) Perceived harmonic structure of chords in three related musical keys. J Exp Psychol Human Percept Perform 8(1):24–36. https://doi.org/10.1037/0096-1523.8.1.24
Kurth E (1990) Musikpsychologie. G. Olms, Hildesheim, 2. nachdruck der ausgabe Berlin 1931 edition. https://doi.org/10.2307/932010
Lakatos S (2000) A common perceptual space for harmonic and percussive timbres. Percept Psychophys 62(7):1426–1439. https://doi.org/10.3758/bf03212144
Lee S, Park SH, Sung K-M (2011) A musical source separation system using a source-filter model and beta-divergence non-negative matrix factorization. In: Audio engineering society conference: 42nd international conference: semantic audio, Jul 2011, pp 216–220
Leman M (2000) Visualization and calculation of the roughness of acoustical musical signals using the synchronization index model (SIM). In: Proceedings of the COST G-6 conference on digital audio effects (DAFx-00). Verona, Dec 2000
Leman M (1995) Music and schema theory. Cognitive foundations of systematic musicology, Springer, Berlin
Leman M, Vermeulen V, De Voogdt L, Moelants D, Lesaffre M (2005) Prediction of musical affect using a combination of acoustic structural cues. J New Music Res 34(1):39–67. https://doi.org/10.1080/09298210500123978
Lerdahl F (2001) Tonal pitch space. Oxford University Press, Oxford. https://doi.org/10.1093/acprof:oso/9780195178296.001.0001
Lerdahl F, Jackendoff R (1983) A generative theory of tonal music. MIT Press, Cambridge
Levinit DJ (2004a) Instrument (and vocal) recording tips and tricks. In: Greenbaum K, Barzel R (eds) Audio anecdotes, vol I. A K Peters, Natick, pp 147–158
Levinit DJ (2004b) How recordings are made I: analog and digital tape-based recording. In: Greenbaum K, Barzel R (eds) Audio anecdotes, vol II. A K Peters, Natick, pp 3–14
Lewald J (2006) Auditives Orientieren im Raum uns seine Störung. In: Karnath H-O, Thier P (eds) Neuropsychologie. Springer, 2. aktualisierte und erweitere edition, pp 185–196. https://doi.org/10.1007/3-540-28449-4_18
Lind S (2009) An interactive trichord space based on measures 18–23 of clermont pépin's toccate no. 3. Music Theory Online 15(1). http://www.mtosmt.org/issues/mto.09.15.1/mto.09.15.1.lind.html
Logan B (2002) Content-based playlist generation: exploratory experiments. In: Proceedings of the 2nd international conference on music information retrieval, Paris, p 10
Logan B (2004) Music recommendation from song sets. In: Proceedings of the 5th international conference on music information retrieval, Barcelona, p 10
Maempel H-J (2008) Medien und Klangästhetik. In: Bruhn H, Kopiez R, Lehmann AC (eds) Musikpsychologie. Das neue Handbuch, Rowohlt, Reinbek bei Hamburg, pp 231–252
Magix AG (2012) Mufin vision in mufin player. Your 3d music collection. http://www.mufin.com/us/3d-music/. Accessed 17 May 2013
De Man B, Reiss JD (2017) The mix evaluation data set. In: Proceedings of the 20th international conference on digital audio effects. Edinburgh, Sep 2017, pp 436–442

Martins JO (2011) Interval cycles, affinity spaces, and transpositional networks. In: Agon C, Andreatta M, Assayag G, Amiot E, Bresson J, Mandereau J (eds) Mathematics and computation in music. Third international conference, MCM 2011 Paris, France, 15–17 June 2011, Proceedings. Berlin, Heidelberg. https://doi.org/10.1007/978-3-642-21590-2_10

Mazolla G (1990) Geometrie der Töne. Elemente der mathematischen Musiktheorie, Birkhäuser, Basel

Meyer J (1986) Gedanken zu den originalen Konzertsälen Joseph Haydens. In: Bräm T (ed) Musik und Raum. Eine Sammlung von Beiträgen aus historischer und künstlerischer Sicht zur Bedeutung des Begriffes <Raum> als Klangträger für die Musik. GS-Verlag, Basel, pp 26–38

Meyer J (2003) Wechselbeziehungen zwischen Orgel und Raumakustik. In: Fortschritte der Akustik—DAGA'03. Aachen, pp 518–519

Mores R (2018) Music studio technology. Springer, Berlin, pp 221–258. https://doi.org/10.1007/978-3-662-55004-5_12

de la Motte-Haber H (2000) Raum-Zeit als musikalische Dimension. In: Böhme T, Mehner K (eds) Zeit und Raum in Musik und Bildender Kunst. Böhlau, Cologne, pp 31–37

Mozart M (2015) Your mix sucks. The complete mix methodology from DAW preparation to delivery. Mozart & Friends Limited, Gießen

Myers DG (2008) Psychologie. 2. erweiterte und aktualisierte edition. Springer, Berlin. https://doi.org/10.1007/978-3-642-40782-6

Nagel F, Kopiez R, Grewe O, Altenmüller E (2007) EMuJoy: software for continuous measurement of perceived emotions in music. Behav Res Methods 39(2):283–290. https://doi.org/10.3758/BF03193159

Nauck G (1997) Musik im Raum—Raum in der Musik. Ein Beitrag zur Geschichte der seriellen Musik. Franz Steiner Steiner, Stuttgart. https://doi.org/10.2307/3686862

Nolan C (2003) Combinatorial space in nineteenth- and early twentieth-century music theory. Music Theory Spectr 25(2). https://doi.org/10.1525/mts.2003.25.2.205

Otondo F (2008) Contemporary trends in the use of space in electroacoustic music. Organ Sound 13(1):77–81. https://doi.org/10.1017/s1355771808000095

Owsinski B (2014) The mixing engineer's handbook, 3rd edn. Corse Technology PTR, Boston

Panagakis Y (2009) Constantine Kotropoulos, and Gonzalo R. Arce. Music genre classification via sparse representations of auditory temporal modulations. In: Proceedings of the 17th European signal processing conference, Glasgow, Aug 2009

Park SH, Lee S, Sung K-M (2011) Polyphonic music transcription using weighted cqt and non-negative matrix factorization. In: Audio engineering society conference: 42nd international conference: semantic audio, pp 39–43, Jul 2011

Pulkki V (2004) Spatialization with multiple speakers. In: Greenbaum K, Barzel R (eds) Audio anecdotes, vol II. A K Peters, Natick, pp 159–171

Rauber A, Pampalk E, Merkl D (2002) Using psycho-acoustic models and self-organizing maps to create a hierarchical structuring of music by sound similarity. In: Proceedings of the 3rd international symposium on music information retrieval, Paris

Richard G, Sundaram S, Narayanan S (2013) An overview on perceptually motivated audio indexing and classification. Proc IEEE 101(9):1939–1954. https://doi.org/10.1109/JPROC.2013.2251591

Roeder J (2009) A transformational space structuring the counterpoint in adès's 'auf dem wasser zu singen'. Music Theory Online 15(1). http://www.mtosmt.org/issues/mto.09.15.1/mto.09.15.1.roeder_space.html

Rogers SE (2004) The art and craft of song mixing. In: Greenbaum K, Barzel R (eds) Audio anecdotes, vol II. A K Peters, Natick, pp 29–38

Rosner A, Kostek B (2018) Automatic music genre classification based on musical instrument track separation. J Intell Inf Syst 50(2):363–384. https://doi.org/10.1007/s10844-017-0464-5

Russell JA (1980) A circumplex model of affect. J Pers Soc Psychol 39(6):1161–1178

Révész G (1937) Gibt es einen Hörraum? Theoretisches und Experimentelles zur Frage eines autochthonen Schallraumes nebst einer theorie der Lokalisation. Acta Psychol 3:137–192

Schäfer S (2000) TopoPhonien ein künstlerisches Entwicklungsvorhaben von Sabine Schäfer und Sukandar Kartadinata. In: Anders B, Stange-Elbe J, (eds) Musik im virtuellen Raum. Rasch, Osnabrück, pp 247–256

Schneider A (1989) On concepts of 'tonal space' and the dimensions of sound. In: Sprintge R, Droh R (eds) MusicMedicine. International society for music in medicine IV, international musicmedicine symposium 25–29 Oct 1989. California

Shao B, Wang D, Li T, Ogihara M (2009) Music recommendation based on acoustic features and user access patterns. IEEE Trans Audio, Speech, Lang Process 17(8):1602–1611. https://doi.org/10.1109/tasl.2009.2020893

Shepard RN (1964) Circularity in judgments of relative pitch. J Acoust Soc Am 36(12):2346–2353. https://doi.org/10.1121/1.1919362

Shepard RN (1982) Structural representations of musical pitch. In: Deutsch D, (ed) The psychology of music. Elsevier, pp 343–390. https://doi.org/10.1016/b978-0-12-213562-0.50015-2

Stevenson MH (1968) Acoustical features of a new television studio. In: International acoustics symposium. Australian Acoustical Society, Sydney, pp K1–K8, Sep 1968

Stoianova I (1989) Textur/Klangfarbe und Raum. Zum Problem der Formbildung in der Musik des 20. Jahrhunderts. In: Morawska-Büngeler M (ed) Musik und Raum. Vier Kongressbeiträge und ein Seminarbericht. Mainz, pp 40–59

Sturm BL (2014) A simple method to determine if a music information retrieval system is a 'horse'. IEEE Trans Multimed 16(6):1636–1644. https://doi.org/10.1109/TMM.2014.2330697

Thom R (1983) Paraboles et catastrophe. Flammarion, Paris

Troiviainen P (1997) Optimizing self-organizing timbre maps. two approaches. In: Leman M (ed) Music, gestalt, and computing. Springer, Berlin, pp 337–350. https://doi.org/10.1007/bfb0034124

Tzanetakis G, Cook P (2002) Musical genre classification of audio signals. IEEE Trans Speech Audio Process 10(5):293–302. https://doi.org/10.1109/TSA.2002.800560

von Bismarck G (1974) Sharpness as an attribute of the timbre of steady sounds. Acustica 30:159–172. https://www.ingentaconnect.com/contentone/dav/aaua/1974/00000030/00000003/art00006

von Ehrenfels C (1890) Über Gestaltqualitäten. Vierteljahrsschrift für wissenschaftliche Philosophie 14:249–292. https://doi.org/10.1515/9783035601602.106

Wang B, Plumbley MD (2005) Musical audio stream separation by non-negative matrix factorization. In: DMRN summer conference, p 7

Weinzierl S (2008) Virtuelle Akustik und Klangkunst. In: Fortschritte der Akustik—DAGA'08. Dresden, pp 37–38, Mar 2008. http://pub.dega-akustik.de/DAGA_1999-2008/data/articles/003709.pdf

Weston J, Samy B, Hamel P (2011) Multi-tasking with joint semantic spaces for large-scale music annotation and retrieval. J New Music Res 40(4):337–348. https://doi.org/10.1080/09298215.2011.603834

Wilson A, Fazenda B (2015) 101 mixes: a statistical analysis of mix-variation in a dataset of multi-track music mixes. In: Audio engineering society convention 139, Paper no. 9398, Oct 2015. http://www.aes.org/e-lib/browse.cfm?elib=17955

Wilson A, Fazenda BM (2016) Perception of audio quality in productions of popular music. J Audio Eng Soc 64(1/2):23–34. http://www.aes.org/e-lib/browse.cfm?elib=18102

Yaslan Y, Cataltepe Z (2006). Audio music genre classification using different classifiers and feature selection methods. In: 18th international conference on pattern recognition (ICPR'06), vol 2. Hong Kong, pp 573–576, Aug 2006. https://doi.org/10.1109/ICPR.2006.282

Zagorski-Thomas S (2014) The musicology of record production. Cambridge University Press, Cambridge. https://doi.org/10.1017/cbo9781139871846

Zbikowski LM (2002) Conceptual music. Cognitive structure, theory, and analysis. Oxford University Press, New York

Ziemer T (2017) Source width in music production. methods in stereo, ambisonics, and wave field synthesis. In: Schneider A (ed) Studies in musical acoustics and psychoacoustics, current research in systematic musicology chapter 10, vol 4. Springer, Cham, pp 299–340. https://doi.org/10.1007/978-3-319-47292-8_10

Ziemer T, Yu Y, Tang S (2016) Using psychoacoustic models for sound analysis in music. In: Majumder P, Mitra M, Sankhavara J, Mehta P, (eds) Proceedings of the 8th annual meeting of the forum on information retrieval evaluation, FIRE'16. ACM, New York, NY, USA, pp 1–7, Dec 2016. https://doi.org/10.1145/3015157.3015158

Ziemer T, Schultheis H, Black D, Kikinis R (2018) Psychoacoustical interactive sonification for short range navigation. Acta Acust United Acust 104(6):1075–1093. https://doi.org/10.3813/AAA.919273

Zwicker E (1958) Über psychologische und methodische Grundlagen der Lautheit. Acustica 8(4):237–258. https://www.ingentaconnect.com/contentone/dav/aaua/1958/00000008/a00104s1/art00002

Zwicker E, Fastl H (1999) Psychoacoustics. Facts and models. Second updated edition. Springer, Berlin. https://doi.org/10.1007/978-3-662-09562-1

Chapter 3
Biology of the Auditory System

The evolutionary origin of the auditory system as well as the bilateral nature of the ear confirm that hearing is in many aspects related to space. Its original function is associated with spatial orientation. The auditory system processes sensory input so that a mental representation of the physical outside space can be created at higher stages of the human brain. Such a mental map represents especially spatial locations, extents and relations of sound sources and the environments. In this chapter the functional evolution of the auditory system is traced from sensory hair cells in earliest vertebrates over the lateral line system in fish to the human ear.

3.1 Functional Evolution of the Auditory System

Over 520 million years ago ancestral chordates originated in the oceans. Within a hundred million years the first sensory hair cells arose in these chordates. From these the vertebrate ear evolved about 370 million years ago. Still, they were the last paired sensory receptors to arise.[1] It is under debate whether the auditory system has evolved or derived from the lateral line organ or not.[2] Without any doubt the auditory system is closely related to the mechanosensory lateral line system,[3] which is known to exist in the earliest vertebrates and presumed to be the evolutionary earlier.[4] In fish and amphibians auditory and lateral line system have a multi modal overlap and form the

[1]See e.g. Fritzsch et al. (2010), Mallatt (2009), p. 1201, Manley and Clack (2004), p. 8 and Clack (1993), p. 392.

[2]For a discussions about the "octavolateralis hypothesis" and "acousticolateralis hypothesis", see e.g. Popper et al. (1992), Coombs et al. (1992), Popper and Platt (1993), Ryugo (2011), p. 8, Will and Fritsch (1988), p. 160, Kalmijn (1989), pp. 201f, Jørgensen (1989), p. 115 and pp. 132ff, Manley and Clack (2004), p. 7 and Webb et al. (2008), p. 145.

[3]See Kalmijn (1989), p. 187, Braun and Grande (2008), and Manley and Clack (2004), p. 15.

[4]See e.g. Coombs et al. (1992), p. 267 and Gans (1992), p. 7.

© Springer Nature Switzerland AG 2020

T. Ziemer, *Psychoacoustic Music Sound Field Synthesis*, Current Research in Systematic Musicology 7, https://doi.org/10.1007/978-3-030-23033-3_3

octavolateralis system.[5] There are functional similarities between hair cells of both sensory organs as well as mechanical linkages between ear and lateral line.[6] The lateral line organ of fish is discussed next,[7] followed by a description of the auditory system of fish and finally the human auditory system.[8]

3.1.1 Lateral Line System

The lateral line system can be found in the head, trunk and tail of fishes and amphibians. The end organ of the lateral line system is the neuromast. It is a patch of cells with bundles of hair cells in the center. These hair cells are displacement receptors which shear due to movement of water particles relative to the body, sensing velocity of hydrodynamic flow. The sensory hair cells are surrounded by nonsensory support cells. Stereociles "anatomically polarize" these bundles which means that they have a maximum response along one axis.[9] Figure 3.1 shows two regions of hair cell bundles with different polarizations. Neuromasts can be found in pored lateral line canals and on the skin's surface, sensing accelerations and velocity of water flow, created or reflected by other animals or obstacles. In fact, since fishes have about the same density as their surrounding water, they are accelerated themselves as water displacements arrive. It is the acceleration gradient along their body from which they derive the source of particle displacement that propagates through water as waves. Figure 3.2 shows the distribution of canals in a fish head.

The flow field of swimming fishes can be described as dipole-like[10]; water accumulates in front of the forward moving fish while a low pressure region arises in the rear, where the body displaced water before the forward movement. This pressure gradient accelerates the fluid particles and creates hydrodynamic flow and large particle motion in the near field. These near field effects predominate the flow fields which propagate as waves in the far field. The larger the wavelength the larger the near field and the higher the order of a pole the stronger the domination of near field effects compared to far field propagation.[11] Thus, the lower the frequency, the larger the local flow compared to the propagating particle displacements. Details about the acoustical properties like particle displacement, particle acceleration, nearfield and far field are given in Sect. 5.1.

---

[5] See e.g. Webb et al. (2008), pp. 161ff and Braun and Grande (2008).

[6] See Webb et al. (2008), p. 145 and pp. 161ff.

[7] Mainly based on literature edited by Richard Fay and colleagues, Sheryl Coombs and David H. Evans, particularly Fay et al. (2008), Coombs et al. (1992), Popper and Schilt (2008), Webb et al. (2008), and Braun and Grande (2008), Kalmijn (1989), and Popper and Platt (1993).

[8] Mainly based on Gelfand (1990), Zwicker and Fastl (1999), Young (2007) and Dallos (1978).

[9] See Coombs et al. (1992), p. 268 and Popper and Platt (1993), pp. 101ff.

[10] See e.g. Kalmijn (1989), p. 202.

[11] See e.g. Sand and Bleckmann (2008), pp. 138f.

Fig. 3.1 Scanning electron micrograph showing hair cells on a zebrafish's neuromast. The dashed white line separates two regions with different hair cell orientations. The black arrows indicate the axis of maximum response. From Popper and Platt (1993), p. 102

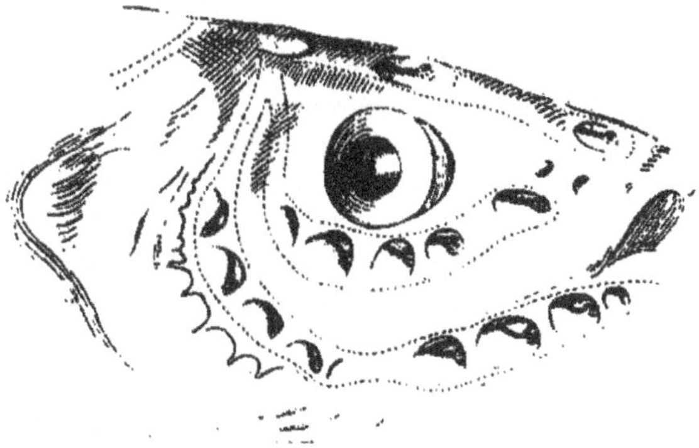

Fig. 3.2 Drawing of a fish's head with removed skin. The canals containing the neuromasts are distributed along lateral lines, naturally covered by the skin. Taken from Dijkgraaf (1989), p. 8

Swimming movements of fishes create hydrodynamic accelerations in a frequency region from almost 0–45 Hz with a dominance of frequencies below 20 Hz in steady swimming. Their tail beat produces frequencies up to 1 kHz.[12] The lateral line detects accelerations within a frequency range of almost 0–100 Hz best and hardly responses to frequencies above 200 Hz, as illustrated in Fig. 3.3. Therefore, it is able to perceive only those frequency components which remain in the near field of fishes. Since the lateral line system is bilaterally distributed and its hair cells' response intensity is direction-dependent, this sensory system provides fishes with a localization capability of sources, reflectors and deflectors of local flow fields. Thus, it delivers a mental representation of the close environment, in the range of several body lengths.[13]

"The primary function of any complex sensory system is to represent the structure of the outside world."[14] Since somatosensoric, i.e., haptic, perception only informs about magnitudes in a striking distance, the lateral line extends the perceived part of

[12]Detailed information about pressure of swimming fish is given in Webb et al. (2008), p. 155, Kalmijn (1989), p. 204 and Schellart and Popper (1992), p. 302.

[13]See Sand and Bleckmann (2008), p. 184 and Popper and Platt (1993), p. 100.

[14]See Braun and Grande (2008), p. 105.

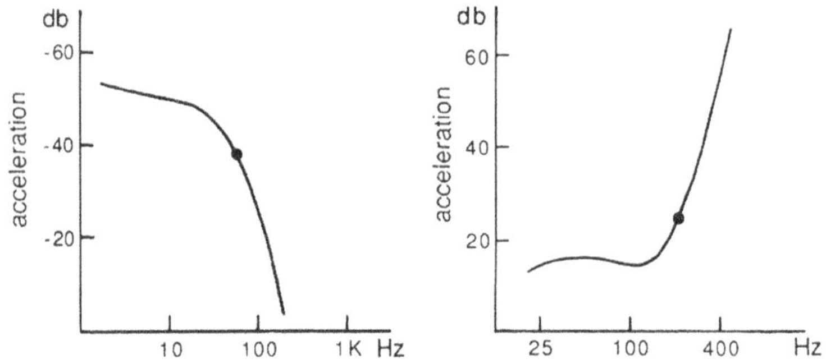

Fig. 3.3 Frequency response of sensory hair cells in the lateral line (left) and auditory system (right) of fish. Figure taken from Kalmijn (1989), p. 199

the outside world. In the literature it is described as "distant touch", "hydrodynamic imaging" or "intermediate between touch and hearing".[15] The term "svenning"[16] is used to describe the perception of the lateral line system, as a counterpart of the term *hearing* for the perception of pressure changes by the auditory system. Svenning serves for predator- and prey detection, group cohesion of schooling fishes, mate attraction, obstacle avoidance and a general awareness of the environment. Furthermore, in prey detection it supports chemoreception and vision, especially in murky water or darkness.[17] Many attempts have been made to mimic the lateral line system by technical means with hydrophone arrays in vessels.[18] Passive sonar systems mimic hearing by recording sound pressure from the environment and estimating their origin. Especially in densely settled and shallow waters the high amount of sound sources and the strong reverberation make it difficult to detect the location of sources by means of sound pressure recordings. In this case the near field effect of swimming objects is leveraged: particle accelerations along swimming objects are strong but especially low frequencies do not radiate to the far field as a sound wave. Rather, they remain in the near field and create little reverberation. Lateral line sensors detect such particle accelerations by means of a hydrophone array and near field methods.

[15]See Popper and Platt (1993), p. 117 or Coombs et al. (1992), p. 280, Webb et al. (2008), p. 156 or Coombs et al. (1992), p. 280.

[16]See e.g. Popper and Schilt (2008), p. 18, Popper and Platt (1993), p. 117.

[17]All functions gathered from Popper and Platt (1993), p. 100 and pp. 117f and Popper and Schilt (2008), p. 18.

[18]Examples of lateral line sensors can be found in Ziemer (2014), Ziemer (2015b), Xu and Mohseni (2017), Ziemer (2015a), Santos et al. (2010).

3.1.2 *Auditory System of Fish*

Auditory scene analysis is considered the primary function of audition.[19] A detailed explanation of auditory scene analysis is given in Sect. 4.5. Basically, it means an identification and discrimination of different items and happenings in the outside world, grouping sounds from one source and localizing it, leading to "a spatial model of the local world".[20] Consequently, detection of predator and prey, rather than communication, is the primary function of hearing in fish. It evolved to enable fish and other animals with a mechanism to perceive part of their environment in greater distance. The tactile sense only detects direct touch, the lateral line only works for near field effects, visual cues only exist in the visual field with enough light and chemical signals propagate slowly and with little directional information and do not travel large distances.[21]

As can be seen in Fig. 3.3, the auditory system in fishes is sensitive to those frequencies of fish's swimming movement which widely propagate as waves. Predators first detect prey by the inner ear which guides them towards the prey. Then the lateral line offers cues precise enough for the predator's final strike or the prey's quick evasion maneuver.[22] Today, many fish have a hearing range from 50 to 1500 Hz and use the auditory system for communication and mate, too.[23] "It is reasonable to suggest that improvement of hearing and refinement of vocalization co-evolved."[24] The swim bladder of fishes resonates with pressure fluctuations and is considered the origin of the middle ear which has a similar function of reinforcing sound from the outside world for the inner ear.[25] However, detection of pressure fluctuations has evolved dozens of times and therefore it is difficult to follow evolution from the fish's to the human's ears.[26] These are described in the following section.

3.2 Human Auditory System

The human auditory system is very sensitive. It perceives pressure fluctuations with no more than 10^{-6} kJ energy at optimum and has a frequency range from 16 Hz to 20 kHz which is about 10 octaves. In contrast, the eye needs one hundred times the

---

[19] See e.g. Braun and Grande (2008), p. 105 or Popper and Schilt (2008), p. 19.

[20] See Fay (1992), p. 229.

[21] See Popper and Schilt (2008), pp. 18–19, Gans (1992), p. 7, Popper and Platt (1993), pp. 123ff.

[22] See Kalmijn (1989), p. 210 and Popper and Platt (1993), p. 117.

[23] See Popper and Schilt (2008), p. 19 about the hearing range and Popper and Platt (1993), p. 116 about acoustical communication of fishes.

[24] See Schellart and Popper (1992), p. 302.

[25] See e.g. Fay et al. (2008), p. 8.

[26] See Braun and Grande (2008), p. 99, Coombs et al. (1992), p. 269, Gans (1992), p. 9 and p. 39, Sterbing-d'Angelo (2009), p. 1286.

amount of energy and has a range of one octave only.[27] The auditory system can be divided into the ear and the auditory pathway. The fundamentals of these two are discussed successively in this section.[28]

3.2.1 Human Ear

The ear can be divided into the outer ear, the middle ear and the inner ear as illustrated in Fig. 3.4. They are followed by the "auditory nerve"[29] and the central auditory pathways discussed subsequently.

The Outer Ear:

The pinna is the visible part of the outer ear. It mainly consists of cartilage, covered with vestigial muscles and skin. It works as a sort of funnel, as well as a filter, collecting frequencies with slightly different intensities for each incidence angle. For example the pinnae create an acoustic wave shadow for high frequencies from the rear but not from the front. The slightly S-shaped ear canal is about 3 cm long and also acts as a filter, resonating widely around 4 kHz. Tiny hairs protect the ear from invasion, wax and oil lubricate the ear and keep debris outside. The eardrum separates the outer ear from the middle ear, being displaced by arriving sound pressure fluctuations.

The Middle Ear:

The eardrum is about 0.074 mm thin, concave outward and elliptical with a diameter of 0.8–1 cm. At the peak the tensor tympani muscle is attached to the first bone in the ossicular chain, the malleus (hammer). Together with the incus (anvil) and the stapes (stirrup), it transfers the displacement of the ear drum to the oval window, the entrance of the inner ear, acting as impedance converter between the ambient air and the perilymph-filled scala vestibuli. The Eustachian tube connects the middle ear to the upper throat region. It allows for a pressure adjustment between middle ear- and ambient pressure. In some situations, e.g. on airplanes and when diving deeper than a few meters, the pressure discrepancy becomes apparent and tends to encourage people to equalize the pressure consciously, by deliberate muscle contraction or general chewing motions. Together, the outer and middle ear gain up the signal by a factor of about 30 by resonance and impedance conversion.

---

[27] See Motte-Haber (1972), p. 26.

[28] The following descriptions and an even much deeper insight into biology, mechanics, neurology, and psychology of the auditory system can be found e.g. in Zwicker and Fastl (1999), Ando (2010), Warren (1982), Hall and Barker (2012), Roederer (2008) and Zatorre and Zarate (2012).

[29] Also referred to as "cochlear nerve" or "eighth cranial nerve", see e.g. Gelfand (1990), p. 33 and Schneider (2018), p. 615. Strictly speaking, the auditory nerve is the auditory branch of the eighth cranial nerve which also includes the vestibular nerve, see, e.g., Herman (2007), p. 592.

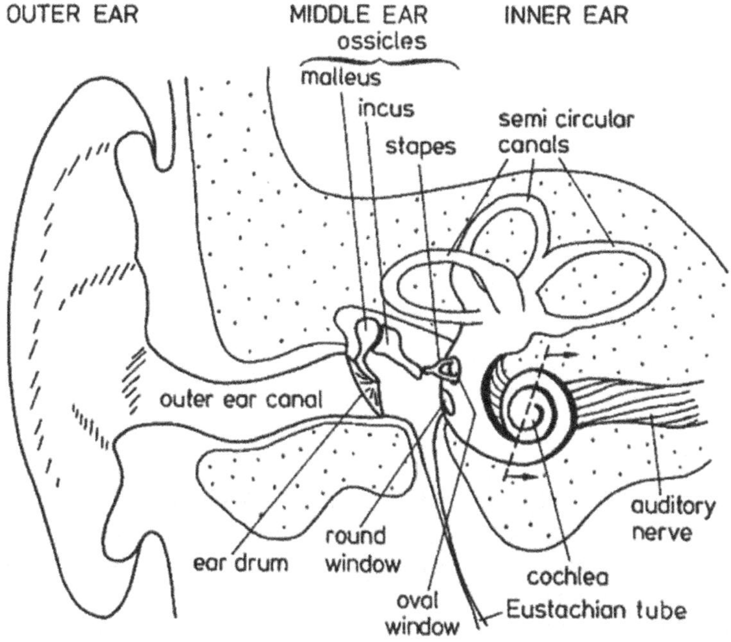

Fig. 3.4 Schematic drawing of the human ear. From Zwicker and Fastl (1999), p. 24

The Inner Ear:

Over the oval window the cochlea receives displacements of the stapedial footplate. The cochlea is a 35 mm long, snail-shaped spiral with $2\frac{3}{4}$ turns, tapering from 9 mm at the base to 5 mm at the apex. It has three chambers or *scalae*: The scala vestibuli begins behind the oval window and contains perilymph. At the apex it is connected with the scala tympani through the helicotrema opening. At its end, the round window membrane equalizes the displacement of the oval window, which is necessary, since the fluids in the cochlea and the surrounding bone are incompressible. The scala media lies between the two scalae. It is filled with endolymph and is separated from scala vestibuli by the Reissner's membrane which is so thin that it has no considerable mechanical influence but separates the fluids, which have very dissimilar ionic properties. From the scala tympany it is separated by the basilar membrane. There are steady positive potentials in the scala media and steady negative potentials in and near the basilar membrane.[30] Figure 3.5 illustrates an uncoiled cochlea with the scalae and membranes.

The end organ of hearing, the organ of Corti, lies on the basilar membrane. It transduces mechanical movements into electrochemical activity over the eighth cranial nerve, connecting the sensory hair cells with the nervous system. It contains a row of about 3,500 inner hair cells and 12,000 outer hair cells, arranged in three rows,

[30] See Thurlow (1971), p. 230.

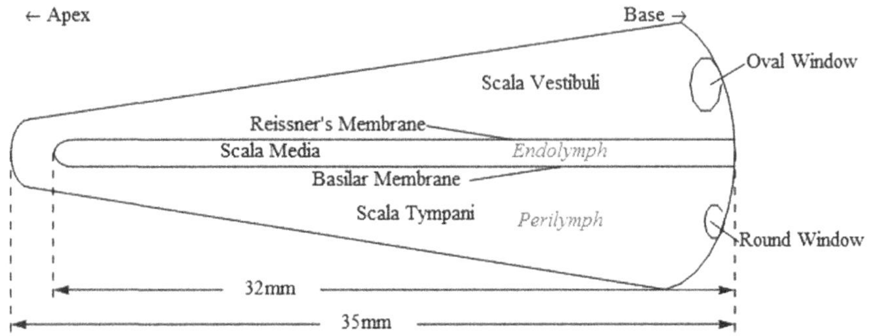

Fig. 3.5 Schematic illustration of an uncoiled cochlea. Scalae vestibuli and tympany connect the oval and round window, being filled with perilymph. The scala media separates those two, being filled with endolymph

surrounded by various supporting cells. About 100–150 sensory hairs, *stereocilia*, lie on top of each outer hair cell, partly being attached to the fibrous tectorial membrane which separates them from the scala media. 40–70 thicker stereocilia lie on each inner hair cell. Tiny lines connect the sensory hairs.

Movements of the oval window induce a traveling wave within the cochlea. This wave propagates inwards, slowly builds up and suddenly collapses after reaching the climax, i.e., its *amplitude* or *peak*. Due to complicated biomechanical effects—mainly the variant tension and width of the basilar membrane—the location of the traveling wave's peak depends on frequency. The frequency-place transformation is known as the "place principle" and means that every frequency has a corresponding area on the tectorial membrane, i.e. frequency is spatially encoded in the cochlea.[31] This principle is illustrated in Fig. 3.6 for a high and a low frequency. The figure shows the envelope of two traveling waves in the cochlea. Positive elongations cause hair cells to shear, which allows neurons to fire. In addition to the peak region both frequencies excite the higher-frequency region at the base. However, they barely excite the lower-frequency region towards the apex. This is the main reason why high frequencies barely mask low ones.[32]

Even though frequency is mostly considered as temporal quantity, having the unit 1/s, already the inner ear encodes it by spatial means. This transformation underlines the importance of space in hearing. With this encoding technique the inner ear performs a sort of frequency analysis—similar to a Fourier transform which will be discussed in Sect. 5.1.3—often referred to as "cochlear filter" or "Ohm's auditory law".[33] The displacement of the tectorial membrane relative to the basilar membrane causes the inner hair cells to shear proportional to positive elongation. Thereby, transduction channels open and close, allowing ion movements and thus a neural firing at the auditory nerve. Furthermore, hair cells show a "microphonic"

[31] See e.g. Zwicker and Fastl (1999), p. 29.

[32] Details on masking are given in Chap. 4.

[33] See e.g. Ando (2010), p. xv and Gelfand (1990), p. 140.

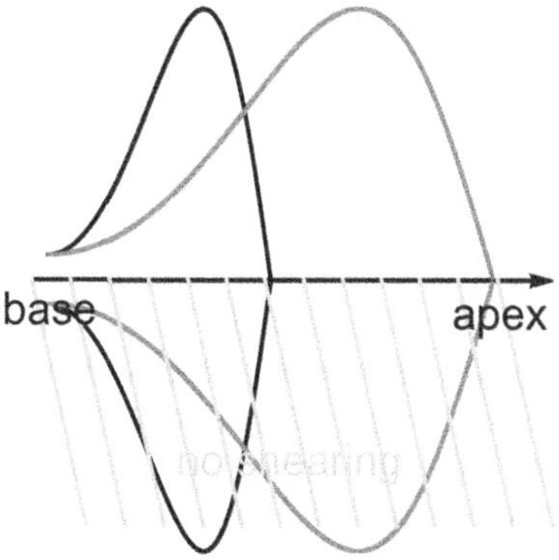

Fig. 3.6 Envelope of a high frequency (black) and a low frequency (gray) traveling wave in the cochlea. The envelopes are left-skewed, i.e., the high-frequency base region is excited stronger than the low-frequency apex region

response.[34] The inner hair cells are primarily displacement receptors, almost sensing magnitudes of the size of an atom. The 3,500 inner hair cells are connected to about 30,000 neurons in the auditory nerve. This connection is the entrance to the auditory pathway. Neurons that lie at the peak of a frequency's traveling wave envelope tend to be tuned to this very frequency. This means that the neuron is most sensitive to excitement in this frequency and responds even at low amplitudes. The sensitivity per frequency is referred to as *tuning curve* and the frequency of highest sensitivity is called *best frequency*.

The Vestibule:

Besides the cochlea the inner ear consists of the equilibrium organ, the vestibule. It contains the two balance organs utricle and saccule. These semi circular canals contain receptors for rotational acceleration and maintaining balance. The close relationship between receptors of alternating pressure and accelerations in the human body is an interesting parallel to the close relationship between the ear and the lateral line organ of fish. The equilibrium organ plays an important role for the sense of proprioception, together with vision and touch. Proprioception provides us with the perception of position, motion and forces of body parts in relation to each other and to the outside world. Of course, this is an important aspect of spatial perception and orientation and is necessary to navigate our body through the physical world. Although not generally considered as a part of proprioception, auditory input certainly gives helpful feedback. It is assumed that micro-movements of the head are unconsciously performed when listeners try to localize sources that are not in their field of view.

[34] See Thurlow (1971), p. 230.

3.2.2 Human Auditory Pathway

As stated earlier in this chapter, the main function of the auditory system is auditory scene analysis. The auditory pathway performs the major preprocessing steps to achieve this. A simplified scheme of the auditory pathway is illustrated in Fig. 3.7. It contains about 6 stations in hierarchic order. The higher the stage, the less automatic and the more centrally controlled the processing is, i.e., the more the processing tends to be affected by motivation, decision, knowledge or consciousness. Due to the hierarchy, processing steps have a certain order. One can roughly say that the higher the order, the longer the processing takes. The sequential, ascending processing is referred to as *feedforward mechanism*. After the inner ear, the auditory pathway mainly consists of the cochlear nuclei, superior olivary complex, nuclei of lateral lemniscus, inferior colliculus, thalamus and the auditory cortex. In fact, we have one auditory pathway in each hemisphere. These have many interconnections.[35]

Most stations of the auditory pathway are nuclei. These are spatially distributed within the brain and contain populations of neurons. Neurons are the processing units in the brain. They are distributed along the auditory pathway and are connected with each other via neural connections or *synapses*. Some synapses are ipsilateral, connecting neurons within one hemisphere. An example for ipsilateral connection is the auditory nerve. Other neural connections are contralateral. For example the cochlear nuclei of both hemispheres exhibit a contralateral connection. The cochlear nuclei provide additional connections to neurons in the superior olivary complexes of both hemispheres which is an example of bilateral synapses. All three examples of neural connections are either towards higher stations along the auditory pathway or the same station of both hemispheres is interconnected before further processing happens at higher stations. Such connections are referred to as *afferent*, *ascending*, or *bottom-up*. They connect lower stages with higher stages. In the other direction, synapses are *efferent*, *descending*, or *top-down*. The lowest possible stage in the hierarchy is the *periphery*. In the case of audition this is typically the ear. Ascending connections are directed towards higher stages along the central nervous system over the brainstem and the midbrain towards the stages of the cortex. Some afferent and efferent connections are indicated in Fig. 3.7. Neurons receive, process, and send data. The receiving and sending of data is also referred to as *projection*. The projected data are electric potentials, also referred to as *neural firing*, or *spikes*. Neurons fire binary. At each point in time they either fire or they do not. They cannot fire with different intensities. After each spike, the nerve cell needs to recharge which takes some time. The maximum firing rate of a neuron lies around 1 kHz. Neurons are most sensitive for excitations with a certain frequency, i.e., their best frequency. They only tend to respond to a rather narrow frequency range around their best frequency at all. An exemplary best frequency of an auditory nerve fiber is plotted in Fig. 3.8. Due to the limited bandwidth of neurons, different frequencies are not processed and

[35]The figure and the description rely largely on the illustrations and explanations in Ryugo (2011), p. 4, Schofield (2011), p. 263, Zwicker and Fastl (1999), p. 60, Ando (2010) p. 43, Cariani and Micheyl (2012), p. 370, Hong et al. (2012), p. 3, and Schneider (2018), pp. 615ff.

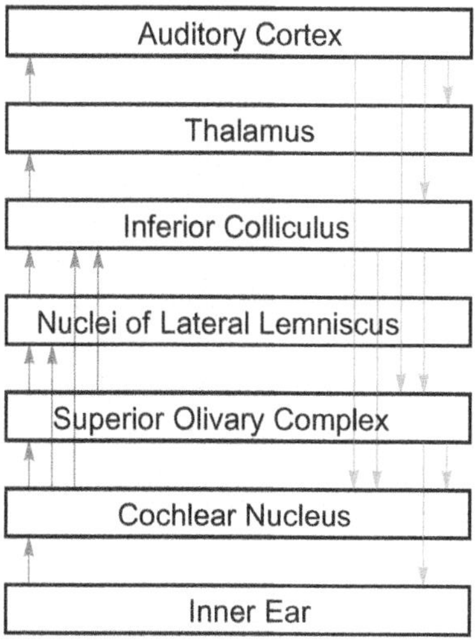

Fig. 3.7 Simplified scheme of the auditory pathway including the 6 stations and some ascending (left) and descending (right) connections. After Ryugo (2011), p. 4

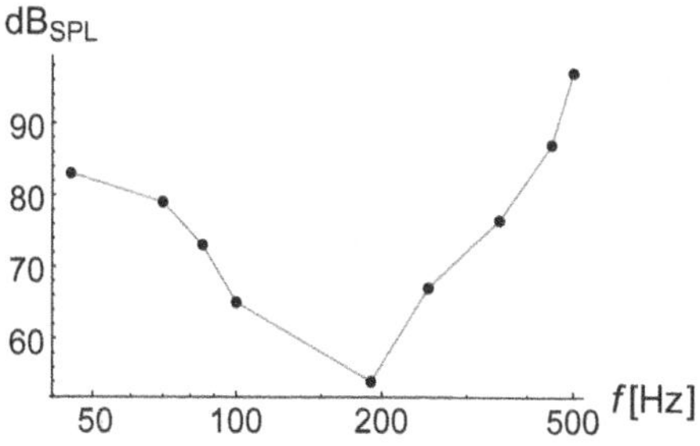

Fig. 3.8 Exemplary frequency-threshold curve for an auditory nerve fiber. At the best frequency a low sound pressure level at the eardrum is sufficient to activate neural firing

transmitted through the same neurons and synapses but rather side-by-side, referred to as *tonotopic* principle. The tonotopic principle starts in the cochlea. Here, the best frequency of each auditory nerve fiber is tuned to the frequency that peaks at this location. One exception that has been found in cat are auditory nerve fibers that are located at the peak region of frequencies higher than 3 kHz.[36] The tonotopic principle is kept throughout practically all stations in the auditory pathway up to the primary auditory cortex. Considering the narrow bandwidth of neurons, it becomes obvious why the cochlear filter transforms the incoming sound into its frequency components. Single neurons in the auditory nerve cannot handle broadband signals, so the cochlea has to divide incoming broadband signals to narrowband portions, process them separately, and integrate them at higher stages.

Even though a lot about the auditory system is known today, details about auditory processing are still under debate. The higher the stage the less is known about its exact functions. Much knowledge has been gained by invasive animal experiments. It not certain whether the human auditory system exhibits the same neural processing. The

[36]See e.g. Nedzelnitsky (1974), pp. 51f.

human auditory pathway is largely examined by noninvasive imaging techniques, which tend to have a high spatial resolution but also a rather high integration time. Other techniques, like electroencephalography, have a high temporal but low spatial resolution.[37] There is large consensus about the ascending auditory pathway in the literature, which will be described in the following. In addition to that, a descending pathway is evident, modulating ascending input. These modulations are gating sensory information, improve the discrimination of signals from noise and enable the switching of attention.[38] Comparatively little is known about the descending auditory pathway. There is no doubt, however, that hearing is an active rather than a passive process. Compared to a passive microphone pair, our pair of ears is active. The interplay of afferents and efferents is responsible for the active processing of sound to provide us with all the auditory information that we finally perceive. So after the treatise of afferents, the fundamentals of efferent processing are given.

3.2.2.1 The Afferent Auditory Pathway

A scheme for the encoding of a frequency with its specific amplitude and phase is illustrated in Fig. 3.9. Inner hair cells are sensitive to elongations along one direction. This leads to a sort of half-wave rectification: the inner hair cells only shear proportionally to an incoming half wave. This is indicated in the upper plots of (a) and (b) for one frequency with two different amplitudes. Each inner hair cell is connected to several neurons in the auditory nerve. The shearing evokes neural firing, typically in phase with the incoming half wave, referred to as "synchronization", "phase locking" or "entrainment".[39] After the cochlear filtering spike trains encode the incoming narrow band portion of the signal. This principle is referred to as *volley principle*.[40] Not all neurons necessarily fire perfectly in phase with each incoming half wave. Some may fire a bit earlier or later, leave out a period or even exhibit spontaneous firing in between. This is referred to as *jitter*. But when summing up the neural activity, the plot may look as indicated in the lower plots of (a). Neural activity peaks in phase with the incoming half wave, so the phase information is kept. Furthermore, the peak-to-peak period encodes the frequency. Now, if the amplitude increases, as indicated in (b), the plot changes in three ways. First, the noise floor will get lower because neurons tend to show less spontaneous activity. Second, the peaks will get higher because more neurons fire in phase with the incoming half wave. And thirdly, the peaks will become narrower because the in-phase firing becomes more precise. So the amplitude-encoding is threefold: peak height, peak width and the degree of spontaneous firing indicate the amplitude. All three observations result from the fact that phase-locking is stronger at high amplitudes. Note that only frequencies below

---

[37] A vast review of the methods of neuroscience is the context of neuromusicology can be found in Neuhaus (2017).

[38] See e.g. Ryugo (2011), p. 4.

[39] See Thurlow (1971), p. 230, Ando (2010), p. xiv or Bader (2015), p. 1054.

[40] See e.g., Opstal (2016), pp. 152f.

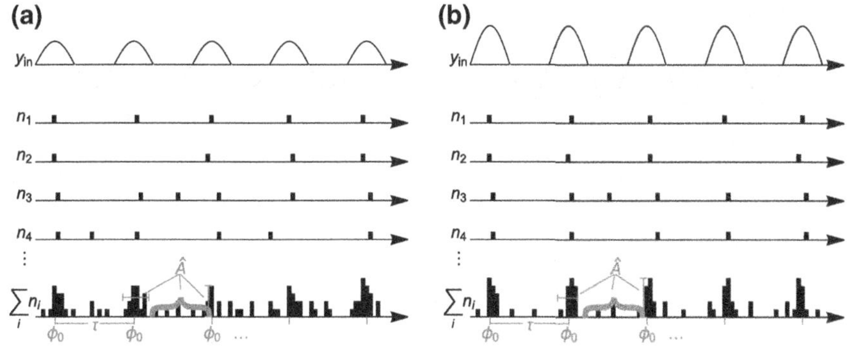

Fig. 3.9 Encoding scheme of frequency ($1/\tau$), amplitude ($\hat{A}$) and phase (ϕ) in the auditory nerve

about 1 kHz can be represented by this principle. At higher frequencies, all neurons will leave out at least one period to recharge, so the phase representation may become a bit blurred. Furthermore, phase locking is weaker at this high repetition rate. At about 4 kHz, the phase locking is so weak and the neurons leave out so many periods, that neither frequency nor phase are represented well. It is mainly the amplitude that is encoded by this temporal encoding mechanism. However, another vague frequency representation is given due to the tonotopic principle and the best frequency of the neurons. If neurons at a specific spatial location in the auditory nerve fire, the incoming wave certainly contains the corresponding frequency. So this is an indicator that a frequency is contained in the incoming wave. However, this information is prone to mistakes. As mentioned earlier in this chapter, low frequencies travel all along the basilar membrane, so they also stimulate the high-frequency region. This is where the best frequency comes into play. Neurons barely respond if they are not attracted with a frequency near their best frequency. Thanks to this redundancy the auditory system is able to distinguish between low frequencies and broadband signals, even though the low traveling wave of a low frequency passes the complete cochlea.

So for a wide frequency region, amplitude, frequency and phase are encoded already on the way to the first station of the ascending auditory pathway. The neural firing can be considered as oscillations that are related to the physical sound wave. They are transferred ipsilaterally to the cochlear nuclei. Here, the incoming frequency, amplitude and phase information are largely preserved. Furthermore, some neurons seem to be sensitive to the onsets of single frequencies. The cochlear nuclei of both hemispheres are connected with each other via the trapezoid body. Due to this connection, a sort of interaural level difference calculation can be performed. Neural activity in the left auditory nerve and cochlear nucleus is stronger than in the right, if the level of the incoming wave is higher in the left ear compared to the right.[41] Interaural level differences are important for horizontal source localization. This will be discussed in detail in Sect. 4.4. Even more important for spatial hearing is the processing that happens in the superior olivary complex. Here, interaural

[41] See Ando (2010), p. 43.

time differences seem to be reflected in neural responses.[42] Furthermore, neurons fire proportionally to the interaural crosscorrelation coefficient and the width of the correlation peak. This binaural processing is important for source localization and the perception of source width as will be demonstrated in detail in Sects. 4.4 and 6.2. So the first stations of the auditory pathway are strongly interconnected to provide us with information for spatial hearing. Interestingly, this happens already in the brainstem. Sound is localized already before it reaches the cortex, i.e., before we consciously identify the sound and perceive its characteristics. The *where* comes before the *what*.

The nuclei of lateral lemniscus are responsible for the routing of data for further processing. The ventral and intermediate dorsal nuclei of the lateral lemniscus are mainly involved in monaural processing and have ipsilateral synaptic connections to the inferior colliculus. The dorsal nuclei of lateral lemniscus exhibit bilateral connections to the inferior colliculus of both hemispheres and support binaural processing. The inferior colliculus is the midbrain center. It receives projections from all earlier stages as well as from some other modalities, such as vision and balance. Again, it is amazing to see that the first multi-modal processing is initiated long before the sound passes the thalamus and reaches the cortex so that it can consciously be perceived, interpreted or related to emotions, memory and experience. In the inferior colliculus neural correlates to the signal auto-correlation have been found. The auto-correlation is important for the sensation of pitch, harmonicity of complex sounds and consonance of intervals. Harmonic complex tones exhibit a regular auto-correlation whose peak-to-peak period indicates the pitch. In the chinchilla, single neurons have been found to respond to mistuned complex tones.[43] Modulations below 50 Hz are important for melody recognition and syllabic structure in speech. Faster modulations contribute to the sensation of pitch and roughness and may indicate inharmonicity or dissonance. Nonlinear distortion products have been found in neural response measurements, indicating that the inferior colliculus not only retrieves but even exaggerates dissonance.[44] So the inferior colliculus is a station that not only extracts information from the incoming signals but starts integrating them.

The inferior colliculus projects information mainly ipsilaterally to the thalamus via its so-called *brachium*. The thalamus relays sensory information from all modalities except smell. It can be considered as a sort of gate. It filters out unimportant information and lets important information pass through to the cortex. The medial geniculate is the main auditory nucleus of the thalamus. The medial geniculate nuclei of both hemispheres exhibit no interconnections, so most contra- and bilateral interactions happen already at lower stages.[45] Information that passes the thalamus is projected to the primary auditory cortex and some other cortical auditory areas. This

---

[42]It is speculated that ITDs are neurally encoded by response latency at in the superior olivary complex and maybe even at earlier stages, see Ando (2010), p. 44.

[43]See Sinex et al. (2002).

[44]See Lee et al. (2015).

[45]See Schofield (2011), p. 264 and Warren (1982), p. 14.

is a premise for conscious perception. In general, one can say that on the way towards higher stages phase locking declines and processing slows down while the number of neurons involved increases.

Heschl's gyrus in the primary auditory cortex seems to be the last station within the auditory pathway that exhibits a tonotopic map. In the primary auditory cortex, frequencies may be partly integrated. It is likely that auditory scene analysis in terms of integration and segregation take shape at this stage.[46] The right Heschl's gyrus seems to be involved in the pitch perception in the case of a missing fundamental and lesion of the right Heschl's gyrus reduces pitch resolution. Pitch perception is assumed to follow onset detection and take longer. As mentioned already in Sect. 2.4, the perception of pitch is two-fold, consisting of a rectilinear height and a cyclic chroma dimension. Height seems to be processed in the posterior auditory cortex and chroma in more anterior regions. In the non-primary auditory cortex fields, neurons tend to be more broadly tuned and some respond to certain intervals of complex tones, to noise or clicks rather than to sinusoidal components. Some neurons seem to indicate spectral bandwidth, note or pitch onsets. It is observable that synchronization with incoming waves and modulations decreases gradually along the ascending pathway. At the stage of the auditory cortex temporal encoding has largely disappeared, temporal integration seems to happen and the neural encoding becomes more complex, performing both serial and parallel processing.[47] In the belt and parabelt regions in the right hemisphere as well as areas along the right superior temporal gyrus, melodyperception seems to take place. Processing in the superior frontal sulcus is assumed to be important for spatial processing. The auditory cortex extends over a large region within the ventral portion of the lateral sulcus and the and superior temporal gyrus. Naturally, the auditory cortex is connected to other cortical regions. It is for example connected to the frontal cortex which is involved the integration of sensory input as well as in anticipation, planning and expectation, working memory and learning. Wernicke's area is a part of the inferior frontal lobe and one of the main areas involved in the comprehension of speech. This involves both heard speech and read words. Projections between the auditory and Broca's area in the premotor cortex are important for the analysis and creation of sound sequences, which is another important aspect of speech. The auditory cortex has synaptic connections to non-auditory cortical brain structures as well. The reticular formation is involved in arousal and the amygdala in alert and emotion. Located closely under the cerebral cortex is the hippocampus. It belongs to the limbic system and plays a major role in working memory functions as well as in consolidation from short-term memory to long-term memory and spatial memory. It is important for orientation and navigation. The basal ganglia are involved steering and switching attention, and are linked to the dopamine prediction-reward system.

---

[46] An attempt to associate aspects of auditory scene analysis to cortical structures can be found in Griffiths et al. (2012).

[47] See Hall and Barker (2012), p. 180, Griffiths et al. (2012), p. 214 and Hong et al. (2012), pp. 6 and 10–12.

A lot of auditory analysis and pre-processing happens already at the early stations of the auditory pathway and is subcortical. It does not require conscious examination. Many processing steps are, however, modulated by descending connections originating e.g. in the superior olivary complex, the inferior colliculus or even the auditory cortex. These efferents are discussed in the following section.

3.2.2.2 The Efferent Auditory Pathway

The discovery of efferents in the auditory system is relatively new, dating back to the 1950s.[48] Until then, the auditory system was commonly assumed to be a receiver and amplification system without central control. Otoacoutic emissions are evidence of efferents: the response of the ear to a sound input contains more energy than can be explained by a purely passive response. Furthermore, spontaneous otoacoustic emissions have been observed in many individuals. This means that the ear can exhibit both neural firing and even motion without the need for sound input.

Efferents were mostly attributed to the known autonomic functions like pupillary tension or heart rate, and to other autonomic or voluntary responses like muscle contraction. Today, efferents tend to be included in observations and explanations of auditory functioning. Efferents modulate sensory input at multiple synaptic stations along the auditory pathway to improve discrimination of signals from noise, balance sensitivity to increase the dynamic range, to enhance or suppress pieces of information or switch attention between aspects of sound. Sensory efferents may project information from the highest stations. Some input modulations are automatic, others are driven by prior knowledge, expectation, attention, motivation or conscious decision. There is evidence that they are either organized as parallel *regional circuits* or *feedback loops*, or they form a *descending chain*, or *branches* that project towards several lower targets within the descending auditory pathway. In this section, some examples of assumed efferent projections and active functions in the auditory system are provided. A much broader overview as well as a deeper insight can be found in the literature.[49]

There are several examples of well-observable afferent influences on signal processing in the auditory system. For example the dynamic range of inner hair cells is assumed to lie only between 40 and 60 dB. However, the auditory system is able to process sound with a dynamic range of more than 100 dB. Efferents play a crucial role to achieve this. The medial and the lateral olivocochlear systems exhibit efferents that change the biomechanical behavior of the cochlea.[50] Their name already implies that the cilia on the outer hair cells in the cochlea can contract. They change their length by up to 5%. This contraction changes the behavior of the traveling wave in the cochlea. It is likely that at a certain sound pressure level a saturation of the inner hair cell motion is reached. The cilia contraction decreases the deflection of

[48] See Rasmussen (1953).

[49] Especially in the chapters of Ryugo et al. (2011).

[50] See Guinan (2011) for a deeper insight.

the balisar membrane. This way higher sound pressure levels are necessary to deflect the hair cells. The saturation occurs at higher amplitudes. The contrary effect is also very likely. The outer hair cells amplify low-intensity sounds and sharpen frequency tuning curves of the auditory nerve over efferent feedback.[51] These feedback loops could account for another 40 dB of dynamic range.

The evolutionary background as well as the primary function of the auditory system provide evidence for the direct linkage between hearing and space. This mental representation of the outside world may be a biological explanation for the association of music and space in music theory, music perception, and composition as discussed in Chap. 2. Psychoacoustic principles describe the relationship between physical sound wave and auditory sound sensation, largely based on biomechnical considerations and modeling of the earliest stages of the ear and the efferent auditory pathway. An overview of psychoacoustical aspects that are related to spatial hearing and psychoacoustic sound field synthesis are discussed in the following chapter.

References

Ando Y (2010) Auditory and visual sensation. Springer, New York, Dordrecht, Heidelberg, London. https://doi.org/10.1007/b13253

Bader R (2015) Phase synchronization in the cochlea at transition from mechanical waves to electrical spikes. Chaos Interdiscip J Nonlinear Sci **25**(10): 103124. https://doi.org/10.1063/1.4932513

Braun CB, Grande T (2008) Evolution of peripheral mechanisms for the enhancement of sound reception. In: Webb JF, Fay RR, Popper AN (eds) Fish bioacoustics, Chap 4, pp 99–144. Springer, New York. https://doi.org/10.1007/978-0-387-73029-5_4

Cariani P, Micheyl C (2012) Toward a theory of information processing in auditory cortex. In: Poeppel D, Overath T, Popper A, Fay R (eds) The human auditory cortex. Springer Handbook of Auditory Research, Chap 13, vol 43, pp 351–390. Springer, New York. https://doi.org/10.1007/978-1-4614-2314-0_13

Clack JA (1993) Homologies in the fossil record. The middle ear as a test case. Acta Biotheor 41(4): 391–409. https://doi.org/10.1007/bf00709373

Coffin A, Kelley M, Manley GA, Popper AN (2004) Evolution of sensory hair cells. In: Manley GA, Fay RR, Popper AN (eds) Evolution of the vertebrate auditory system, pp. 55–94. Springer, New York. https://doi.org/10.1007/978-1-4419-8957-4_3

Coombs S, Janssen J, Montgomery J (1992) Functional and evolutionary implications of peripheral diversity in lateral line systems. In: Webster DB, Popper AN, Fay RR (eds) The evolutionary biology of hearing, Chap. 15, pp. 267–294. Springer, New York. https://doi.org/10.1007/978-1-4612-2784-7_19

de la Motte-Haber H (1972) Musikpsychologie. Hans Gerig, Cologne

Dijkgraaf S (1989) A short personal review of the history of lateral line research. In: Coombs S, Görner P, Münz H (eds) The mechanosensory lateral line. Neurobiology and evolution, pp 7–14. Springer, New York. https://doi.org/10.1007/978-1-4612-3560-6_2

Fay RR, Popper AN, Webb JF (2008) Introduction to fish bioacoustics. In: Webb JF, Fay RR, Popper AN (eds) Fish bioacoustics, Chap 1, pp 1–15. Springer, New York. https://doi.org/10.1007/978-0-387-73029-5_1

[51] See e.g. Coffin et al. (2004), p. 72.

Fay RR (1992) Structure and function in sound discrimination among vertebrates. In: Webster DB, Fay RR, Popper AN (eds) The evolutionary biology of hearing, Chap 14, pp 229–263. Springer, New York. https://doi.org/10.1007/978-1-4612-2784-7_18

Fritzsch B, Eberl D, Beisel K (2010) The role of bHLH genes in ear development and evolution. Revisiting a 10-year-old hypothesis. Cellular Mol Life Sci 67: 3089–3099. https://doi.org/10.1007/s00018-010-0403-x

Gans C (1992) An overview of the evolutionary biology of hearing. In: The evolutionary biology of hearing, Chap 1, pp 3–13. Springer, New York. https://doi.org/10.1007/978-1-4612-2784-7_1

Gelfand SA (1990) Hearing: An Introduction to Psychological and Physiological Acoustics, 2nd edn. CRC Press, New York and Basel

Griffiths TD, Micheyl C, Overath T (2012) Auditory object analysis. In: Poeppel D, Overath T, Popper AN, Fay RR (eds) The human auditory cortex. Springer Handbook of Auditory Research, Chap 8, pp 199–223, vol 43. Springer, New York. https://doi.org/10.1007/978-1-4614-2314-0_8

Guinan JG (2011) Physiology of the medial and lateral olivpcochlear system. Audit Vestib Efferents. https://doi.org/10.1007/978-1-4419-7070-1_3

Hall D, Barker D (2012) Coding of basic acoustical and perceptual components of sound in human auditory cortex. In: Poeppel D, Overath T, Popper AN, Fay RR, (eds) The human auditory cortex. Springer Handbook of Auditory Research, Chap 7, vol 43, pp 165–197. Springer, New York. https://doi.org/10.1007/978-1-4614-2314-0_7

Herman IP (2007) Sound, speech, and hearing, pp 555–628. Springer, Heidelberg. https://doi.org/10.1007/978-3-540-29604-1_10

Jørgensen JM (1989) Evolution of octavolateralis sensory cells. In: Coombs S, Görner P, Münz H (eds) The mechanosensory lateral line. Neurobiology and evolution, Chap 6, pp 115–145 (1989). Springer, New York. https://doi.org/10.1007/978-1-4612-3560-6_6

Kalmijn AJ (1989) Functional evolution of lateral line and inner ear sensory systems. In: Coombs S, Görner P, Münz H (eds) The mechanosensory lateral line. Neurobiology and evolution, Chap 9, pp 187–215. Springer, New York. https://doi.org/10.1007/978-1-4612-3560-6_9

Lee KM, Skoe E, Kraus N, Ashley R (2015) Neural transformation of dissonant intervals in the auditory brainstem. Music Perception Interdiscip J 32(5):445–459. https://doi.org/10.1525/mp.2015.32.5.445

Lee SY, Yeo SG, Seok Min Hong (2012) The anatomy, physiology and disorders of the auditory cortex. In: Elhilali M (ed) Auditory Cortex: Anatomy. Functions, and Disorders, Physiology - Laboratory and Clinical Research, Chapter I. Nova Science, New York, pp 1–26

Mallatt J (2009) Evolution and phylogeny of chordates. In Binder MD, Hirokawa N, Windhorst U (eds) Encyclopedia of neuroscience, pp 1201–1208. Springer, Heidelberg. https://doi.org/10.1007/978-3-540-29678-2_3116

Manley GA, Clack JA (2004) An outline of the evolution of vertebrate hearing organs. In: Manley GA, Popper AN, Fay RR (eds) Evolution of the vertebrate auditory system, pp 1–26. Springer, New York. https://doi.org/10.1007/978-1-4419-8957-4_1

Nedzelnitsky V (1974) Measurements of sound pressure in the cochlea of anesthetized cats. In: Zwicker E, Terhardt E (eds) Facts and models in hearing, pp 45–55. Springer, Berlin. https://doi.org/10.1007/978-3-642-65902-7

Neuhaus C (2017) Methods in neuromusicology: principles, trends, examples and the pros and cons. In: Schneider A (ed) Studies in musical acoustics and psychoacoustics. Current research in systematic musicoogy, Chap 11, vol 4, pp 341–374. Springer, Cham. https://doi.org/10.1007/978-3-319-47292-8_11

Dallos P (1978) Biophysics of the cochlea. In: Carterette EC, Friedman MP (eds) Handbook of perception, vol IV. Hearing, pp 125–162. Academic Press, New York. https://doi.org/10.1016/b978-0-12-161904-6.50011-7

Popper AN, Platt C (1993) Inner ear and lateral line. In: Evans DH (ed) The physiology of fishes, Chap 4, pp 99–136. Springer, Boca Raton

Popper AN, Platt C, Edds PL (1992) Evolution of the vertebrate inner ear. An overview of ideas. In: Webster DB, Fay RR, Popper AN (eds) The evolutionary biology of hearing, Chap 4, pp 49–57. Springer, New York. https://doi.org/10.1007/978-1-4612-2784-7_4

Popper AN, Schilt CR (2008) Hearing and acoustic behaviour. Basic and applied considerations. In: Webb JF, Fay RR, Popper AN (eds) Fish bioacoustics, Chap 2, pp 17–48. Springer, New York. https://doi.org/10.1007/978-0-387-73029-5_2

Rasmussen GL (1953) Further observations of the efferent cochlear bundle. 99:61–74. https://doi.org/10.1002/cne.900990105

Roederer JG (2008) The physics and psychophysics of music, fourth edn. New York. https://doi.org/10.1007/978-0-387-09474-8

Ryugo DK, Fay RR, Popper AN (eds) (2011) Auditory and vestibular efferents. Springer, New York. https://doi.org/10.1007/978-1-4419-7070-1

Ryugo DK (2011) Introduction to efferent systems. In: Ryugo DK, Fay RR, Popper AN (eds) Auditory and vestibular efferents. Springer Handbook of Auditory Research, pp 1–15. Springer, New York, Dordrecht, Heidelberg, London. https://doi.org/10.1007/978-1-4419-7070-1_1

Sand O, Bleckmann H (2008) Orientation to auditory and lateral line stimuli. In: Webb JF, Fay RR, Popper AN (eds) Fish bioacoustics, Chap 6, pp 183–231. Springer, New York. https://doi.org/10.1007/978-0-387-73029-5_6

Santos P, Felisberto P, Jesus SM (2010) Vector sensor arrays in underwater acoustic applications. In: Camarinha-Matos LM, Pereira P, Ribeiro L (eds) Emerging trends in technological innovation, pp 316–323. Springer, Heidelberg. https://link.springer.com/chapter/10.1007/978-3-642-11628-5_34, https://doi.org/10.1007/978-3-642-11628-5_34

Schellart NAM, Popper AN (1992) Functional aspects of the evolution of the auditory system of actinopterygian fish. In: Webster DB, Fay RR, Popper AN (eds) The evolutionary biology of hearing, pp 295–322. Springer, New York. https://doi.org/10.1007/978-1-4612-2784-7_20

Schneider A (2018) Pitch and pitch perception, pp 605–685. Springer, Heidelberg. https://doi.org/10.1007/978-3-662-55004-5_31

Schofield BR (2011) Central descending auditory pathways. In: Ryugo DK, Fay RR, Popper AN (eds) Auditory and vestibular efferents. Springer Handbook of Auditory Research, Chap 9, pp 261–290. Springer, New York, Dordrecht, Heidelberg, London. https://doi.org/10.1007/978-1-4419-7070-1

Sinex DG, Sabes JH, Li H (2002) Responses of inferior colliculus neurons to harmonic and mistuned complex tones. Hearing Res 168(1–2):150–162. https://doi.org/10.1016/S0378-5955(02)00366-0. A collection of papers presented at the symposium on the inferior colliculus: from past to future

Sterbing-d'Angelo SJ (2009) Evolution of the auditory system. In: Binder MD, Hirokawa N, Windhorst U (eds) Encyclopedia of neuroscience, pp 1286–1288. Springer, Heidelberg. https://doi.org/10.1007/978-3-540-29678-2_3144

Thurlow WR (1971) Audition. In: Kling JW, Riggs LA (eds) Woodworth & Schlosberg's experimental psychology, Third American edn, pp 223–271, London

van Opstal J (2016) The auditory nerve. In: The auditory system and human sound-localization behavior, pp 147–169. Academic Press, San Diego. https://doi.org/10.1016/B978-0-12-801529-2.00006-4

Warren RM (1982) Auditory Perception: A New Synthesis. Pergamon General Psychology Series. Pergamon Press, New York, Oxford, Toronto, Sydney, Paris, Frankfurt

Webb JF, Montgomery JC, Mogdans J (2008) Bioacoustics and the lateral line system of fishes. In: Webb JF, Fay RR, Popper AN (eds) Fish bioacoustics, Chap 5, pp 145–182. Springer, New York. https://doi.org/10.1007/978-0-387-73029-5_5

Will U, Fritsch B (1988) The eighth nerve of amphibians. Peripheral and central distribution. In: Fritsch B, Ryan MJ, Wilczynski W, Hetherington TE, Walkowiak W (eds) The evolution of the amphibian auditory system, pp 159–183. Springer, New York

Xu Y, Mohseni K (2017) A pressure sensory system inspired by the fish lateral line: hydrodynamic force estimation and wall detection. IEEE J Ocean Eng 42(3):532–543. https://doi.org/10.1109/JOE.2016.2613440

Young ED (2007) Physiological acoustics. In: Rossing TD (ed) Springer handbook of acoustics, pp 429–457. Springer, New York. https://doi.org/10.1007/978-0-387-30425-0_12
Zatorre RJ, Zarate JM (2012) Cortical processing of music. In: Poeppel D, Overath T, Popper AN, Fay RR (eds) The human auditory cortex. Springer Handbook of Auditory Research, Chap 10, vol 43, pp 261–294. Springer, New York. https://doi.org/10.1007/978-1-4614-2314-0_10
Ziemer T (2014) Towards a lateral line sensor to supplement sonar in shallow water. In: American Society of Mechanical Engineering (ASME) (ed) ASME 2014 33rd international conference on ocean, offshore and arctic engineering, Ocean Space Utilization; Professor Emeritus J. Randolph Paulling Honoring symposium on ocean technology, OMAE2014–23624, vol 7, San Francisco, CA, June 2014. https://doi.org/10.1115/OMAE2014-23624
Ziemer T (2015a) Localizing swimming objects in noisy environments by applying nearfield acoustic holography and minimum energy method. In: American Society of Mechanical Engineering (ASME) (ed) ASME 2014 34th international conference on ocean, offshore and arctic engineering (OMAE), Ocean space utilization, vol 6 OMAE2015–41733, St. John's, June 2015. https://doi.org/10.1115/OMAE2015-41733
Ziemer T (2015b) Simulating the lateral line with low-frequency nearfield acoustic holography based on a vector hydrophone array for short-range navigation in littoral waters. J Acoust Soc Am 138(3):2015b. https://doi.org/10.1121/1.4933959
Zwicker E, Fastl H (1999) Psychoacoustics. Facts and models, Second updated edn. Springer, Heidelberg. https://doi.org/10.1007/978-3-662-09562-1

Chapter 4
Psychoacoustics

The main function of the auditory system is auditory scene analysis. This psychological representation of the physical world relies on several psychoacoustic mechanisms which are outlined in this chapter. Psychoacoustics is the translation of physical sound input to auditory perception. However, perception is difficult to generalize as it depends not only on sound properties but also on situational context and the individual. Therefore, most psychoacoustic considerations tend to be restricted to the translation of sound properties to auditory sensation. Auditory sensation is assumed to be inter-subjective, depending less on situation and the individual experience, preference, and state of mind. At the same time it is the basis of auditory perception. Perception is a result of filtering, analysis, segregation and integration of physical signal input to sensory organs. Certain absolute and relative thresholds limit the region of sound which is psychologically processed. Signals which do not surpass these thresholds are neglected for auditory processing and perception. Therefore, these thresholds are discussed next.[1] Many of the filtering processes are based on the spatial representation of sound within the cochlea, the critical bands, which are explained subsequently, followed by an associated psychoacoustic phenomenon, namely masking. Sound source localization and other aspects of spatial hearing are also performed separately for each critical band. Auditory scene analysis explains how those sounds that are not filtered out are grouped and mentally represented.

4.1 Thresholds and Just Noticeable Differences

As already mentioned in the previous chapter, the human auditory system is sensitive to pressure fluctuations with a rate from about 16 Hz to 20 kHz. That is if these fluctuations exceed at least the *threshold in quiet* or *hearing threshold* which is dependent on frequency. It is $p_{ref} = 2 \times 10^{-5}$ Pa at a frequency of 1 kHz, slightly

[1]Mainly based on Zwicker and Fastl (1999).

© Springer Nature Switzerland AG 2020

T. Ziemer, *Psychoacoustic Music Sound Field Synthesis*, Current Research in Systematic Musicology 7, https://doi.org/10.1007/978-3-030-23033-3_4

less for frequencies around 3 kHz and up to 2×10^{-2} Pa at the limits of the audible frequency range. The threshold in quiet can be approximated[2] by

$$p_{\min}(f) = 3.64 \left(\frac{f}{\text{kHz}}\right)^{-0.8} - 6.5e^{-0.6\left(\frac{f}{\text{kHz}}-3.3\right)^2} + 10^{-3}\left(\frac{f}{\text{kHz}}\right)^4 \text{ dB}. \quad (4.1)$$

The *threshold of pain* starts at $p_{\max} = 20$ Pa. These thresholds are the limits of the *hearing area*. Music encompasses a large part of this area. As the range of audible pressure amplitudes is 10^6, amplitudes of sound pressure are usually not denoted absolute and linear but, adapted to the auditory perception, relative and logarithmic. Via

$$\text{dB} \equiv 20 \lg \frac{p}{p_0} \quad (4.2)$$

the logarithmic relative Sound Pressure Level (SPL) in dB can be derived from the decadic logarithm lg, the sound pressure p and a reference sound pressure value p_0. It can be calculated back to an absolute, linear value by

$$p = p_0\, 10^{\frac{\text{dB}}{20}}. \quad (4.3)$$

If the base value p_0 is unknown a relative, linear value is calculable. In this logarithmic scale the range of audible sound pressure levels is

$$20 \lg \frac{p_{\max}}{p_{\min}} = 20 \lg \frac{20}{10^{-5}} = 120 \text{ dB}. \quad (4.4)$$

In this work, if not explicitly denoted differently, units in dB are referred to the sound pressure level, i.e. "1 dB" means "1dB_{SPL}".

The hearing area is illustrated in Fig. 4.1. The solid line is the threshold in quiet. It is lowest in the most sensitive frequency region around 3 kHz and increases a lot towards the lowest and highest audible octaves, i.e., below 40 Hz and above 10 kHz. The actual hearing area is quite individual. It has been reported that subjects in experiments were able to hear frequencies up to no less than 28 kHz.[3] For subjects who frequently listen to loud the threshold in quiet drastically increases in the sensitive frequency region between 3 and 9 kHz. The area of music does not include extreme cases like impulsive sounds in music which can cover a much wider frequency range and even higher sound pressure levels.

Just noticeable differences (JNDs) are thresholds of change of certain physical parameters. The JND in change of sound pressure roughly lies around 0.8 dB, being lowest around 1 kHz at high sound pressure levels.[4] The JND in SPL of successive tones is generally lower. It lies between 0.3 and 1.4 dB depending on frequency and

---

[2]This formula can be found, e.g., in Terhardt et al. (1982), p. 682, Lin and Abdulla (2015), p. 24 and Kostek (2005), p. 10.

[3]See Ashihara (2007) for an overview of experiments.

[4]See Bruhn (2002a), pp. 667ff and Zwicker and Fastl (1999), pp. 175ff.

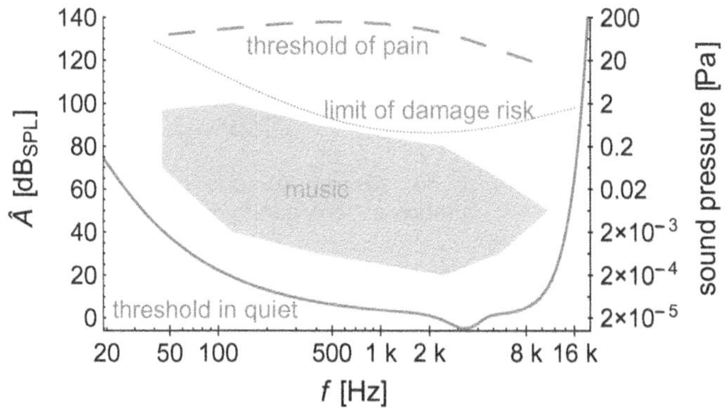

Fig. 4.1 Threshold of audibility and pain. After Zwicker and Fastl (1999), p. 17

Fig. 4.2 Just noticeable difference (JND) in sound pressure level for three different frequencies. After Backus (1969), p. 86

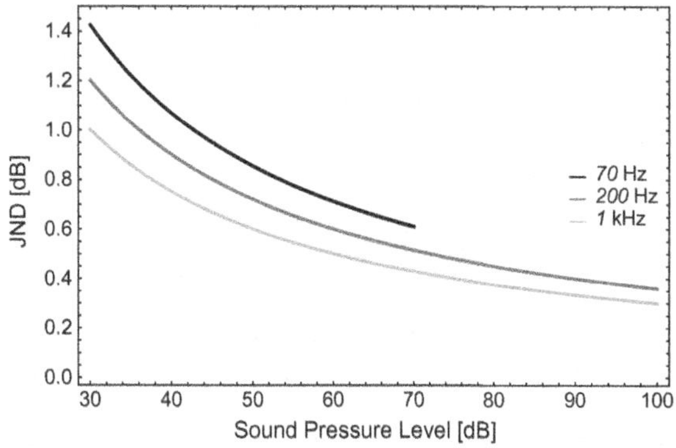

Fig. 4.3 Just noticeable variation in sound pressure level for different levels of white noise (WN) and a 1 kHz-tone. From Zwicker and Fastl (1999), p. 176

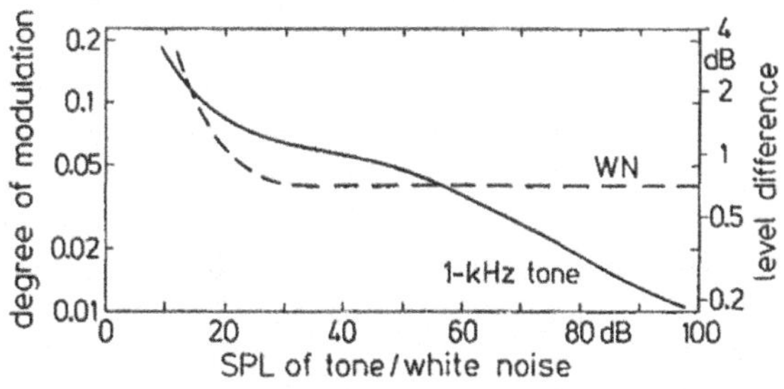

absolute SPL as can be seen in Fig. 4.2. These values are valid for durations of the presented sounds of 200 ms and more. For shorter tone bursts the JND can be up to four times larger as demonstrated in Fig. 4.4. The JND of both continuous and successive sounds is larger for white noise at most sound pressure levels. The JND in sound pressure modulation for a 1 kHz-tone and white noise are illustrated in Fig. 4.3.

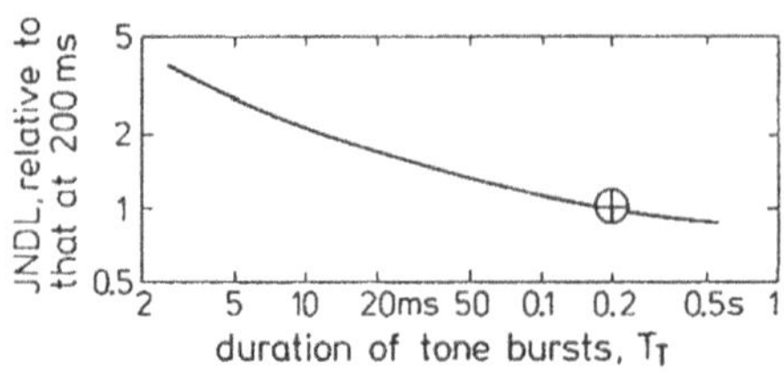

Fig. 4.4 Just noticeable difference in sound pressure level of successive tone bursts over signal duration relative to a duration of 200 ms the of a 1 kHz-tone for different modulation frequencies and different sound pressure levels. From Zwicker and Fastl (1999), p. 181

These 200 ms are an *integration time* of the auditory system which not only affects the perception of amplitude differences but also masking thresholds and the source motion detection which are discussed in subsequent sections.[5] One temporal threshold of the auditory system is about $T = 50$ ms, becoming less for increasing frequencies.[6] Successive acoustical events happening quicker than that are not discriminated but perceived as one sound or noise. This explains the lowest audible frequency of around $\frac{1}{T} = 20$ Hz. 50 ms is the time it takes for pitch perception to build up.[7] However, in an experiment the duration necessary to discriminate alternating from simultaneous complex tones has been found to lie below that threshold. This has been tested by Ziemer et al. (2007) with two complex tones with triangular wave form in the interval of a fifth in three frequency regions.[8] The complex tones are attached at their peaks or zero crossings to ensure that neither an impulsive sound nor silence occurs between them. Subjects were asked to judge whether the presented tones were perceived as clearly simultaneous (1), tend to be simultaneous or alternating (2 and 3), or clearly alternating (4). The arithmetic means are plotted in Fig. 4.5. However, as we have an ordinal scale of measurement the median values just above and below 2.5 are considered as threshold between perceived simultaneity and alternation. With fundamental frequencies of about 1 and 1.5 kHz, i.e. in the most sensitive frequency region, a duration between 7 and 11.2 ms of each tone is sufficient to recognize that they are presented in succession and not simultaneously. At higher frequencies a duration of 11.2 and 13.2 ms is necessary. For very low frequencies even simultaneous tones were perceived as alternating but with a large deviation. This is not surprising since both fundamentals fall into the same critical band and are therefore hardly perceived as individual tones as will be discussed extensively in the subsequent section. Only at a duration of 30 ms and more subjects identified the alternating tones as clearly as for the other frequency regions with a small deviation. Here, the threshold lies between 15.2 and 30.3 ms. The test has been conducted with 33 musicology students, most of them trained musicians, who may

[5] See also Zwicker and Fastl (1999), pp. 83f, Middlebrooks and Green (1991), pp. 150f and Grantham (1986), and Sect. 4.3.

[6] See e.g. Bruhn (2002a), p. 669.

[7] See e.g. Bader (2013), p. 325.

[8] The experiment is described in detail in Ziemer et al. (2007).

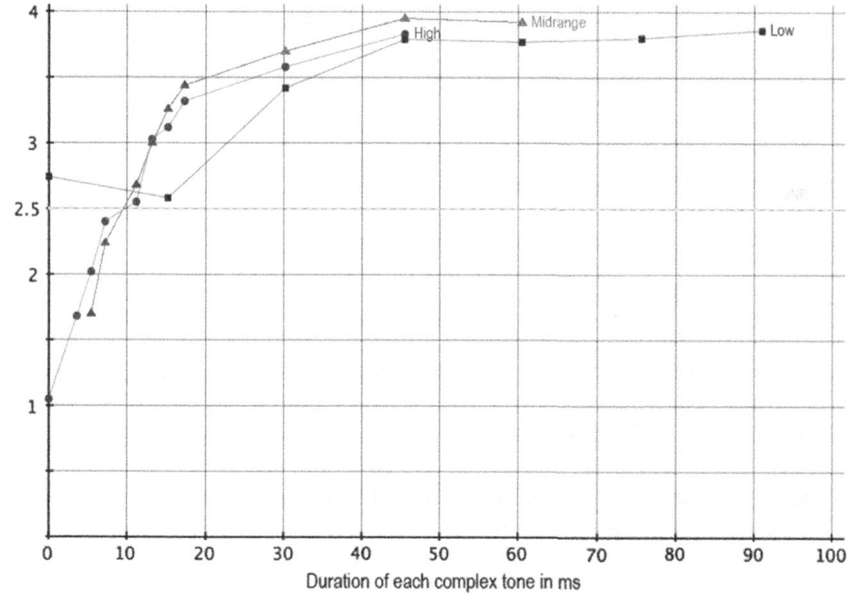

Fig. 4.5 Just noticeable difference in temporal order for low (33 and 49.5 Hz), midrange (1056 and 1584 Hz) and high (5280 and 7920 Hz) sounds with triangular waveform. From Ziemer et al. (2007), p. 23

have lower thresholds than an average person. The monaural temporal resolution of the auditory system is about 2 ms.[9] Auditory events need a duration of more than 2 to 5 ms to be perceived as having a timbre, rather than being a simple 'click'.[10] The temporal binaural resolution is even better by several orders of magnitude. Interaural arrival time differences of several μs are sufficient for sound source localization as will be discussed below in Sect. 4.4.

The JND of phase is difficult to quantify. Zwicker and Fastl (1999) deal with this subject and essentially give the following quantitative and qualitative statements[11]: Changes of phase result in instantaneous frequency changes and a change of envelope. Experiments have been carried out with complex tones consisting of three frequencies with equal amplitudes. Here, the just noticeable difference in phase change of one tone relative to the others has been found to lie around 10° at best conditions, 20° at worse laboratory conditions and as much as 60° in everyday-environments like the living room. It is audible apparently due to the change in envelope. As demonstrated in Sect. 3.2.2.1, phase is encoded in the auditory nerve up to a frequency of about 1 kHz. The phase relations of frequencies within one critical band play an important role in the perception of beating and roughness. Frequencies that lie further apart are

[9]Cf. Zwicker and Fastl (1999), p. 293.

[10]See Bader (2013), p. 324.

[11]These data are given in Zwicker and Fastl (1999), pp. 188–191.

processed alongside until they are integrated at higher stages of the auditory pathway. The higher this stage the more likely the phase information has been lost on the way. That is why phase relations are well-audible in some sounds, like impulsive attack transients, but inaudible in harmonic stationary sounds.

Thresholds of frequency discrimination, masking and sound source position exist as well and are discussed in the following sections.

4.2 Critical Bands

People perceive their environment through sensory organs. Information is represented as mental map which does not necessarily resemble the physical relations. A selection process reduces the amount of physical stimuli that are perceived and eventually allocated in the mental map. Filtering, like absolute thresholds in amplitude and frequency, masking and just noticeable differences of pitch and loudness arise in consequence of physical constraints of our auditory system or selective processes in the brain.

Sound pressure is processed in time-frequency domain by the auditory system, as already anticipated in Sect. 2.5 and described in more detail in Sect. 3.2. Furthermore, frequencies that do not have a corresponding resonance area in the cochlea are filtered out by that cochlear filter mechanism. This is the case for frequencies below 16 Hz and above 20 kHz. The resonance of a traveling wave excites many hair cells. Data on the envelope of such a traveling waves range from a decrease of 6 to 12–15 dB per octave on the lower frequency side and 20 to 70–95 dB per octave on the higher frequency side.[12] An additional filter mechanism narrows the excited area by heavily amplifying the highest firing rate, referred to as "cochlear amplifier".[13] Also, in the cochlear nucleus—which is directly connected to the auditory nerve—and at higher levels of the auditory pathway, sensitivity to narrow frequency bands has been observed.[14] How exactly this filtering is accomplished by the auditory system is unknown. Hypotheses regarding the motility of hair cells as well as efferent controlling mechanisms have been discussed as explanation for the extremely fine frequency selectivity of the auditory system.[15] The minimum sound pressure necessary to activate a cochlear amplification is the absolute threshold. The higher the frequency the closer the traveling wave peak lies to the oval window at the base. Consequently, the traveling wave of a low frequency passes the area of higher frequencies and evokes cochlear activity. It resonates in a certain region which is narrowed by a cochlear amplifier. Behind the resonance region the traveling wave collapses. The traveling wave of a frequency has to surpass the envelope of simultaneous traveling waves to

[12] See Klinke (1970) p. 318.

[13] See e.g. Luce (1993), p. 74, and Schneider (2018), p. 618. The effect is illustrated in Fig. 4.6.

[14] As discussed in Thurlow (1971), p. 230, and in Sect. 3.2.2.

[15] See Hellbrück (1993), pp. 101ff.

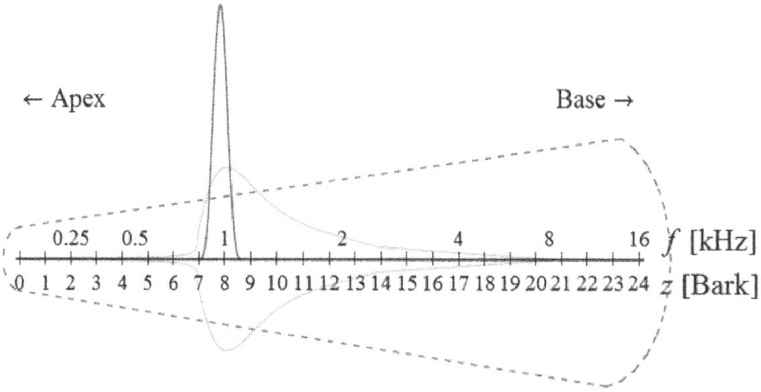

Fig. 4.6 Schematic diagram of a rolled-out cochlea (dashed contour) with the envelope of a traveling wave induced by a frequency of 1 kHz (light gray). At its peak the neural firing is amplified (dark gray curve) by a cochlear mechanism. The abscissa illustrates the linear alignment of frequencies in Bark in contrast to the logarithmic distribution in Hertz

protrude and thus be audible. Figure 4.6 schematically illustrates the envelope of a traveling wave and the cochlear amplification.

The region, affected by the cochlear amplifier is the so-called *critical band*. The width of a critical band is about 1.3 mm and roughly includes 160 hair cells. Within one critical band approximately 25 JNDs in pitch can be discriminated. Frequencies that simultaneously fall into the same critical band cannot be identified individually. They create a common sound impression. Depending on their interval, it is the impression of one single note, beats or roughness. Only frequencies from different frequency bands can be heard as different tones with a certain interval. The Bark scale divides the cochlea into 25 fixed equal areas z representing critical bands as can be seen in Fig. 4.6. Every frequency can be transferred to its corresponding position on the cochlea by

$$z[\text{Bark}] = 13 \arctan\left(0.76 \frac{f}{\text{kHz}}\right) + 3.5 \arctan\left(\frac{f}{7.5\,\text{kHz}}\right)^2 . \tag{4.5}$$

The width of one Bark, the critical band width $\Delta f_{\text{critical}}$, can be calculated as

$$\Delta f_{\text{critical}} = 25 + 75\left(1 + 1.4\left(\frac{f}{\text{kHz}}\right)^2\right)^{0.69} . \tag{4.6}$$

Table 4.1 lists the Bark scale and the corresponding lower and upper boundary frequency f_l and f_u, the mean frequency f_{mean} and the critical band width.[16] The mean frequency does not represent the arithmetic mean of lower and upper frequency but anatomically the mean position within one critical band on the basilar membrane.

[16]Cf. Zwicker and Fastl (1999), p. 159.

Table 4.1 Bark scale and corresponding frequencies

z [Bark]	f_l [Hz]	f_u [Hz]	$\Delta f_{\text{critical}}$ [Hz]	f_{mean} [Hz]
0	0	100	100	50
1	100	200	100	150
2	200	300	100	250
3	300	400	100	350
4	400	510	110	450
5	510	630	120	570
6	630	770	140	700
7	770	920	150	840
8	920	1080	160	1000
9	1080	1270	190	1170
10	1270	1480	210	1370
11	1480	1720	240	1600
12	1720	2000	280	1850
13	2000	2320	320	2150
14	2320	2700	380	2500
15	2700	3150	450	2900
16	3150	3700	550	3400
17	3700	4400	700	4000
18	4400	5300	900	4800
19	5300	6400	1100	5800
20	6400	7700	1300	7000
21	7700	9500	1800	8500
22	9500	12000	2500	1050
23	12000	15500	3500	13500
24	15500			

Up to a center frequency of 500 Hz, the critical band width is about 100 Hz, as can be seen in Fig. 4.7. From there on it is approximately 20% of the frequency. Certainly, the Bark scale does not describe the whole complex nature of critical bands. In fact, critical bands have no fixed positions but are areas around a center frequency. They are dynamic and adapt to the incoming signal.

As discussed in Sect. 3.2, the auditory system does not process sound as a whole but in frequency bands.[17] At higher stages of the auditory pathway they are integrated to draw a meaningful mental map of sound objects and their relations. This is demonstrated in the following sections in the context of masking, spatial hearing, and auditory scene analysis.

[17]See also Blauert (1974), p. 173, Allen (2008), pp. 28ff and Kostek (2005), p. 9.

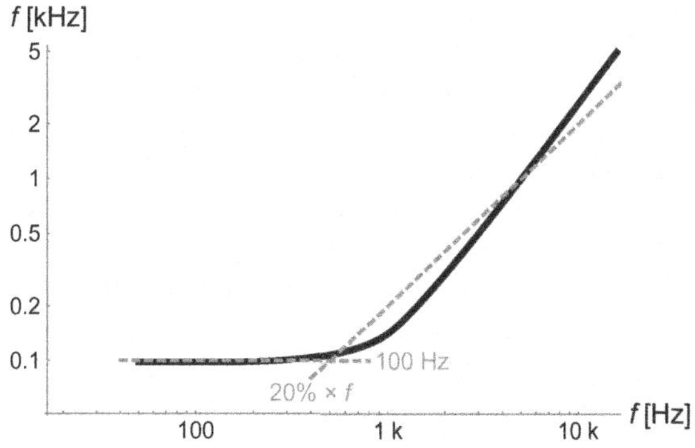

Fig. 4.7 Plot of the critical band width over frequency. After Zwicker and Fastl (1999), p. 158

4.3 Masking

Masking is an everyday life phenomenon. Imagine you are at home, put on your headphones, and choose a comfortable volume for your music. When leaving your apartment the music becomes suddenly less distinct as you open the doors. While approaching the nearest busy street, the music becomes less and less audible. The music is masked more and more by the traffic noise. You either have to turn up the volume or get back to your apartment to keep listening to the music. In psychoacoustical terms the voice is the *maskee*, *test signal* or *probe signal*, the street noise is the *masker*.[18] The intensity level at which the music becomes inaudible due to the traffic noise is called *masked threshold*, *masking!masking threshold* or *masking pattern*.[19] A signal has to surpass this threshold to become audible.

In the following, the basic masking phenomenons are explained by the results of listening tests with stationary masking sounds, meeting the same ear as the test signal (monaural/ipsilateral masking). Subsequently, the coaction of several simultaneous maskers is illuminated, as well as temporal effects during the onset of a masker (overshoot phenomenon) and the time period right before and after on- and offset of a masker (pre- and post-masking). Considerations of the interaction of subsequent maskers (temporal masking patterns) close the monaural subsection. Accordingly, simultaneous and temporal masking are described for the bilateral/contralateral case (central masking), followed by a short recapitulation of the results.

4.3.1 Monaural Masking

For the investigation of monaural masking, masker and maskee are presented to the same ear via headphones. This method ensures that the signals don't enter the other

[18] See Gelfand (1990), p. 353.

[19] According to Gelfand (1990) or, respectively Fastl (1977), p. 317.

ear by sound propagation, deflection or reflection. In natural listening situations this is hardly the case. Therefore, it is inadequate to simply transfer the results of monaural listening tests to everyday life. For this purpose binaural masking has to be taken into account, too. Furthermore, dynamic signals, interaction with the environment, individual listening experience and the ability of the auditory system to adapt to situations play a crucial role in auditory perception as a whole and in masking in particular. Monaural masking experiments with passive listeners and artificial, quasi-stationary signals have led to results that are relatively accurate and reliable. Although not universally applicable for dynamic, binaural signals as experienced in music listening, the experiment results served as a basis for masking-based compression of music, speech and other audio signals.

4.3.1.1 Simultaneous Masking

As described in Sect. 4.2, the biomechanics of the inner ear lead to traveling waves which peak in a certain frequency-dependent area. This traveling wave excites the hair cells along the basilar membrane to neuronal fire. Amplitudes of frequencies that do not protrude the envelope of simultaneous traveling waves do not undergo a cochlear amplification. They are masked. Thus, a sound increases the absolute threshold to a masked threshold. The amplitudes of simultaneous sounds need to surpass this masked threshold to be audible.

Sound waves enter the cochlea at the oval window, i.e., at the base. Here, traveling waves are excited. They build up until they reach their peak region. Behind the peak the wave collapses quickly. The peak of high frequencies lies near the base. The peak of low frequencies lies closer to the apex. Hence, traveling waves of low frequencies pass the peak region of higher frequencies, but not vice versa. Consequently, low frequencies excite the whole bandwidth, i.e., hair cells along the whole basilar membrane fire. To ensure that a low frequency sounds like a low tone instead of a broadband sound, the auditory system detects the peak region and amplifies the neural firing at the corresponding hair cells. At the same time the neural activity at the other hair cells remains comparably low for three reasons: First, the elongation is much lower than at the peak region, so the hair cells are deflected less, which causes less neural activity. Second, the frequency of the traveling wave does not coincide with the best frequency of these neurons, so their response is relatively weak. Third, due to the mismatch of traveling wave frequency and best frequency, the excitation seems to be actively suppressed by the auditory system. Only if a peak protrudes from the envelope of this low-frequency tone, it will be amplifies as well. Otherwise, it is masked. This behavior is reflected in the different masking thresholds per masker frequency and amplitude, as illustrated in Fig. 4.8.

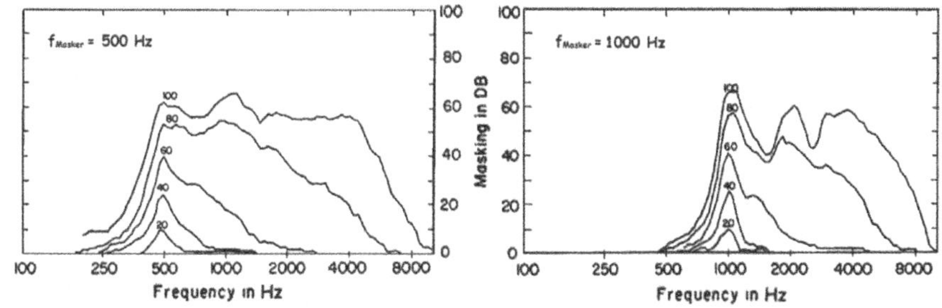

Fig. 4.8 Masking patterns for a 500 Hz-masker and a 1 kHz-masker with five different amplitudes (indicated by the numbers near the lines). A second frequency has to surpass this threshold to be perceivable for a listener. Reproduced from Ehmer (1959, p. 1117), with the permission of the Acoustical Society of America

4.3.1.2 Simultaneous Maskers

Several simultaneous maskers can create a joint masked threshold as illustrated in Fig. 4.9. The effect from maskers within the same critical band adds up and the masked threshold increases. Maskers in other frequency regions create their own masking threshold. The joint masked threshold from frequencies of different frequency bands is not the sum of all contributing masking patterns. Nor does the masking threshold equal themasking pattern of the nearest or strongest masker. Nonlinear effects have been observed and the presence of a second masker has the potential both to increase, but even to suppress, some masking effect of the first.

Two simultaneous tones with similar frequencies $f1$ and $f2$ and amplitudes create the impressions of one tone with a slowly modulated amplitude (beat), namely their mean frequency $\frac{(f1+f2)}{2}$ with a beat rate of their frequency distance $|f1 - f2|$. The reason for that is that the areas on the basilar membrane, excited by the similar frequencies highly overlap. That means they evoke firing of the same neurons with a firing rate proportional to their envelope.[20] Beat frequencies between 15 and 300 Hz lead to the impression of roughness, occurring together with combination tones, caused by nonlinearities in the auditory system which are not fully understood.[21] Larger frequency differences are perceived as two different tones.

Interaction between tones, like beats, roughness, combination tones or fusion can affect masking. For example beats between two tones can be heard even if one of the tones' amplitude is below absolute threshold; then the tone becomes indirectly audible. The same applies for combination tones of a masker and a maskee. Furthermore, in some listening tests complex sounds were reported to not create a joint masking pattern but one masker leads to improvement or attenuation of another masker's masking effect. This phenomenon is called *suppression*. Its appearance

[20] See Gelfand (1990), pp. 406f.

[21] See e.g. Zwicker and Fastl (1999), p. 33. An examination of nonlinearities in the auditory system and roughness can be found e.g. in Zwicker and Fastl (1999), pp. 50ff and 257ff.

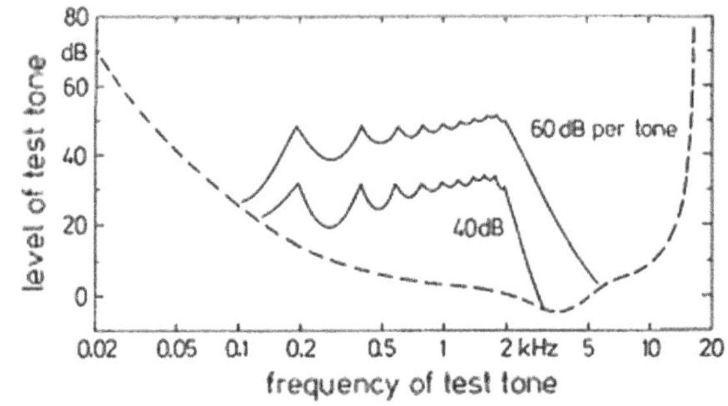

Fig. 4.9 Joint masking pattern of a 200 Hz-tone with the first nine harmonics with random phase and equal amplitudes of 40 and 60 dB. The dashed line shows the absolute threshold. From Zwicker and Fastl (1999), p. 71

and magnitude strongly differs between different investigations and seems to be quite vague and subjective. To avoid these effects, narrow banded noise is used for the investigation of masking effects. Their amplitudes are specified as spectrum level (SL)[22]:

$$dB_{SL} = dB_{overall} - 10\,\lg\,(\text{Bandwidth})\,. \tag{4.7}$$

However, in natural listening situations those effects do occur and may attenuate masking.

4.3.1.3 Overshoot Phenomenon

At the onset of a masker, its masking effect lies considerably higher than during stationary state. The increased masked threshold decreases within 50 ms to the magnitude of the masked threshold during stationary state. This phenomenon is called *overshoot phenomenon* and leads to an increase of the masked threshold of up to 26 dB for broadband signals and between 3 and 8 dB for a sinusoidal masker. However, it is not observed for narrow band noise. Furthermore, dependent on phase relation between masker and maskee or spectral distribution of the masker, the overshoot varies within a range of up to 10 dB. As for the suppression phenomenon, it strongly varies interindividually and in some investigations no overshoot could be produced at all.

4.3.1.4 Pre-masking

The term pre-masking or backwards-masking describes the masking effect right before the onset of a masker. This effect is caused by the different processing times for sounds with different magnitudes in the auditory system. The processing of soft tones takes more time than that of loud ones. Thus, the processing time of a soft sound followed by a slightly delayed loud sound largely overlap and are therefore perceived

[22] See Gelfand (1990), p. 356.

simultaneously. A possible reason is that in nature loud sounds may indicate danger and therefore have higher priority. Pre-masking is effective for a duration of about 20 ms before the masker's onset.[23] The temporal envelope of the masked threshold is only slightly dependent on the absolute amplitude. However, the duration of the masker has quite a big influence. Longer signals cause a flattened and longer masked threshold up to a length of 10 ms for broadband noise, 20 ms for critical-band-wide noise and 50 ms for a sinusoidal sound.[24] Listening test results show a wider variance for pre-masking than for simultaneous masking.[25]

4.3.1.5 Post-masking

After a masker's offset the masked threshold remains constant for about 5 ms before descending towards absolute threshold. The reason for this 5 ms-sustain is the inertia of the auditory system; the oscillation of the excited hair cells attenuates. Furthermore, the integration time of the filter mechanisms of the auditory system is 2.5 ms. Around 5 ms the sounds of masker and maskee "smear".[26] As for pre-masking, the masking effect of post-masking is dependent on duration of the masker rather than its absolute amplitude. It is weaker than in case of pre-masking but lasts up to 200 ms. Figure 4.10 illustrates the masked threshold for different masker durations. It needs to be mentioned that, as for pre-masking, different listening tests led to different results.[27] Figure 4.11 schematically illustrates the whole masking threshold of monaural pre-, simultaneous, and post-masking.

4.3.1.6 Temporal Masking Patterns

As complex sounds create a joint spectral masking pattern, successive masking sounds create a joint temporal masking pattern. Between consecutive masking sounds, it sometimes lies higher than the sum of overlapping pre- and post-masking. Even the simultaneous masking threshold is increased. These observations are illustrated in Fig. 4.12.[28] Sometimes, however, the joint masking pattern is quite similar to the sum of pre- and post-masking. It is expected that nonlinear legalities apply for the detection of single elements from spectrally and temporally complex sounds, leading to interindividually different masking patterns. This would explain the high variance of performance data compared to nonsequential masking experiments.

---

[23]See Zwicker, p. 82. According to Gelfand (1990), pre-masking effects up to 100 ms were observed, see Gelfand (1990), p. 374.

[24]According to Brandter (2007), p. 126.

[25]See Brandter (2007), p. 125.

[26]Cf. Brandter (2007), p. 120.

[27]Cf. Brandter (2007), p. 120.

[28]Discussed in detail in Fastl (1977), Fastl (1979) and Gelfand (1990), p. 375.

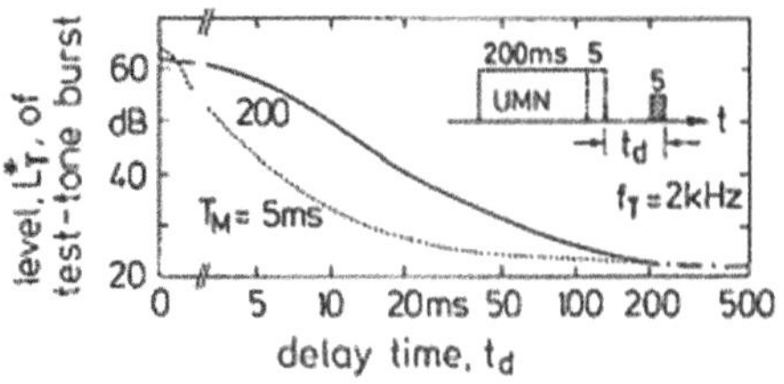

Fig. 4.10 Temporal development of the masked threshold for a 2 kHz masker with different durations (solid line = 200 ms, dashed line = 5 ms). For masker durations up to 200 ms it applies: The shorter the signal the steeper the temporal decline in masking threshold. From Zwicker and Fastl (1999), p. 84

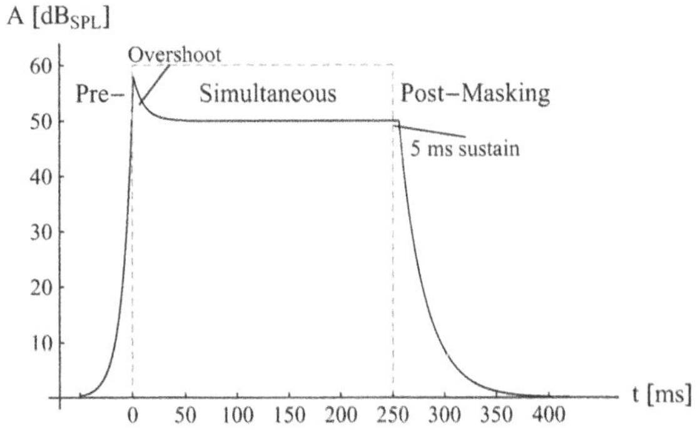

Fig. 4.11 Schematic illustration of a temporal masking pattern including pre-masking, overshoot phenomenon, simultaneous masking, a 5 ms-sustain and post-masking for a masker of 60 dB$_{SPL}$

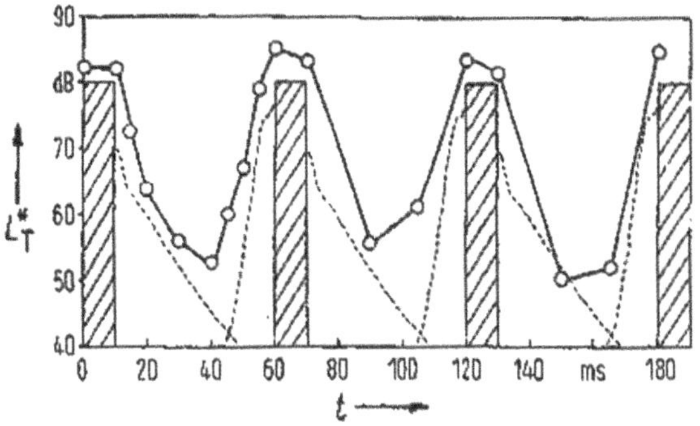

Fig. 4.12 Temporal masking pattern of a succession of critical band wide noise. The hatched bars indicate the durations of the 70 dB loud maskers, the solid line connects the examined masked thresholds which are indicated as circles. The dashed lines represent the pre- and post-masking thresholds as expected from research results with single critical band wide noise. Reproduced from Fastl (1977, p. 329), with the permission of Deutscher Apotheker Verlag

4.3.2 Binaural Masking

Binaural masking, or *central masking*, describes the masking effect which arises when masker and maskee are presented to opposing ears. As the term indicates, the masking effect in central masking presumably emerges in the central nervous system. The masked threshold typically lies 1–2 dB higher than absolute threshold and is almost independent of the masker amplitude. It becomes remarkably higher when both sounds have a simultaneous on- and offset. In this case it is likely that they are both integrated into the same auditory stream, as will be described in more detail in Sect. 4.5. Basically, it means that they are perceived as belonging together and are therefore processed as one. Figure 4.13 compares the masking patterns of both cases. As binaural masking happens at higher stages of the auditory pathway, the masking effect is much weaker.

4.3.2.1 Temporal masking

The overshoot phenomenon also occurs in central masking and lasts up to 200 ms.[29] Central pre-masking is weaker than monaural pre-masking. Yet the difference is smaller than for simultaneous and post-masking. Figure 4.14 shows a comparison of pre- and post-masking for monotic and dichotic performance.

4.4 Spatial Hearing

The ability of people to localize sound sources is a well researched topic in psychoacoustics. The literature of Blauert[30] is considered a standard work summarizing the state of research in the seventies—with the postscript in the mid-eighties—especially derived from listening tests. Supplemented by Blauert and Braasch (2008), Dickreiter (1987), Webers (2003) and Strube (1985), the most important results are expounded in the following subsection.

After presenting the general testing conditions and a short clarification of terms, findings are shown, structured in four separate domains:

1. Localization in the horizontal plane
2. Localization in the median plane
3. Distance hearing
4. Localization in case of various sound sources.

---

[29]See Gelfand (1990), pp. 369f.

[30]See Blauert (1974) and Blauert (1985), translated in Blauert (1997).

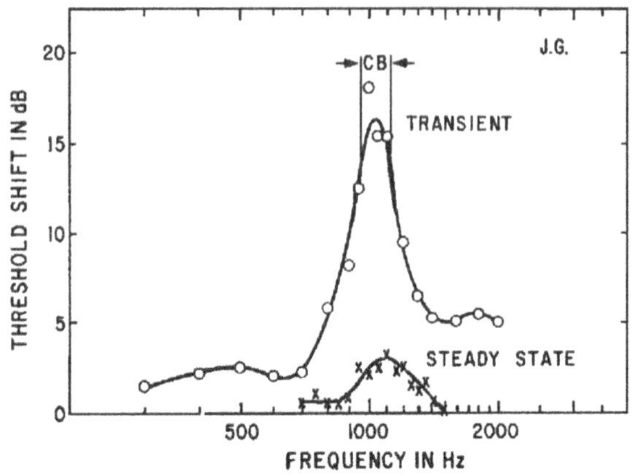

Fig. 4.13 Central masking pattern for a 1 kHz tone burst masker with a duration of 250 ms and maskees of different frequencies and a duration of 10 ms. Closer to the masker onset (TRANSIENT) the masking threshold is much higher compared to later maskee onsets (STEADY STATE). In both cases the masked threshold is far below monaural masking. Reproduced from Zwislocki et al. (1968, p. 1268), with the permission of the Acoustical Society of America

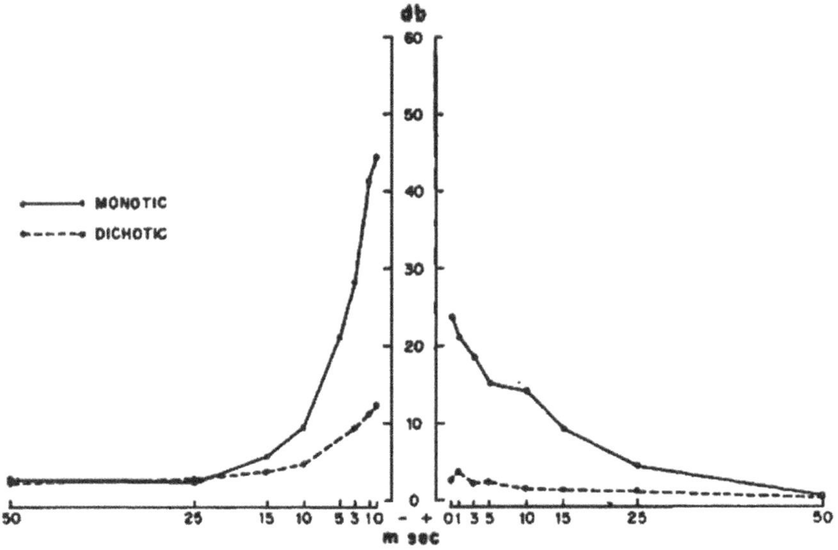

Fig. 4.14 Comparison of temporal pre- and post-masking patterns for monaural (solid lines) and binaural signals (dashed lines). The masker is a 50 ms broad-band noise at $70dB_{SL}$, test signals are 10ms-lasting 1 kHz-tone bursts. Reproduced from Elliott (1962, p. 1112), with the permission of the Acoustical Society of America

4.4.1 Research Conditions and Definition of Terms

The auditory system uses a variety of analytical processes of the acoustical signals impinging the eardrum for spatial hearing. These usually proceed preconscious: Sensations are assigned to a direction and distance in a bottom-up-processing.[31] In case of familiarity with the signal or the acoustical environment top-down-processes support the localization.[32] Parameters gathered for localization are *interaural time difference* (ITD) and *Interaural Level Differences* (ILDs) and an individual filtering of the signal depending on incidence angle. The *Head Related Transfer Function* (HRTF) quantifies the changes a sound undergoes from its origin to the eardrums of a listener as resulting from sound propagation and filtering, caused by reflections in the pinna, diffractions around and acoustic shadow behind head, torso, shoulders, etc. Sound playback via headphones often leads to a localization inside the head. Hence, it is only spoken of *lateralization* in this case. Localization parameters and accuracy immensely distinguish between spatial planes, therefore they are illuminated individually.

People tend to consult visual cues additionally to hearing for the localization of sound sources, as will be discussed in detail in Sect. 4.5. In the listening tests this is avoided by blindfolding the subjects. Also, bone conduction is neglected because its auditory threshold is more than 40 dB higher than the threshold of the eardrums.

The following tests take place in free field rooms exclusively, such as the room illustrated in Fig. 4.15. The probands sit in the center of a hemispherical loudspeaker array with fixed heads or in a darkened free field room and judge the assumed origin direction of a sound verbally or by pointing.

Typical test stimuli are pure tones, narrow band and broad band noise and occasionally speech. Furthermore, Gaussian impulses are used. These are pure tones multiplied by a Gaussian function which gives the continuous tones the envelope of the Gaussian bell curve function resulting in a small spectral widening and imprecise note on- and offsets. If nothing else is alluded, one signal at a time is concerned. The distance of the source is so big that the wave fronts reaching the listener can be considered as being plane. Subjects are people without hearing loss.

For sound source localization it is meaningful to use a head-related spherical coordinate system with the head as the origin. The horizontal and median plane can be seen in Fig. 4.16. The vertical plane could be considered as a combination of the other planes. The area below the height of the head is ignored because it is comparatively small and a total surrounding of a subject with loudspeakers is difficult to arrange.

The actual direction/distance is denoted *sound event direction/distance*. *Auditory event* is the position of the source as assumed by the subject. The *localization blur* is magnitude at which 50% of the subjects recognize a change of the sound event

[31] See Blauert (1997), p. 409.

[32] See Bruhn (2002b), p. 444, definitions of bottom-up- and top-down-processing see e.g. Myers (2008), p. 214.

Fig. 4.15 Free field room of the University of Göttingen during a test arrangement with 65 loudspeakers. Reproduced from Meyer et al. (1965, p. 340), with the permission of Deutscher Apotheker Verlag

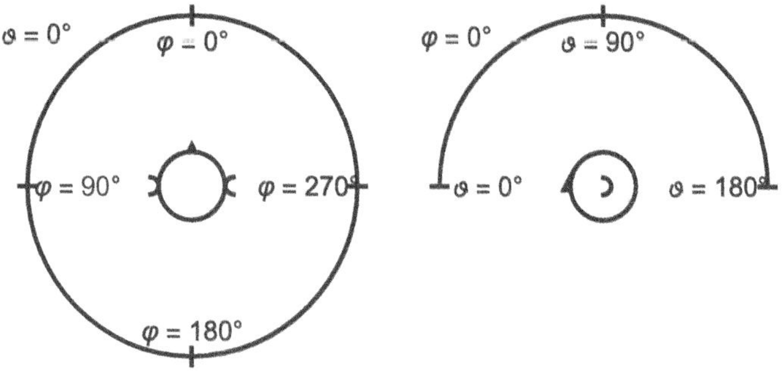

Fig. 4.16 The horizontal (left) and median listening plane (right). After figures in Blauert (1974)

location. The blur is assumed to come from a somewhat vague auditory localization capability. Sometimes, it is put into relation with the perception of source width.

4.4.2 Horizontal Plane

People can localize sound sources more precisely and robustly than in the median plane. Audio signals with certain properties are ideal for localization. Other signals may cause confusion or systematic localization errors.

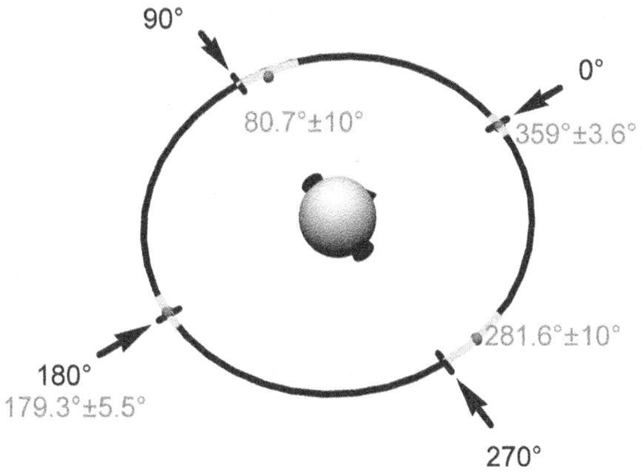

Fig. 4.17 Auditory event directions (spheres) and localization blurs (gray curves) in the cases of fixed sound events (arrows) in the horizontal plane. After Blauert (1997), p. 41, with data taken from Haustein and Schirmer (1970) and Preibisch-Effenberger (1966)

Localization capability:

Mechanisms to localize sound sources are superior in the horizontal plane compared to the median plane. Horizontal location is especially based on binaural signal comparisons. The localization capability is best in the frontal area with an average accuracy of 1° and a localization blur of about ±3.6°. This localization blur can be considered as the JND in position.[33] The localization blur is signal-dependent. It is largest in a frequency range between about 1 or 1.5 kHz and 3–4 kHz. Interestingly, this is exactly the frequency region in which we exhibit the lowest threshold in quiet and the largest dynamic range. It is also a formant region of many vowels.

Towards the sides the deviations and the localization blur increase distinctly. Completely lateral signals are typically estimated too frontal. Here, the JND in position lies between 12 and 18°.[34] Especially for unfamiliar and narrow band signals the auditory event direction be axial symmetric to the actual sound event direction. This effect is known as *localization inversion* and is illustrated in Fig. 4.18. A typical mistake in localization is a "front-back reversal", also called "front-to-rear confusion", especially for low frequencies. In the figure, a sound coming from the front-right is heard as coming from the rear-right and a rear-left incidence is heard as front-left source. At higher frequencies, the HRTF—mostly due to wave shadow behind the pinna—yield audible spectral differences between sources from the front and the rear, which inform subjects on the source position, if they are familiar with the sound.[35]

The coherence between ILD and lateralization of sounds through dichotic headphones is almost linear but with quite quite a localization blur especially off the center as can be seen in Fig. 4.19.

However, the auditory event angle per ILD is dependent on frequency. For broadband signals a level difference above 11 dB leads to a total lateral impression, narrow banded signals need larger differences. The coherence between interaural phase dif-

[33] See e.g. Webers (2003), p. 120.

[34] See Webers (2003), p. 120.

[35] See Kling and Riggs (1971), p. 351 and Blauert (1997), p. 360.

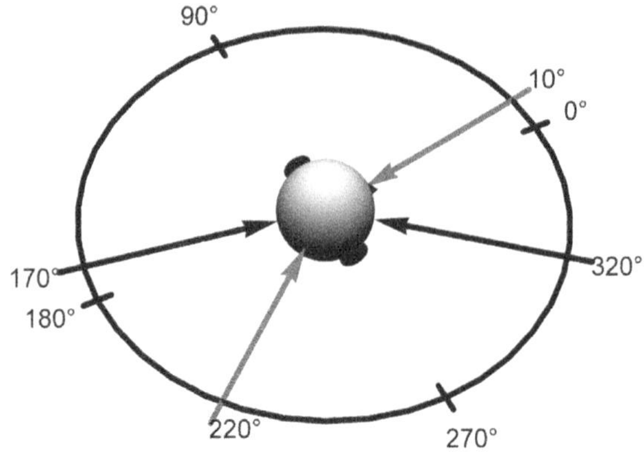

Fig. 4.18 Examples of localization inversions in the horizontal plane, after Blauert (1974), p. 35

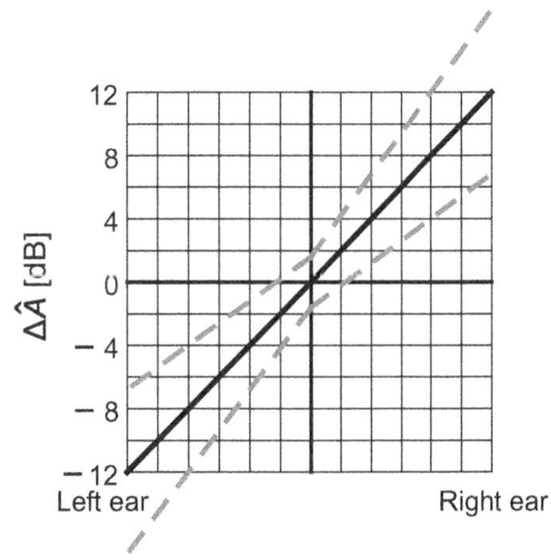

Fig. 4.19 Lateralization (black line) and lateralization blur (region within the dashed lines) per interaural level difference ($\Delta\hat{A}$ in dB). After Blauert (1997), p. 158

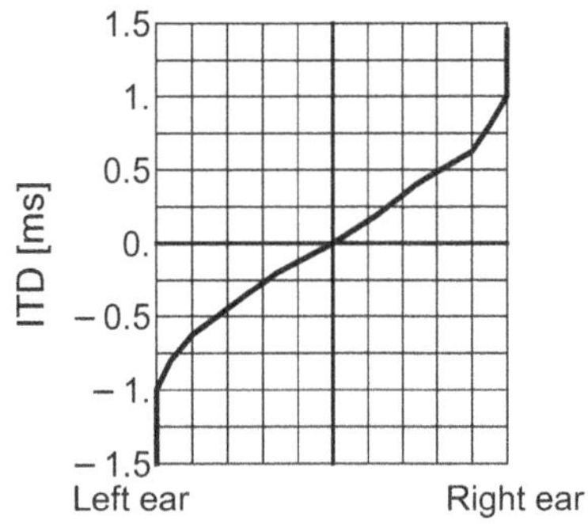

Fig. 4.20 Lateralization per ITD according to data from Blauert (1997), p. 144

ference and lateralization is also relatively linear in the range from 0 to 80% of a completely lateral angle until approximately 640 μs arrival time difference. This can be seen in Fig. 4.20.

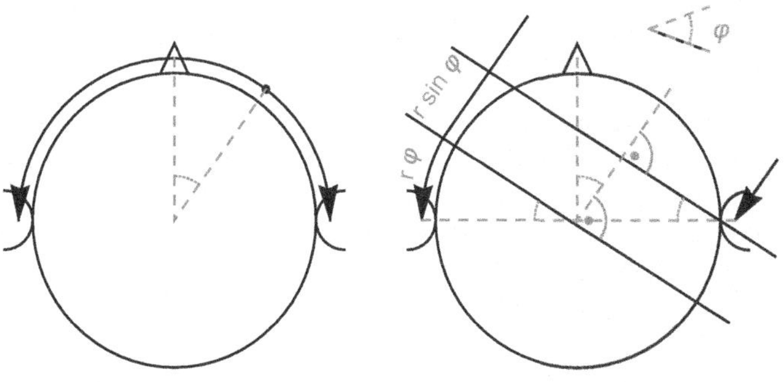

Fig. 4.21 Binaural distance difference for a source in the near and the far field. After Kling and Riggs (1971), p. 351

ITD above 640 μs lead to hardly any increase of the auditory event angle, probably because a source actually placed at 90° leads to an ITD of roughly 640 μs, assuming a head-radius of 8.5 cm. It has to be taken into account that completely lateral signals are localized 10° too frontal, as discussed above and shown in Fig. 4.17. According to Myers (2008), the JND in ITD lies at about 27 μs whereas (Zwicker and Fastl 1999) consider it to lie at about 50 μs, mentioning that individual values between 30 and 200 μs have been measured. Kling and Riggs (1971) even state that values up to 300 μs are possible.[36] Kling and Riggs (1971) quantified the relationship between incidence angle of a source and the length of the path of its propagating wave to the two ears for a head in the near field and in the far field of the source, considering diffraction of a wave around a sphere representing the listener's head.[37] This formulation is illustrated in Fig. 4.21. Dividing this path difference by sound velocity yields the relationship between source angle and ITD[38]:

$$
\begin{aligned}
ITD_{\mathrm{NF}} &= \frac{2r\varphi}{c} \\
ITD_{\mathrm{FF}} &= \frac{r\left(\sin\varphi + \varphi\right)}{c}
\end{aligned}
\tag{4.8}
$$

Here, the subscripts NF and FF denote the near field and the far field, r is the radius of the sphere, c is the sound velocity and φ is the azimuth angle of the source in a head-related coordinate system where $-\frac{\pi}{2} \leq \varphi \leq \frac{\pi}{2}$. For a source in the far field, this formulation can be extended to cover sources beyond the horizontal plane[39]:

---

[36] See Myers (2008), p. 240, Zwicker and Fastl (1999), pp. 293f and Kling and Riggs (1971), p. 355.

[37] See Kling and Riggs (1971), p. 351.

[38] See Kling and Riggs (1971), p. 352.

[39] See Larcher and Jot (1999).

$$ITD_{\mathrm{FF}} = \frac{r\left(\arcsin\left(\cos\varphi\sin\vartheta\right) + \sin\varphi\cos\vartheta\right)}{c} \tag{4.9}$$

$-\frac{\pi}{2} \leq \vartheta \leq \frac{\pi}{2}$ is the elevation angle.

Contradictory interaural attributes might compensate each other via *trading* or the signal is perceived as two signals from different directions where low frequencies seem to arrive from the direction suggested by phase difference, whereas high frequencies seem to arrive from the direction derived from ILD.

Demands on the signal:

Very low frequencies reveal barely no evaluable level differences due to a negligible wave shadow behind the head. Likewise, the ITD yields no detectable phase differences. It is these minor cues that make it difficult for the auditory system to localize low frequencies. Here, the ITD—especially of onsets, transient sounds, short signals and the envelope of sound—play a central role.[40] Low to medium frequencies may show an evaluable phase difference due to ITD which becomes a dominant localization cue in that frequency region. In the range from about 1.5 kHz to 3 kHz the localization capability is poor despite the high sensitivity to volume in this region. Here, on the one hand, the frequencies are too high for unambiguous phase relations. Furthermore, the auditory neurons are not capable of firing rapidly enough to display the phase difference at higher frequencies.[41] On the other hand, the wave lengths are too large to create noticeable level differences by acoustic wave shadow. For higher frequencies, filtering by head, hair, pinna and shoulders cause ILD and make it the dominant aspect which leads to a proper localization even for stationary sounds. Furthermore, the human auditory system is capable of detecting ITD of the envelope of high frequencies but it is unknown whether these envelope delays deliver reliable localization cues.[42] The spectrum resulting from the individual HRTF shows prominent peaks and notches between 3 and 14 kHz which support localization. Still, localization of high pass noise above 16 kHz is imprecise because ILD are the only evaluable cues.[43] Thus, a good localization demands a large bandwidth, transients and distinct sound envelopes. However, the bandwidth of many musical instruments is so large that both aspects, ITD and ILD, occur in combination.[44] For front/back localization the direction-dependent filtering, HRTF, delivers the only valuable auditory cues. The HRTF is very individual, depending on size and shape of head, trunk, pinna and torso.

[40] See e.g. Kling and Riggs (1971), pp. 350ff and Morikawa and Hirashara (2010), p. 419.

[41] As discussed in Sect. 3.2.2, see also Hall (2008), p. 343, Davis (2007), p. 750 and Ross et al. (2007).

[42] See e.g. Middlebrooks and Green (1991), pp. 142f.

[43] See Morikawa and Hirashara (2010), p. 419.

[44] The so-called "duplex theory", see e.g. Bruhn and Michel (2002), p. 651.

4.4.3 *Median Plane*

In the median plane we assume only minor interaural differences. Sources in the experiments have rather omnidirectional radiation properties, so the signals reaching both ears should be about the same. Consequently, monaural cues dominate localization in the median plane.

Localization capability:

People's localization capability in the median plane is distinctly worse than in the horizontal plane. Because the human's head is relatively symmetric, the signals from a source in that plane reaching both eardrums are quite similar; monaural signal features are dominant. Results of a localization test with speech are illustrated in Fig. 4.22.

The localization capability is best in the frontal area. The frontal signal in height of head is localized correctly. The divergence of 6° for the elevation angle of 36° is relatively small compared to signals from overhead and behind. The localization blur even in the frontal area is ±10°. As in the horizontal plane, sound events from 90° are estimated too frontal. Also, localization inversion appears with the vertical axis. The minimal localization blur for unfamiliar speech is twice as much as for familiar speech. For white noise the minimal localization blur in the frontal area is only 2°. For narrow band signals, the auditory event angle is dependent on frequency and almost independent from the actual source position. Figure 4.23 schematically shows the auditory event course per center frequency of narrow band noise with a width of one to two thirds. The pathway also holds for complex signals, if the particular frequency region is dominant. Note, that this path is very rough. In fact, it is neither as smooth nor as continuous as depicted in this figure. Furthermore, it describes a general observation, which is not inter-individually true.

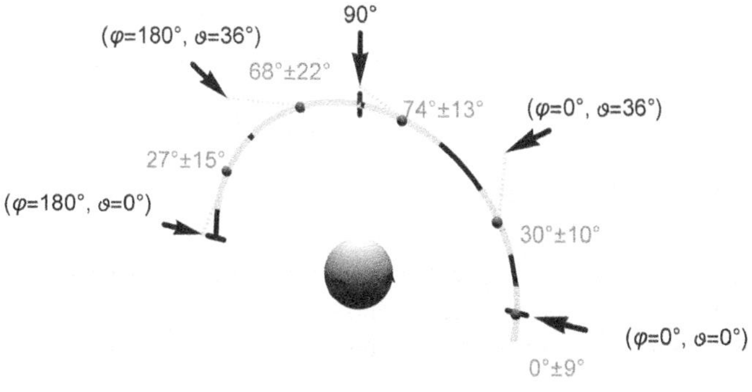

Fig. 4.22 Localization (spheres) and localization blur (gray curves) in the median plane for speech of a known speaker. The gashed gray lines connect the associated sound event and auditory event. After Blauert (1997), p. 44

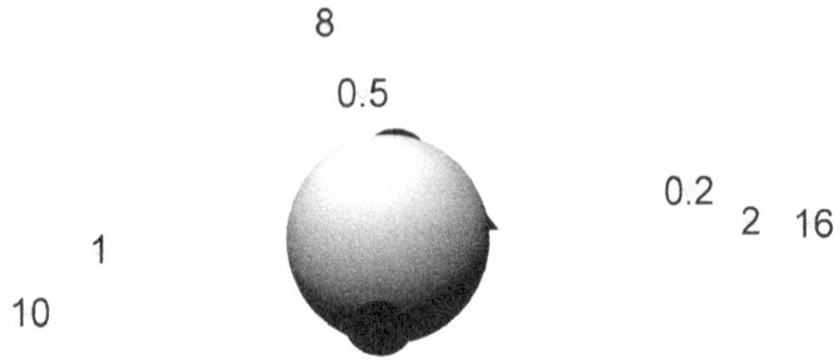

Fig. 4.23 Schematic pathway of the auditory event direction for narrow band noise of variable center frequencies from arbitrary directions in the median plane. After Blauert (1974), p. 36

Demands on the Signal:

In the median plane the frequency range plays an important role for the localization of narrow band noises. This can be adapted to pure tones, and to sounds and broadband noises, if a narrow frequency range protrudes. Impulse-containing, short signals are often localized in the rear area. For known sound signals the HRTF delivers an evaluable directional clue. From listening test and data analysis by means of principal component analysis—a method to reduce a number of variables from a large set of variables—(Martens 1987) derived that subjects estimate the elevation of a source with the help of five components whereas for horizontal localization one or—according to Sodnik et al. (2006)—two components, namely ILD and ITD are sufficient.[45] For example notches, band-reject-, band-pass- and high-pass-filter characteristics were reported to correlate with elevations in the median plane.[46]

Demands on the signal for a good recognition of the elevation is a wide spectrum with frequencies above 7 kHz. A correct front-back-localization is given if the signal contains a bottom cutoff frequency between 2 and 8 kHz. In the median plane learning effects are noticeable.

A languor of the audition effectuates that people localize rapidly moving sources diffuse. The determination of a direction is perceived as integral over a time interval of approximately 172 ms in the horizontal plane and 233 ms in the median plane. Contradictory directivity cues by ILD, ITD and monaural filtering lead to a localization determined by ITD, as long as low frequencies are contained in the signal.[47]

4.4.4 Distance Hearing

Experiments on distance hearing mostly concentrate on the front. Consequently, no interaural cues are considered. In an anechoic environment, distance localization is rather coarse, especially for unfamiliar sounds.

[45] See Martens (1987) and Sodnik et al. (2006).

[46] See Middlebrooks and Green (1991), p. 145.

[47] See Verheijen (1997), p. 8.

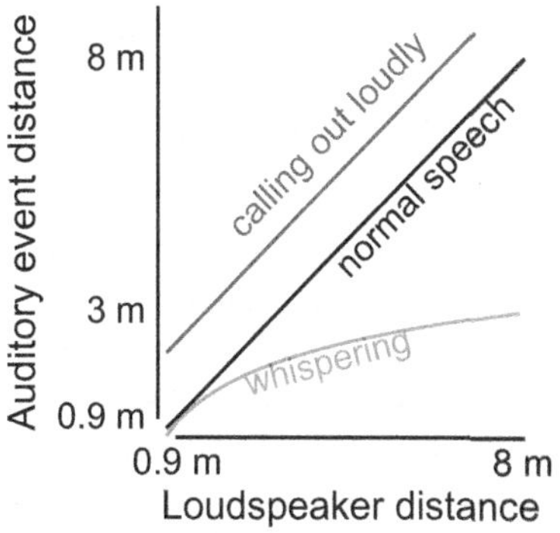

Fig. 4.24 Auditory event distance for different types of speech presented via loudspeaker in front of a listener. After Blauert (1997), p. 46

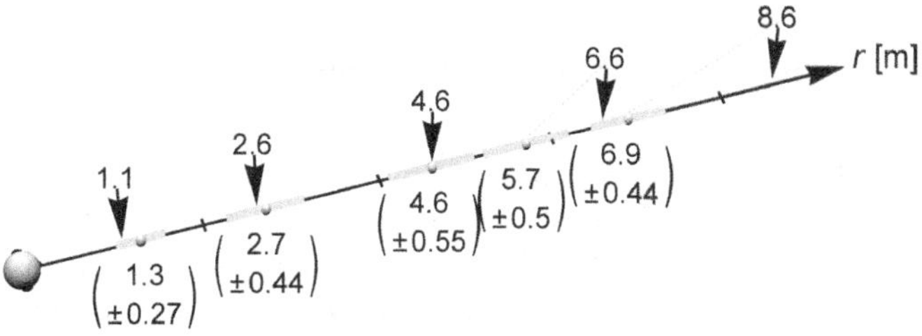

Fig. 4.25 Sound source and auditory event distance for Bangs with approximately 70 Phon. The dashed gray lines connect the related sound event and auditory event. After Blauert (1997), p. 47

Localization capability:

Since during the attempts about distance hearing some sources are placed very close to the subject, the wave fronts reaching the ears cannot be considered as being plane. For distance estimation the familiarity with the signal is of great importance. As illustrated in Fig. 4.24 only common speech is localized correctly.

The distance of frontal noises is underestimated for distances above 5 m. The localization blur is about 0.5 m except for immediate head proximity where it is smaller (Fig. 4.25).

For close unfamiliar signals the auditory event lies too close up to a localization inside the head or the impression of a source directly behind the head. The perceived spectrum from familiar sounds helps to localize distances between 0.25 and 15 m since the pressure level decays evenly whereas perceived loudness does not because contours of equal loudness change over the overall amplitude.[48] In a natural environment the pressure level relation between direct sound and first reflections denote the distance. For distances further than about 15 m high frequencies are damped stronger

[48]Detailed information on the *inverse distance law* which describes the sound pressure decay is given in Sect. 5.1.6. Details on contours of equal loudness can be found e.g. in Zwicker and Fastl (1999), pp. 203ff.

than lower ones, because the distance between areas of excess and reduced pressure are smaller in cases of small wavelengths. Thus, more acoustic energy transfers to heat exchange. Therefore, signals from afar sound more dull.

Demands on the signal:

When a listener is familiar with the signal, loudness and spectrum are important distance clues. The distance of a completely unknown sound can't be localized under laboratory conditions. Even for more familiar sounds distance hearing is imprecise. Screaming is localized too far away, whispering too close by. Best localization accuracy can be observed for distances between 1 and 5 m. The localization ability notably improves when the sound is heard in a natural, known environment with its specific spatial acoustics. Then, arrival time- and loudness difference between direct sound and early reflections give applicable distance cues.

4.4.5 Localization of Various Sound Sources

If a source is located in the median plane, its sound will reach both ears simultaneously. Amplitudes of contained frequencies will be similar at both ears. Only the radiation characteristics of the source and maybe slight anatomic dissymmetry create little interaural level and phase differences. For any other location, the sound will reach the ears at different points in time and especially high frequencies will arrive with different amplitudes due to the inverse distance law and due to a wave-shadow. The auditory system uses these interaural cues to localize sound sources. If several sources are present, the auditory system needs to identify what portion of sound belong together and find the location of all or at least of one or some important sounds. This organization of sound—i.e. the integration of sounds that belong together and their segregation from each other—is referred to as *auditory scene analysis* and will be addressed in the following section, Sect. 4.5. If the auditory system correctly interprets the ear signals as a result of different traveling paths from one source to the two ears, the above mentioned localization mechanisms are valid and pretty robust. However, this interpretation by the auditory system is prone to mistakes. The auditory system may interpret signals coming from different sources as belonging to the same source, having only one source location. According to the idea of "summing localization" this perceived position is exactly that position at which a sound source would have to be located to create similar ITD and ILD.[49] This position does not necessarily have to coincide with one of the actual source positions. If such a position is found, the localization is distinct. If not, a wide or diffuse sound source is perceived. Many stereophonic sound systems make use of this principle to generate *phantom sources* or *diffuse sound fields* as will be discussed in Chap. 7. Two loudspeaker signals which are identical except for a slight level difference can create the impression of a source location somewhere between the two loudspeakers.

---

[49]See e.g. Strube (1985), p. 69.

Similar, but somewhat incoherent signals can provoke the impression of a wide or diffuse sound field. Theile (1980) criticized the theory of summing localization as a result of simple comparison between localization cues of a superimposed sound field with that of a single source. In his "association model" he expands the idea of summing localization by auditory mechanisms which later became part of the auditory scene analysis principles as formulated and extensively discussed by Bregman (1990) to explain the general psychological treatment of sound.[50] However, in the case of several sources present, another localization effect can occur. The auditory system can distinguish direct sound from reflections to a certain degree. The "precedence effect", "Haas-Effekt" or "Law of the first wavefront"[51] indicates that sound events are localized solely in the direction of the first wave front arriving at the ears, even if similar but later arriving sounds are much louder. Even if a sound arriving with a delay of 5–30 ms is 10 dB louder than the first arriving wave front it won't affect the localization.[52] The effect occurs especially but not exclusively with transient signals, particularly at onsets. A frontal sound is localized correctly, even if lateral reflections reach the ears, since the first wavefront was already crucial for the localization. Premise is that the first and second arriving signal fuse, i.e. that they are integrated into one auditory stream as will be discussed in the upcoming section. From a time delay of about 50 ms on, auditory event and echo are perceived individually as it is a typical threshold of the auditory system.[53] The precedence effect can last for seconds and more.[54] Zurek and Saberi (2003) found evidence that the precedence effect does not fully suppress other localization cues that follow the onset. Rather an interaural cross correlation after the onset can stabilize or adjust the auditory event position.[55]

As described in Sect. 4.3, the sound of one source can completely or partially mask the sound of another source. The masking threshold at the listener's position is independent of spatial distribution of the sources. What counts are the sound pressure levels at the ears. As people have two ears, the masking effect may be reduced if the sound is masked at one ear only.

4.5 Auditory Scene Analysis

Bregman (1990) published an extensive elaboration regarding the perceptual organization of sound, called "auditory scene analysis". He demonstrates it on the basis of laboratory experiments with artificial sounds, such as sinusoidal tones or noise,

[50] See Theile (1980) and Bregman (1990).

[51] See David jr. (1988), p. 159, Friedrich (2008), p. 39, Hall (2008), p. 469 and Blauert (1997), p. 411.

[52] See e.g. Dickreiter (1978), p. 77 and Friesecke (2007), p. 139.

[53] See e.g. Blauert (1974), p. 180 or Strube (1985), p. 68.

[54] According to Blauert (1974), p. 224.

[55] See Zurek and Saberi (2003).

as well as from listening experience and experiments with music. This section summarizes the essence of his work.[56]

Whatever we hear is our perceived part of the acoustic outside world: the auditory scene. Single units in the auditory scene are called *auditory streams*, a counterpart to the *visual object*. In natural listening situations sounds from different acoustic sources overlap in time and spectrum and the sound pressures reaching the ears are always the sum of all propagated sounds and their reflections. The task of the auditory system is to analyze these complex sounds to be able to identify what parts belong together (*integration*) as well as to discriminate between different streams (*segregation*). This *grouping* is the attempt of the auditory system to create a mental representation of the world in which every stream is derived from the same environmental entity or *physical happening*. Such a categorical perception is crucial for a proper understanding of and orientation in the outside world. Auditory scene analysis is not an explanation of how exactly this is accomplished by the auditory system by means of biological, biochemical, physiological or neurological functionality. Rather it describes organization patterns which can be observed in the perception of the acoustic environment, most of which are primitive, innate, pre-attentive bottom-up grouping processes, whereas higher levels of grouping are schema-based, attention-directing top-down processes, according to our knowledge of attributes and behavior of familiar sounds.[57]

If components have arisen from the same physical event they naturally have many more things in common than could be explained by chance, e.g. timing, frequency, and differential effect on our two ears. There exists not one exclusive parameter which determines auditory scene since there is no law of nature from which we could derive a concept for an adequate auditory scene analysis. Rather a complex system of certain principles is used for this task, many of which are known from Gestalt psychology. The necessity for this redundancy can be easily explained. One could think that the location of a sound source is a proper parameter to distinguish different physical happenings since only one thing can be on one position at a time. But firstly, our localization capability is quite weak e.g. in the lateral region and especially in the vertical dimension, as discussed previously in this chapter. Secondly, it is crucial to understand an echo as reflection of a direct sound and assign it to the same physical happening, even if locations are different. Thirdly, a correct localization from a mixture of sounds from various sources already requires a correct grouping. And finally, we would not be able to distinguish several sounds from a monophonic presentation if location was the only parameter considered. This deficiency is found for every single parameter, such as similarity in pitch, timbre or temporal change, or proximity in onset or spectrum. Among others, organization principles base on the named parameters and are described more extensively in the following, subdivided into three categories, namely:

[56]Particularly based on Bregman (1990).

[57]See Bregman (1990), pp. 38f, 137, 204, 395, and 451.

1. Properties of streams and their elements
2. Primitive grouping principles
3. Schema-based grouping principles.

4.5.1 Properties of Streams and Their Elements

Several principles concern the properties of auditory streams and their elements. These are predominant and can be considered as framework for auditory scene analysis. Given examples refer to grouping principles which are explained in detail later in this subsection.

Emergence:

Integration into streams takes a certain time which can vary. But still there is no arising or fading in of streams. Grouping takes place spontaneously, even when controlled by the listener's attention. This principle is visualized in Fig. 4.26. The person on the right does not take form bit by bit. His body parts become recognizable after the person as a whole is perceived. Although the raised arm and the legs are barely recognizable, we imagine him as a person and not as a distribution of body parts. A man is more than just the sum of his body parts. In German, this principle is called "Übersummenhaftigkeit".

Fig. 4.26 Demonstration of emergence. One can recognize the person standing on the right although his legs are missing. The original photo is presented in Sect. 6.2.1

Simultaneous and sequential grouping:

All sounds created from a source last for a certain time and undergo some changes. Therefore, it is necessary to group sounds that arrive simultaneously at the listener (*simultaneous/vertical grouping*) and sounds arriving at different points in time (*sequential/horizontal grouping*). While simultaneous grouping is necessary to discriminate different auditory streams from the sum of arriving sound, sequential grouping is needed to keep track of streams and to trace back continuous or successive sounds to the same physical happening. Simultaneous and sequential grouping are not independent of one another. Principles can affect both and auditory streams typically consist of both.[58]

Units:

Auditory streams are units which can be embedded in larger streams and form a "higher order unit".[59] Being perceived as one object does not mean we cannot differentiate between single parts of the object. Not only can we imagine a person on the right hand side in Fig. 4.26 but also can we locate his head, legs, and so on. Imagine a person wandering through a room. Although we know the person has a head, arms and legs, these single body parts are not considered separately, since a person cannot walk away, leaving the head where he started. One note, played on the piano, will certainly be integrated into one stream. But still, at least with some training, it is possible to hear out some single frequencies from its spectrum.

The old-plus-new-heuristic:

If a part of current sound can be interpreted as continuation of sound before, it will be integrated and then the remaining part is analyzed for grouping. It is also referred to as "wrap up all your garbage in the same bundle" heuristic.[60] Unattended elements can still be grouped within a stream. It even makes it easier to reject them as a group.

The principle of belongingness:

The principle of belongingness forces *exclusive allocation* of sound parameters. That means every aspect of sound is always exclusively part of one stream at a time. It takes up to four seconds to establish a stream and this stream lasts until there is evidence for a more authentic new grouping for several seconds. "This conservatism prevents the system from oscillating widely among perceptions of various numbers of streams in a complex environment."[61] Sudden happenings, distractions or change of attention or concentration can reset the scene analysis. In vision, the principle of belongingness can easily be seen in the example illustrated in Fig. 4.27. Here, the contours can either be seen as forming three violins or two busts; they cannot be considered as belonging to both objects at the same time. Additional cues can force one specific grouping and impair another.

---

[58]This section gives an overview of grouping principles. However, a broader overview concerning factors influencing sequential streaming is given in Moore and Gockel (2002).

[59]See Bregman (1990), p. 72.

[60]Bregman (1990), p. 450.

[61]Bregman (1990), p. 130.

Fig. 4.27 Illustration of the principle of belongingness. In the picture on top either a number of violins or two persons standing shoulder on shoulder can be seen at a time. Additional cues can force a specific grouping (bottom), like the complete violins (left) or additional facial features

Although a part of sound belongs to a stream it does not have to be considered as totally different from another stream. "There are levels of perceptual belongingness intermediate between 'the same thing' and 'unrelated things'."[62] As units can form higher order units, they may reveal relationships.

Retroactive effects:

In the auditory stream segregation process retroactive effects can occur. Two tones starting at the same time may fuse, which means they are integrated into the same stream. But when one of the tones stops earlier, the two get reconsidered as different tones.

4.5.2 Primitive Grouping Principles

Primitive grouping principles do not premise attention, knowledge or experience regarding sources of sound. They are typically suggesting grouping patterns based on proximity or similarity of temporal or spectral aspects of sound. The following grouping principles subsume occurring phenomenons:

1. Harmonicity
2. Timbre
3. Proximity
4. Common Fate

[62] See Bregman (1990), p. 204.

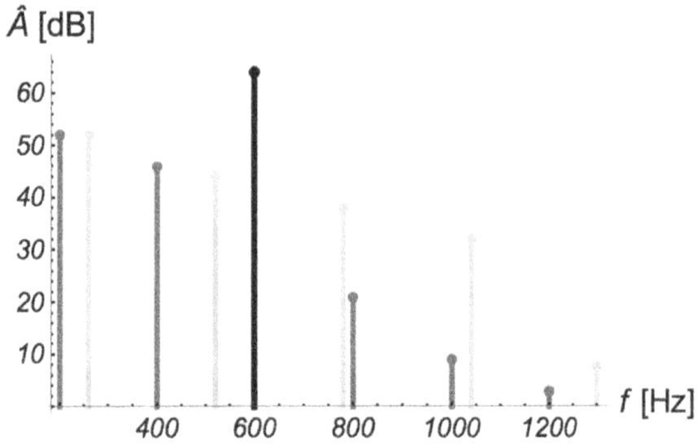

Fig. 4.28 Illustration of the principle of harmonicity. Two harmonic series are encoded with different gray levels. The frequency plotted in black protrudes from the series due to its high amplitude. It may thus be perceived as a third auditory stream, especially if its temporal behavior is not in agreement with the rest of the harmonic series

5. Synchrony
6. Continuity
7. Trajectories
8. Closure
9. Spatial Location
10. Comparison with other Senses

Harmonicity:

If simultaneous tones are harmonics of a common fundamental, they fuse, which means they are likely to integrate into one stream. In an inharmonic series the auditory system makes a "best fit"[63] guess, which results in a less strong integration. Yet, very loud harmonic and inharmonic spectral components can protrude and segregate from a steam. An example is given in Fig. 4.28.

Timbre:

A common harmonicity will also group successive sounds into the same stream. Similar timbre, spectral balance or auditory brightness and simplicity of the behavior of the harmonics and the attack support this integration, even when frequency relations of tones in successive sounds change. Note that timbre is already a quality of a stream. It is a result of spectral grouping. Thus, stream segregation based on timbre is especially a matter of temporal grouping.

Proximity of a succession of frequencies:

Resembling sounds are grouped. Especially in fast sequences short movements in frequency are preferred for a grouping.[64] Bregman (1990) shows this in a listening test in which three high notes 4, 5 and 6 are interlocking with three low notes 1, 2 and 3. Although the actual sequence is 1-4-2-5-3-6, the "apparent motion"[65] is one 1-2-3- and one 4-5-6-sequence. The actual sequence, which was jumping between

[63]Bregman (1990), p. 236.

[64]Referred to as "Körte's law", see Bregman (1990), p. 22.

[65]Bregman (1990), p. 21.

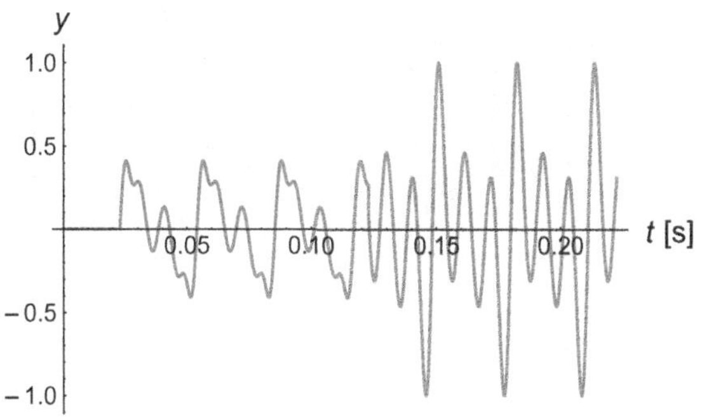

Fig. 4.29 Illustration of the principle of synchrony. Five frequencies start at the same time and create a harmonic sound. After about three periods another partial with a much higher amplitude starts and protrudes visually and audible

high and low notes is segregated into two streams, one with high and one with low notes. Faster sequences and larger frequency distances between the high and low notes increase the grouping strength. Accurate judgments about the order of notes in quick sequences is only possible for notes within one stream. This experiment even works with missing fundamentals and when every note is presented randomly to one ear only, as long as the other ear is simultaneously stimulated e. g. by noise. Spectral edges, spectral balance and frequency proximity play a central role for this grouping.[66] Also can a pure tone integrate with complex tones if it is similar to a harmonic.

Common fate:

If different parts of sound change in the same way, with a common ratio, they are integrated in the same stream, especially concerning frequency- and amplitude-modulations. This is even true for micromodulations and periodicity of beats. On the other hand, changing the frequency, amplitude or phase of one partial only, will segregate it from the stream. Also, echo suppression can be explained by common fate. Echoes are a slightly changed repetition of the direct sound and therefore integrate into the same stream, as long as their temporal distance is not too large. Thus, they manly have an amplifying effect and they can modify the perceived source width. Only when the integration time of the auditory system of 50 ms is exceeded, the echo is segregated from the direct sound. In this case, the echo may reduce the clarity and intelligibility of the direct sound.

Synchrony:

Synchrony of tones, especially synchronous onset, leads to an integration into one stream, particularly if attack and decay of higher harmonics and the corresponding degree of spectral fluctuations coincide. This fusion happens with harmonic and inharmonic sounds. Synchronous changes of frequency, amplitude or spatial direction impose an integration. An example of synchronous onsets is illustrated in Fig. 4.29. The figure actually shows the time series of Fig. 4.28.

[66]See Bregman (1990), p. 76, 90, and 93.

Fig. 4.30 Illustration of the principle of good continuation by three slightly changed versions of beamed eighth notes

Continuity:

A continuous, smooth change leads to a better grouping than sudden changes. A tone changing pitch over time (gliding) is likely to be integrated into one stream, whereas a sudden change of pitch will be perceived as two successive tones, which can support a segregation. In vision this is referred to as good continuation which is demonstrated in Fig. 4.30. In version 1, the irregular figure is perceived as unitary form due to the smooth continuation and our prior knowledge of a beamed pair of eighth notes. The perception occurs despite the fact that the notes are separated by a line which seems to belong to the white rectangle. In version 2, the notes can still be recognized but seem discontinuous through a shift of the part on the right hand side of the vertical line. In version 3, in addition to a vertical shift, the slope of the right part of the shape is differed. This further dissociated the two halves. Of course, knowledge about scores may also contribute to the impression of a correct and two incorrect versions. Grouping principles based on knowledge are referred to as *schema based* grouping principles, which is treated later in this section.

Trajectories:

Crossing lines in vision are not perceived as equal angles being tangent to each other. The contrary holds in auditory perception. Sequences crossing in pitch are perceived as sequences which converge, meet, and diverge again. "When a falling sequence of tones crossed a rising sequence, the listeners tended not to be able to follow either sequence across the crossing point."[67]

Closure:

Masked sounds seem to continue, even if they are physically not present. A repeated note with short silence between the notes is heard as a sequence. But when the gaps between the notes are filled with a masking sound, the repeated note is perceived as one continuous sound, interrupted by the masking signal. We continue to follow a stream even when it is masked or over. Figure 4.31 illustrates this principle in vision and hearing. The lines with periodic incline and decline, interrupted by gaps, are perceived as distinct angles, though one can think of it as being connected. Filling up

[67] See Bregman (1990), p. 447.

Fig. 4.31 Illustration of the principle of closure in vision and hearing. A tone, systematically gliding in pitch, interrupted by silence, is represented by an interrupted zigzag line. When the silence is filled up with noise (bars), the pitch-gliding tone seems to be continuous, as seems the zigzag line. After Bregman (1990), p. 28

the gaps with stripes leads to the impression of a continuous zigzag line, being partly covered by the stripes. The same counts for sound. Tones with periodically ascending and descending pitch glides, interrupted by silence, are perceived as single, though related, events. Filling up the silent parts with masking noise leads to the perception of one continuous tone, occasionally masked by the noise.

Spatial location:

Sounds that are perceived to originate in the same spatial location tend to be grouped to one auditory stream. Each frequency band is treated separately, therefore sound from each band must be localized according to the localization mechanisms explained in Sect. 4.4. However, this does not mean that we cannot distinguish between several sounds just because they come from the same location. If this was the case, listening to monophonic music would be very difficult. Of course, we can follow the melodic line of a single instrument in a music ensemble, even in a monophonic playback. But this source separation becomes easier if the instruments do not share a common source location. Especially if the instruments have a similar timbre and are good in tune and timing, they tend to be integrated in one stream because so many scene analysis principles suggest the same integration. A different spatial location can, however, facilitate a stream segregation. This is why audio engineers tend to pan instruments to different locations if they want them to be perceived as individual instruments and not as an ensemble. To increase the separation, they are slightly detuned or delayed. This procedure has been discussed already in Sect. 2.3 already. Obviously, the motivation behind this procedure is to create several segregation cues for the auditory system.

Vice versa, if sounds are already integrated into one auditory stream—be it due to a similar timbre, pitch, synchrony, common fate etc.—they may be localized at the same position. This position may be dominated by the precedence effect, i.e. the position of the first arriving wavefront as discussed in Sect. 4.4.5. Or it is the region where most partials are localized, except very high and very low frequencies and the region between 1.5 and 3 kHz where localization is ambiguous. It is hypothesized that the auditory system performs a sort of trading when different aspects of sound indicate contradictory grouping. The earlier a grouping information is extracted along the auditory pathway, the heavier this information is weighted for the final grouping decision. When, for example, ILDs point to another source location than ITDs, the perceived source location is a compromise between both locations. Here, ILDs may have a slightly stronger weighting because this information is extracted at an earlier stage in the auditory pathway. Further information, like the analysis of harmonic series or group modulations, is included in the trading process. This could either foster an integration of streams, so that one sound is perceived as coming from the detected source location. Or this additional information fosters a segregation of streams. In this case, two sources may be heard. This is likely, if this information is in agreement with the derived ILD and ITD information. For example, if frequencies whose ILDs point at a certain location are also a harmonic series, they are even more likely to be integrated into one auditory stream. If those frequencies, whose ITDs point at another location do not contribute to this harmonic series, but exhibit synchronous amplitude or frequency modulations, they are likely to be integrated into another auditory stream. These two streams are then segregated from another. They are heard as two sources at two different locations. Contradictory to the heavier weighting of earlier derived aspects of auditory sensation is the observation that harmonic structure and common onset time dominate auditory stream integration over common location, duration or loudness.[68]

Other senses:

Comparison with other senses, such as vision, balance or touch, can influence grouping. We believe a sound is coming from the source suggested by the visual perception, e.g. by similar temporal patterns, especially change in intensity of motion or a corresponding alteration of vertical position and pitch. This principle is known as "ventriloquism effect".[69] Typically, vision is even slightly dominant over hearing. Sense of touch can also influence grouping. Wind from behind, together with a tumbling sound of a wooden wall indicates that there was a physical happening in the rear.

4.5.3 Schema-Based Grouping Principles

Already infants between 1.5 and 3.5 months show evidence of auditory stream segregation.[70] But a six-month old child can locate sound sources only with an accuracy of approximately 15°, even though the physiological development of spectral resolution is completed. Furthermore, they need more obvious cues.[71] From that one

---

[68]See e.g. Cariani and Micheyl (2012), p. 360.

[69]See e.g. Bregman (1990), p. 183 or Schmidhuber et al. (2011).

[70]See e.g. Bregman (1990), p. 405.

[71]See Werner (2012), pp. 4ff.

can assume that further improvements in localization are based on experience which may also be the case for other organization cues.

Learned patterns, like diatonic sequences, can lead to auditory stream segregation. E.g. a non-diatonic note in a diatonic sequence "pops out", in other words segregates, since it does not fit into a learned pattern.[72] Intention of a listener can prefer a desired way of grouping within certain limits. For example hearing a sequence of tones as one or as two separate streams can be chosen by will, as long as the tempo is not too fast and intervals are not too small for a segregation or too large for an integration. A similar phenomenon is the ability to concentrate on certain aspects of sound, like hearing out a particular tone or instrument, and therefore reorganize the auditory scene where necessary. It is easier to segregate a part of a sequence against the grouping, forced by the primitive principles, than to integrate something that would be segregated by primitive grouping.

4.5.4 Organization Based on Auditory Scene Analysis Principles

The more principles suggest the same way of grouping the stronger the grouping gets. In some cases the different principles will lead to a particular scene analysis with distinct integrations into single auditory streams and a clear segregation between them. According to the psychologist Garner (1974), "[...] pattern goodness is correlated with pattern redundancy."[73] However, in many cases the grouping resulting from one or more principles will conflict with the grouping gathered from others. In these cases principles can predominate others, forcing their particular organization preference, as already illustrated in the vase-face Fig. 4.27. Sometimes this leads to insecurity about the grouping. In the worst case conflicting principles may even lead to a total confusion. Bregman (1990) speaks of a "competition among alternative organizations".[74]

Although grouping principles are based on certain parameters of the sound, an auditory stream can have one group-value for a particular parameter which differs from the values of the single components of the group. When sounds from different locations are integrated into one stream due to dominance of other principles—like harmonicity, timbre, common fate, synchrony etc.—they are likely to obtain one common group location. "The auditory system seems to want to hear all the parts of one sound as coming from the same location, and so when other cues favor the fusion of components, discrepant locations for these components may be ignored. It is as if the auditory system wanted to tell a nice, consistent story about the sound."[75]

[72]See Bregman (1990), p. 136.

[73]See Garner (1974), p. 10.

[74]See Bregman (1990), p. 165.

[75]See Bregman (1990), pp. 305f.

An auditory stream can obtain qualities which the single elements within the stream do not have. According to von Ehrenfels (1890), the whole can be more than the sum.[76] Furthermore, relations are clearer between elements within one stream than between those of two different streams, e.g. intervals between notes of a chord from one instrument are easier to identify than between instruments with different timbres and locations. Also, dissonance between frequencies of the same stream are perceived much stronger than between frequencies of different streams. It is harder to tap or count along with a metronome if the clicks fall into different streams, e.g. the location of the click sound alters or the single clicks strongly differ in spectrum. Often a perception of temporal overlap between elements of different streams arises for up to 50 ms of silence between them. This shows the uncertainty concerning a comparison between elements from different streams. It is hard to hear out a melody from a musical piece when the single notes are elements of different streams or when they are integrated into one stream together with distracting other notes. Although all notes of the melody are physically heard, it is almost impossible to recognize it in this case. Bregman (1990) calls this phenomenon "sequential blanking" or "recognition masking".[77]

On the other hand, sounds integrated in a stream can "[...] lose some of their original salience when they are incorporated into a larger unit."[78] For example it is not easy to distinguish all partials of a complex sound or all notes of a 4- or 5-note chord. In a fast sequence of four notes, integrated into one stream, subjects were not able to tell the order of the second and third note. "[...] [T]he phrasing of a sequence of notes makes some of their pitches less distinct."[79] The localization of a source is based on grouping those components of complex, interfering sounds, which can be associated to the same auditory event.[80] Consequently, a subjectively secure localization of a source can suppress the perception of timbre, which explains the inaudibility of the comb filter effect in stereo playback.[81] Despite this loss for details, a formation of separate streams allows for a comparison of global properties.

As grouping principles can conflict, there are also cases in which principles concerning the properties of auditory streams are violated. E.g. when prime and fifth of a triad are presented to one ear, while the third is presented to the other, many listeners perceive a full chord in one ear and the single note in the other.[82] A paradox in grouping is that timbre is an important parameter for sequential grouping, even though timbre is already a result of spectral grouping.

If a part of sound cannot be integrated into a stream it is likely to be neglected for further perception; it will be overheard. This natural selection is necessary to reduce

---

[76] This gestalt quality is known as "Übersummenhaftigkeit", see von Ehrenfels (1890), pp. 249ff.

[77] See Bregman (1990), p. 172.

[78] See Bregman (1990), p. 140.

[79] See Bregman (1990), p. 475.

[80] See Theile (1980), especially p. 24.

[81] See Ono et al. (2002), p. 2 and Theile (1980), p. 12. Details on the comb filter effect in stereo playback are given in Sect. 7.2.2.

[82] See Bregman (1990), p. 693.

the masses of information from the environment to an amount we can handle, to avoid a sensory overload. This reduction of information may be misinterpreted as a deficiency. But it is auditory scene analysis which provides us with reliable information about the acoustical outside world. It is the basis of our understanding of the auditory world as interpreted from all superimposed acoustical signals that confront us. Computational auditory scene analysis has arisen from the idea to compute this impressive capability by imitating mechanisms of the auditory system. This approach might have the potential to assign parts of sound to their physical happening, thus to identify musical instruments from an orchestral recording or recognize speech in a noisy environment.[83]

4.5.4.1 Auditory Illusions

If parameters suggest dissent groupings, this exacerbates the scene analysis and can lead to a diffuseness and even auditory illusions.

Such an illusion may originate in a conflict of visual and auditory information. Bregman (1990) experienced that a "baba" sound on a "gaga"-saying face sounds similar to "baba" but with an open mouth, even for him, who had conscious knowledge of the "trick". The pronunciation can *sound* different, because of a conflicting influence of *vision* which is too strong to ignore.[84] This effect has already been described by McGurk and McDonald (1976), who showed a video of a "ga"-saying speaker while playing a "ba"-sound to subjects.[85] They reported to hear a "da"-sound, a syllable which they considered intermediate between those two. Therefore, this effect is sometimes referred to as "McGurk effect".[86] Donnadieu (2007) summarizes results from several listening tests which showed an influence of visual cues on auditory perception.[87] For example visual information was found to affect loudness perception. Furthermore, it greatly influenced the judgment of how "bowy" or "plucky" an instrument sounded in a listening test with hybrid sounds, intermediate between bowed and plucked. Videos of an instrumentalist plucking a string had a significant influence that could not be observed in the same setup with a video simply presenting the word "plucked". This finding suggests that knowledge—and even the belief of knowledge— about the creation of a sound influences its perceived quality. This ecological way of perceiving may be the cause for the free grouping of similar sounds by subjects based on creation mechanism and resonator type of instruments in timbre studies as described in Sect. 2.1. Another illusion which is based on mismatch of visual and auditory information is the ventriloquism effect as discussed earlier in this chapter.

[83] Concepts, algorithms and the state of research are extensively illuminated in Rosenthal and Okuno (1998) and Wang and Brown (2006).

[84] See Bregman (1990), pp. 183f.

[85] See McGurk and McDonald (1976).

[86] See e.g. Donnadieu (2007), p. 305.

[87] See Donnadieu (2007), pp. 305f.

Bregman (1990) furthermore describes how an oboe-sound was played with two loudspeakers. One loudspeaker played the even partials, one played the odd, both with the same micromodulations in frequency. The perceived sound was an oboe-sound with the original pitch located somewhere in between the two speakers. When the frequency fluctuations were changed independently on each speaker, the perceived sound split up into two streams; the odd and even partials were identified as different sounds emanating from two different locations. The sounds even had a different pitch, since the odd harmonics are multiples of the first and the even harmonics are multiples of the second partial of the original oboe sound. In this example grouping determined perceived pitch and location.[88]

The "Shepard illusion" is also an auditory illusion which can be traced back to scene analysis principles[89]: In a complex tone, all adjacent partials have the same base interval, e.g. a frequency ratio of 1:2. A symmetric spectral envelope ascends from the low frequencies towards the central frequency and descends towards higher frequencies, converging towards 0 at the limits of the audible frequency range. Typically, a Gaussian bell curve is chosen as envelope, but triangles and other axially symmetric curves serve the purpose as well. If now all partials of this sound slowly climb up while keeping their interval constant, the highest frequency will be faded out while a new lowest frequency fades in. As soon as all frequencies have increased by the base interval, the sound exactly equals its initial state again. During this cycle, the spectral centroid barely varies. This *Shepard tone* is illustrated in Fig. 4.32. In fact, it is a complex tone and not a sinusoidal tone. Listening to such a tone creates the impression of an infinitely increasing pitch despite the cyclic repetition. Precisely speaking, the sound exhibits an infinitely clockwise-cycling change of chroma, while stays rather constant.[90]

4.5.5 Auditory Scene Analysis in Composition

Properties like rhythm or harmony are more than the sum of successive or simultaneous sounds. It is auditory scene analysis that actually creates these properties which are important qualities of music. Simultaneous grouping creates timbre and chords, sequential grouping creates melody. At cadences, stream segregation is reduced in favor of integration.

Composers work a lot with grouping, consciously or unconsciously. Ensembles are meant to have a global sound due to integration. Counterpoints in baroque music are a good example for segregated streams. Two simultaneous melodic lines are

---

[88]See Bregman (1990), pp. 256ff.

[89]See Shepard (1964), Burns (1981) and Leman (1995), pp. 23ff for details on pitch perception and the Shepard illusion.

[90]Ziemer and Schultheis (2018a), Ziemer et al. (2018) and Ziemer and Schultheis (2018b) explain how the Shepard tone is created and how it is perceived by passive listeners and in an interactive task.

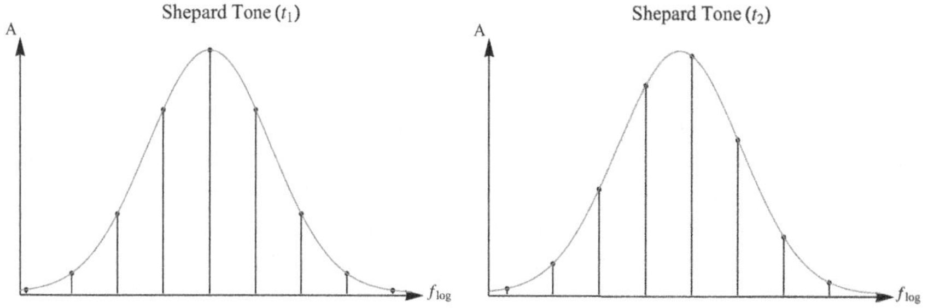

Fig. 4.32 Spectra of a Shepard tone at two different points in time. Although all partials increase in frequency, the spectral centroid stays nearly unchanged. As one partial leaves the envelope at the higher frequency end, a new partial enters at the lower frequency end. This creates the impression of an infinitely rising pitch

perceived distinctly, because they follow grouping principles. These lead to a strong integration of each line by similar pitch regions, timbre, location and a fast tempo. Equally, there is a distinct segregation between them. The melodies consist of smalls steps, to keep them together. They have different locations, timbres, pitch regions and rhythms, and no crossing trajectories. No parallel movements in octave or fifth are accomplished, probably because they would enforce integration. When one line moves up, the other tends to move down to avoid common fate. Compound melodic lines, also known from baroque music, use rapid alternations of high and low notes on the same instrument to create interesting melodies, "ambiguous between one and two streams", similar to virtual polyphony.[91] As Bregman (1990) states, "[...] music often tries to fool the auditory system into hearing fictional streams."[92] In classical music, the flute was used to enrich the spectrum of the violin without being perceived as distinct instrument by playing along one octave higher than the violin and therefore being integrated into the same stream. Solo instruments are segregated from the orchestral sound by the use of slight mistune, vibrato, rubato, by playing slightly sharp or at a different spatial location. Furthermore, groups of instruments are perceived as on instrumental section with an increased loudness, richness in sound and an enlarged source area.

4.6 Usability of Psychoacoustic Knowledge for Audio Systems

The physical nature of the sound in our environment is highly complex. The auditory system supplies numerous mechanisms to adequately transfer the physical signals into psychologically interpretable representations. Not all aspects of physical sound

[91] See Bregman (1990), pp. 496ff, 677f, 457f and p. 464.

[92] See Bregman (1990), p. 457.

stimuli actually contribute to this mental representation. Some magnitudes lie below thresholds, some changes below just noticeable differences. Thus, absolute thresholds and masking thresholds determine the audible frequency- and level region to be reproduced by an audio system. Just noticeable differences as well as integration times tell about the necessary spatial and temporal precision for the reproduction of amplitude, amplitude change, phase change, source direction etc.

Therefore, psychoacoustic knowledge can be used to reduce the amount of data to be recorded, processed, and reproduced without audible effects. This is commonly done in application. For example microphones for musical recordings tend to record only the audible frequency range. All digital systems make use of the temporal and dynamic resolution capacity of the ears by sampling continuous sound into time-discrete values with an appropriate resolution in time and dynamics. The audio-cd reduces continuous sound pressure changes to 44100 discrete sound pressure states per second and codes the dynamic range with 16 bit which allows 2^{16} possible values.

Bader (2013) discusses the approach to efficiently code sounds by using gammatones, imitating the nerve cell output of the auditory system.[93] Masking can be considered as a threshold increase caused by a loud sound. Experiments on simultaneous masking led to quite accurate and valid results, concerning the relationship between masker frequency and amplitude and the resulting masking pattern. Temporal masking shows larger variance and less reliability. Furthermore, interaction between maskers in different frequency regions as well as between masker and maskee has been observed but not fully understood. The existence of binaural masking is an evidence that another masking mechanism exists besides the masking originated in cochlear processes. One sound that exclusively reaches one ear can still mask a sound that reaches exclusively the other ear. Here, processes at higher stages of the auditory pathway cause the masking effect. Here, efferents may play a crucial role. Of course, pure tones, critical band wide noise, white noise and Gaussian sound bursts are not the kind of sound typically faced in a natural listening situation, in communication or musical performance. The same holds for pure monotic or dichotic performance. The masking effects occurring in natural listening contexts may be some sort of mixture of both monaural and binaural masking. Still, investigations of masking led to an understanding of the phenomenon—temporally and spectrally—which gave rise to psychoacoustic audio compression methods quantified in technical applications such as AAC, AC-3 and MP3.[94] It is also qualified for an implementation in a spatial analysis and synthesis system for musical instruments as will be discussed in Chap. 9.

Conventional audio systems are mainly based on psychoacoustic methods to recreate a natural auditory impression rather than aiming at recreating all physical quantities. Spatial localization of sound sources can be accomplished by the auditory system with a high precision especially concerning the horizontal angle in the frontal directions. Due to this fact, early stereophonic audio systems concentrated on sound playback in this region. Further developments added sounds from the rear directions but rarely involved lateral sound, distance or the third dimension, since distance

[93] See Bader (2013), pp. 632ff.

[94] See e.g. Lerch (2008), pp. 872ff. Extensive discussion about MP3 can be found in Ruckert (2005).

hearing and localization capability at the sides and in the median plane are weak. Auditory streams obtain a group value for location and source width. Therefore it can be sufficient to reproduce only some of the acoustical properties to maintain the original auditory scene. The theory of summing localization is used in stereo systems to create the impression of one sound source at any position between two loudspeakers by playing systematically manipulated signals via two loudspeakers. History, development, and functionality of conventional stereophonic audio systems are discussed in Chap. 7.

The theory of sound field synthesis is discussed in Chap. 8. It aims at physically recreating all sound properties, as discussed in Chap. 5, in a large listening area. At first glance, sound field synthesis seems to make applications of psychoacoustic methods superfluous. But when it comes to actual implementation, psychoacoustic considerations are essential as will become clear in Sect. 8.3. Many researchers predict that the future of audio systems lies in psychoacoustics.[95]

References

Allen JB (2008) Nonlinear cochlear signal processing and masking in speech perception. In: Benesty J, Mohan Sondhi M, Huang Y, (eds) Springer handbook of speech processing, chapter 03. Springer, Berlin, pp 27–60. https://doi.org/10.1007/978-3-540-49127-9_3

Ashihara K (2007) Hearing thresholds for pure tones above 16 kHz. J Acoust Soc Am 122(3):EL52–EL57. https://doi.org/10.1121/1.2761883

Backus J (1969) The acoustical foundations of music. W. W. Norton & Co., New York. https://doi.org/10.2307/843219

Bader R (2013) Nonlinearities and synchronization in musical acoustics and music psychology. Springer, Berlin. https://doi.org/10.1007/978-3-642-36098-5

Blauert J (1974) Räumliches Hören. Hirzel, Stuttgart

Blauert J (1985) Räumliches Hören. Nachschrift-Neue Ergebnisse und Trends seit 1972. Hirzel, Stuttgart

Blauert J (1997) Spatial hearing. The pychophysics of human sound source localization, revised edn. MIT Press, Cambridge

Blauert J (2008) 3-d-Lautsprecher-Wiedergabemethoden. In: Fortschritte der Akustik—DAGA'08. Dresden, Mar 2008, pp 25–26

Blauert J, Braasch J (2008) Räumliches Hören. In: Weinzierl S, (ed) Handbuch der Audiotechnik, chapter 3. Springer, Berlin, pp. 87–122. https://doi.org/10.1007/978-3-540-34301-1_3

Brandter C (2007) Ein systematischer Ansatz zur Evaluation von Lautheitsmodellen. Uni-Edition, Berlin

Bregman SA (1990) Auditory scene analysis. MIT Press, Cambridge

Bruhn H (2002a) Verarbeitung einzelner Schallereignisse. In: Bruhn H, Oerter R, Rösing H (eds) Musikpsychologie. Ein Handbuch, 4th edn. Rowohlt, Reinbek bei Hamburg, pp 666–670

Bruhn H. (2002b) Tonpsychologie—Gehörpsychologie—Musikpsychologie. In: Bruhn H, Oerter R, Rösing H (eds) Musikpsychologie. Ein Handbuch, 4th edn. Rowohlt, Reinbek bei Hamburg, pp 439–451

Bruhn H, Michel D (2002) Hören im Raum. In: Bruhn H, Oerter R, Rösing H (eds) Musikpsychologie. Ein Handbuch, 4th edn. Rowohlt, Reinbek bei Hamburg, pp 650–655

[95] See e.g. Blauert (2008), Fastl (2010) and Spors et al. (2013).

Burns EM (1981) Circularity in relative pitch judgements for inharmonic complex tones: the shepard demonstration revisited, again. Percept Psychophys 30(5):467–472. https://doi.org/10.3758/bf03204843

Cariani P, Micheyl C (2012) Toward a theory of information processing in auditory cortex. In: Poeppel D, Overath T, Popper AN, Fay RR (eds) The human auditory cortex. Springer handbook of auditory research, vol 43, chapter 13. Springer, New York, pp 351–390. https://doi.org/10.1007/978-1-4614-2314-0_13

David EE Jr (1988) Aufzeichnung und Wiedergabe von Klängen. In: Winkler K (ed) Die Physik der Musikinstrumente. Spektrum der Wissenschaft, Heidelberg, pp 150–160

Davis MF (2007) Audio and electroacoustics. In: Rossing TD (eds) Springer handbook of acoustics, chapter 18. Springer, New York, pp 743–781. https://doi.org/10.1007/978-0-387-30425-0_18

Dickreiter M (1978) Handbuch der Tonstudiotechnik, vol 1, 2nd edn. In: De Gruyter M et al

Dickreiter M (1987) Handbuch der Tonstudiotechnik, vol 1, 5 völlig neu bearbeitete und ergänzte edition. In: De Gruyter M, et al

Donnadieu S (2007) Mental representation of the timbre of complex sounds. In: Beauchamp JW (ed) Analysis, synthesis, and perception, chapter 8. Springer, New York, pp 271–319. https://doi.org/10.1007/978-0-387-32576-7_8

Ehmer RH (1959) Masking patterns of tones. J Acoust Soc Am 31(8):1115–1120. https://doi.org/10.1121/1.1907836

Elliott LL (1962) Backward masking: monotic and dichotic conditions. J Acoust Soc Am 34(8):1108–1115. https://doi.org/10.1121/1.1918253

Fastl H (1977) Temporal masking effects: II. critical band noise masker. Acustica 36:317–331. https://www.ingentaconnect.com/contentone/dav/aaua/1977/00000036/00000005/art00003

Fastl H (1979) Temporal masking effects: III. pure tone masker. Acustica 43:282–294. https://www.ingentaconnect.com/contentone/dav/aaua/1979/00000043/00000005/art00004

Fastl H (2010) Praktische Anwendungen der Psychoakustik. In: Fortschritte der Akustik—DAGA'10. Berlin, pp 5–10

Friedrich H (2008) Tontechnik für Mediengestalter. Töne hören—Technik verstehen—Medien gestalten. Springer, Berlin

Friesecke A (2007) Die Audio-Enzyklopädie. Ein Nachschlagewerk für Tontechniker, K G Saur, Munich

Garner WR (1974) The processing of information and structure. Lawrence Erlbaum, New York

Gelfand SA (1990) Hearing. An introduction to psychological and physiological acoustics, 2nd edn. Informa. New York and Basel

Grantham DW (1986) Detection and discrimination of simulated motion of auditory targets in the horizontal plane. J Acoust Soc Am 79(6):1939–1949. https://doi.org/10.1121/1.393201

Hall DE (2008) Musikalische Akustik. Ein Handbuch, Schott, Mainz

Haustein BG, Schirmer W (1970) Messeinrichtung zur Untersuchung des Richtungslokalisationsvermögens. Hochfrequenztechnik und Elektroakustik 79:96–101

Hellbrück J (1993) Hören. Physiologie, Psychologie und Pathologie. Hogrefe, Göttingen

Kling JW, Riggs LA (eds) (1971) Woodworth & Schlossberg's experimental psychology, 3 edn. Holt, Rinehart and Winston, New York

Klinke R (1970) Neurophysiological basis of hearing. Mechanisms of the inner ear. In: Grüsser O-J, Klinke R (eds) Pattern recognition in biological and technical systems. Proceedings of the 4th congress of the Deutsche Gesellschaft für Kybernetik held at Berlin, 6–9 Apr 1970. https://doi.org/10.1007/978-3-642-65175-5_29

Kostek B (2005) Perception-based data processing in acoustics. Springer, Berlin. https://doi.org/10.1007/b135397

Larcher V, Jot J-M (1999) Techniques d'interpolation de filtres audio-numériques. application á la reproduction spatiale des sons sur écouteurs. In: Congrès Français d'Acoustique, Marseille, France, Marseille

Leman M (1995) Music and Schema theory. Cognitive foundations of systematic musicology. Springer, Berlin

Lerch A (2008) Bitdatenreduktion. In: Weinzierl S (ed) Handbuch der Audiotechnik, chapter 16. Springer, Berlin, pp 849–884. https://doi.org/10.1007/978-3-540-34301-1_16

Lin Y, Abdulla WH (2015) Audio watermark. Springer, Cham. https://doi.org/10.1007/978-3-319-07974-5

Luce RD (1993) Sound and hearing. A conceptual introduction. Lawrence Erlbaum, Hillsdale. https://doi.org/10.4324/9781315799520

Martens WL (1987) Principal components analysis and resynthesis of spectral cues to perceived directions. In: Proceedings of the international computer music conference. San Francisco, pp 274–281

McGurk H, McDonald J (1976) Hearing lips and seing voices. Nature 264:746–748. https://doi.org/10.1038/264746a0

Meyer E, Burgtorf W, Damaske P (1965) Eine Apparatur zur elektroakustischen Nachbildung von Schallfeldern. Subjektive Hörwirkungen beim Übergang Kohärenz–Inkohärenz. Acustica 15:339–344. https://www.ingentaconnect.com/contentone/dav/aaua/1965/00000015/a00101s1/art00005

Middlebrooks JC, Green DM (1991) Sound localization by human listener. Annu Rev Psychol 42:135–159. https://doi.org/10.1146/annurev.ps.42.020191.001031

Moore BCJ, Gockel H (2002) Factors influencing sequential stream segregation. Acta Acust United Ac 88:320–332. https://www.ingentaconnect.com/contentone/dav/aaua/2002/00000088/00000003/art00004

Morikawa D, Hirashara T (2010) Signal frequency necessary for horizontal sound localization. Acoust Sci Tech 31(6):417–419

Myers DG (2008) Psychologie, 2. erweiterte und aktualisierte edition. Springer, Berlin. https://doi.org/10.1007/978-3-642-40782-6

Ono K, Pulkki V, Karjalainen M (2002) Binaural modeling of multiple sound source perception. coloration of wideband sound. In: Audio engineering society convention 112, Munich, May 2002

Preibisch-Effenberger R (1966) Die Schallokalisationsfähigkeit des Menschen und ihre Audioetaudiom Verwendung zur klinischen Diagnostik. PhD thesis, Technical University of Dresden, Dresden

Rosenthal DF, Okuno HG (1998) Computational auditory scene analysis. Lawrence Erlbaum, Mahwah

Ross B, Tremblay KL, Picton TW (2007) Physiological detection of interaural phase differences. J Acoust Soc Am 121(2):1017–1027. https://doi.org/10.1121/1.2404915

Ruckert M (2005) Understanding MP3. Syntax, semantics, mathematics and algorithms. GWV, Wiesbaden

Schmidhuber M, Völk F, Fastl H (2011) Psychoakustische Experimente zum Einfluss des Ventriloquismuseffekts auf Richtungsunterschiedsschwellen (minimum audible angles) in der Horizontalebene. In: Fortschritte der Akustik—DAGA'11. Düsseldorf, pp 577–578

Schneider A (2018) Pitch and pitch perception. Springer Berlin, pp 605–685. https://doi.org/10.1007/978-3-662-55004-5_31

Shepard RN (1964) Circularity in judgments of relative pitch. J Acoust Soc Am 36(12):2346–2353. https://doi.org/10.1121/1.1919362

Sodnik J, Susnik R, Tomazic S (2006) Principal components of non-individualized head related transfer functions significant for azimuth perception. Acta Acust United Acust 92:312–319. https://www.ingentaconnect.com/contentone/dav/aaua/2006/00000092/00000002/art00013

Spors S, Wierstorf H, Raake A, Melchior F, Frank M, Zotter F (2013) Spatial sound with loudspeakers and its perception: a review of the current state. Proc IEEE 101(9):1920–1938. https://doi.org/10.1109/JPROC.2013.2264784

Strube G (1985) Lokalisation von Schallereignissen. In: Bruhn H, Oerter R, Rösing H (eds) Musikpsychologie. Ein Handbuch in Schlüsselbegriffen. Urban & Schwarzenberg, Munich, pp 65–69

Terhardt E, Stoll G, Seewann M (1982) Algorithm for extraction of pitch and pitch salience from complex tonal signals. J Acoust Soc Am 71(3). https://doi.org/10.1121/1.387544

Theile G (1980) Über die Lokalisation im überlagerten Schallfeld. PhD thesis, University of Technology Berlin

Thurlow WR (1971) Audition. In: Kling JW, Riggs LA (eds) Woodworth & Schlosberg's experimental psychology, Third American edition. London, pp 223–271

Verheijen, E (1997) Sound reproduction by wave field synthesis. PhD thesis, Delft University of Technology, Delft

von Ehrenfels C (1890) Über Gestaltqualitäten. Vierteljahrsschrift für wissenschaftliche Philosophie 14:249–292. https://doi.org/10.1515/9783035601602.106

Wang D, Brown GJ (2006) Computational auditory scene analysis. IEEE Press, Hoboken. https://doi.org/10.1109/9780470043387

Webers J (2003) Handbuch der Tonstudiotechnik. Analoges und Digitales Audio Recording bei Fernsehen, Film und Rundfunk. Franzis, Poing, 8. neu bearbeitete edition

Werner LA (2012) Overview and issues in human auditory development. In: Werner LA, Fay RR, Popper AN (eds) Springer handbook of auditory research, chapter 01. Springer, New York, pp 1–18. https://doi.org/10.1007/978-1-4614-1421-6_1

Ziemer T, Schultheis H (2018a) Psychoacoustic auditory display for navigation: an auditory assistance system for spatial orientation tasks. J Multimodal User Interfaces. https://doi.org/10.1007/s12193-018-0282-2 (Special Issue: Interactive Sonification)

Ziemer T, Schultheis H (2018b) A psychoacoustic auditory display for navigation. In: 24th international conference on auditory displays (ICAD2018), Houghton, MI, June 2018b. https://doi.org/10.21785/icad2018.007

Ziemer T, Below M, Krautwald P, Schade J, Obermöller H (2007) Ein Technical Report zum thema der 'just noticeable differences' (JNDs) zeitlicher unterschiede in musikalischen Signalen. http://www.systmuwi.de/Pdf/Technical%20Reports/Technical%20Report-JND,%20Below,%20Ziemer,%20etc.pdf. Accessed 11 Feb 2013

Ziemer T, Schultheis H, Black D, Kikinis R (2018) Psychoacoustical interactive sonification for short range navigation. Acta Acust United Acust 104(6):1075–1093. https://doi.org/10.3813/AAA.919273

Zurek Patrick M, Kourosh S (2003) Lateralization of two-transient stimuli. Percept Psychophys 65(1):95–106. https://doi.org/10.3758/bf03194786

Zwicker E, Fastl H (1999) Psychoacoustics. Facts and models, second updated edition. Springer, Berlin. https://doi.org/10.1007/978-3-662-09562-1

Zwislocki JJ, Buining E, Glantz J (1968) Frequency distribution of central masking. J Acoust Soc Am 43(6):1267–1271. https://doi.org/10.1121/1.1910978

Chapter 5
Spatial Sound of Musical Instruments

To reach a listener, the sound of musical instruments has to travel, typically through air. Thus, the next section deals with the basic physical principles of sound propagation. This leads to a better understanding of spatial attributes of sound, such as propagation and directivity patterns of musical instruments, which are discussed subsequently. These spatial attributes strongly contribute to the individual sound character of musical instruments. Therefore, many methods have been developed to investigate the radiation characteristics of musical instruments and to represent it in ways that allow for qualitative and quantitative statements. A discussion of these methods completes this chapter.

5.1 Wave Equation and Solutions

In this section the physical fundamentals of sound in air are illuminated.[1] They are the basis of acoustics in the free field and describe sound propagation e.g. of musical instruments and loudspeakers.

5.1.1 Homogeneous Wave Equation

Euler's equation of motion

$$\rho_0 \frac{\partial \boldsymbol{v}(\mathbf{x}, t)}{\partial t} = -\nabla p(\mathbf{x}, t) \tag{5.1}$$

[1]As described in Ziemer (2011, 2018), mainly based on Pierce (2007), Williams (1999), Morse and Ingard (1986), Rabenstein et al. (2006) and Ahrens (2012).

© Springer Nature Switzerland AG 2020

T. Ziemer, *Psychoacoustic Music Sound Field Synthesis*, Current Research in Systematic Musicology 7, https://doi.org/10.1007/978-3-030-23033-3_5

is the first base equation of the wave field. It explains the flow of frictionless fluids by means of time t, direction vector $\mathbf{x}$, particle velocity vector $\boldsymbol{v}$, pressure p, ambient density ρ_0 and nabla operator ∇. In Cartesian coordinates the following is valid:

$$
\begin{aligned}
\mathbf{x} &= \begin{bmatrix} x \\ y \\ z \end{bmatrix} \\
\boldsymbol{v} &= \begin{bmatrix} u(x) \\ v(y) \\ w(z) \end{bmatrix} \\
\nabla &\equiv \frac{\partial}{\partial \mathbf{x}} = \frac{\partial}{\partial x} + \frac{\partial}{\partial y} + \frac{\partial}{\partial z}
\end{aligned}
\tag{5.2}
$$

The second base equation of the wave field is the continuity equation (conservation of mass)

$$
c^2 \rho_0 \nabla \boldsymbol{v}(\mathbf{x}, t) + \frac{\partial p(\mathbf{x}, t)}{\partial t} = 0. \tag{5.3}
$$

with propagation velocity c. Differentiating Eq. 5.3 with respect to time and replacing the velocity term by the right side of the equation of motion, Eq. 5.1, yields the homogeneous wave equation for pressure

$$
\nabla^2 p(\mathbf{x}, t) - \frac{1}{c^2} \frac{\partial^2 p(\mathbf{x}, t)}{\partial t^2} = 0. \tag{5.4}
$$

By differentiating the continuity equation with respect to $\mathbf{x}$ and the equation of motion with respect to t yields the homogeneous wave equation for velocity

$$
\nabla^2 \boldsymbol{v}(\mathbf{x}, t) - \frac{1}{c^2} \frac{\partial^2 \boldsymbol{v}(\mathbf{x}, t)}{\partial t^2} = 0. \tag{5.5}
$$

5.1.2 Wave Field

The sound field magnitudes sound pressure p and sound velocity $\boldsymbol{v}$ are perturbations of the state of equilibrium which propagate as waves. c is the sound propagation velocity and ∇^2 is the Laplace operator

$$
\nabla^2 \equiv \frac{\partial^2}{\partial^2 \mathbf{x}} = \frac{\partial^2}{\partial x^2} + \frac{\partial^2}{\partial y^2} + \frac{\partial^2}{\partial z^2}. \tag{5.6}
$$

Solutions of the wave equation, Eqs. 5.4 and 5.5, are called *sound field* or *wave field*. Note, that although these two equations look very similar, the dependent variables

p and $\boldsymbol{v}$ are not equal. Their relationship is described by the equation of motion, Eq. 5.1. The equations assume the following conditions[2]:

1. The propagation medium is homogeneous
2. The medium is quiescent and vortex free
3. State changes are adiabatic, i.e. no heat interchange between areas of low pressure and areas of high pressure due to the rapid movement of the particles
4. Pressure and density perturbations are small compared to static pressure and density
5. Relationships in the medium are subject to linear differential equations
6. The medium exhibits no viscosity
7. The medium is source-free

In some cases these conditions may over-simplify the actual physics of the observed system. In air, for example, high frequencies exhibit a stronger amplitude decay than low frequencies due to heat transfer from regions of high pressure to regions of low pressure. This high-frequency attenuation for long travel paths becomes audible at distances over say 50 m or so. Inside the tube of brass instruments, the pressure perturbations are not small compared to static pressure and so a nonlinear wave propagation can be observed.[3]

5.1.3 Homogeneous Helmholtz Equation

The pressure term is transformable via Fourier transform

$$P\left(\mathbf{x},\omega\right)=\int_{t=-\infty}^{\infty}p\left(\mathbf{x},t\right)e^{\imath\omega t}\mathrm{d}t \tag{5.7}$$

from the time domain to the frequency domain and back via inverse Fourier transform

$$p\left(\mathbf{x},t\right)=\frac{1}{2\pi}\int_{\omega=-\infty}^{\infty}P\left(\mathbf{x},\omega\right)e^{-\imath\omega t}\mathrm{d}\omega\,. \tag{5.8}$$

e is Euler's number ($e\approx 2.718\ldots$), $\imath=\sqrt{-1}$ is the imaginary unit, $\omega=2\pi f$ is the angular frequency and f the frequency. The wave equation in the frequency domain reads

$$\nabla^2 P\left(\mathbf{x},\omega\right)+k^2 P\left(\mathbf{x},\omega\right)=0 \tag{5.9}$$

with *wave number* or *spatial frequency* $k=\frac{\omega}{c}=\frac{2\pi}{\lambda}$ and wave length λ and is called *Helmholtz equation*. Since the Fourier transform is an integral over time,

---

[2]See Mechel (2008), pp. 5f, Teutsch (2007), Wöhe (1984), Pierce (2007), p. 36 and Baalman (2008), p. 23.

[3]See e.g. Hirschberg et al. (1996), describing shock-waves in brass instruments.

the Helmholtz equation is only valid for stationary signals, i.e. periodic vibrations, and not for transients.[4]

5.1.4 Plane Waves

A general solution of the wave equation is d'Alembert's solution:

$$p(\mathbf{x}, t) = f(\mathbf{x} - ct) + \tilde{f}(\mathbf{x} + ct) \tag{5.10}$$

The first term describes the propagation of a pressure state in **x** direction, the second a propagation in the opposite direction. For waves the principle of superposition applies, i.e. they interfere without affecting each other. Assuming the second term to be 0, only one wave in **x** direction remains. Other directions can simply be added. One possible solution function $f(\mathbf{x} - ct)$ is the function of a plane wave:

$$\begin{aligned} p(\mathbf{x}, t) &= A(\omega)\ e^{-\imath(k\mathbf{x} - \omega t)}, \text{ or, respectively} \\ P(\mathbf{x}, \omega) &= A(\omega)\ e^{\imath k\mathbf{x}} \end{aligned} \tag{5.11}$$

Here $A(\omega)$ is an arbitrary complex amplitude in the form $\hat{A}e^{\imath\phi}$ whose absolute value is the amplitude $\hat{A}$ and whose argument is the phase ϕ of a frequency contained in the signal. $k^2 = k_x^2 + k_y^2 + k_z^2$ is the squared wave number in direction **x**, $\lambda^2 = \lambda_x^2 + \lambda_y^2 + \lambda_z^2$, the wave length in **x**-direction, respectively. A plane wave propagates in direction **x** whereat phase changes with respect to location. k_x, k_y and k_z are called "trace wavenumbers",[5] λ_x, λ_y and λ_z are *trace wavelengths*. They are projections to the spatial axes. The "wavefronts"[6] are infinite planes of equal pressure perpendicular to vector **x**. Although derived from an equation which assumes a source-free medium, plane waves are a good approximation of very far sources. Here, the wave front curvature and the amplitude decay are small compared to a proximate source.

For a wave with non-negative k, two formulations for k_y point out two different sorts of wave[7]:

$$k_y = \begin{cases} \pm\sqrt{k^2 - k_x^2 - k_z^2}, & k^2 \geq k_x^2 + k_z^2 \\ \pm\imath\sqrt{-k^2 + k_x^2 + k_z^2}, & k^2 \leq k_x^2 + k_z^2 \end{cases} \tag{5.12}$$

In the first case all components are real, indicating a propagating plane wave. In the second case k_y is imaginary, leading to an *evanescent wave*. Inserting the second case in Eq. 5.11 yields:

[4] See e.g. Meyer et al. (2001), p. 2.

[5] See e.g. Williams (1999), p. 21.

[6] See Williams (1999), p. 22.

[7] See Ahrens (2012), p. 23.

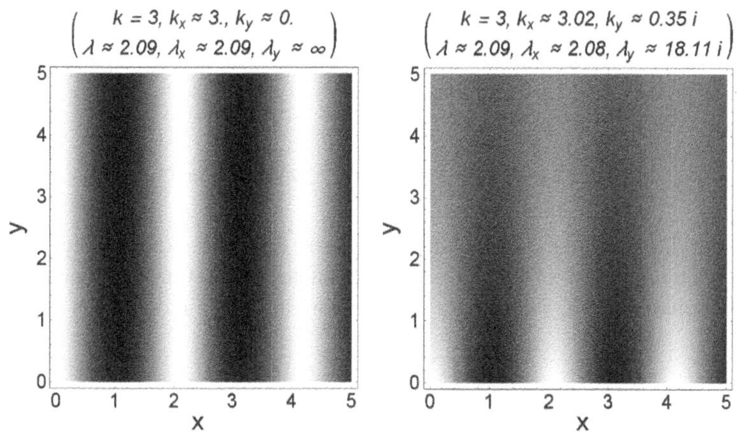

Fig. 5.1 Two dimensional visualization of a propagating plane wave (left) and an evanescent wave (right) propagating along the x-axis. After Ziemer (2018), p. 332. A video can be found on https://tinyurl.com/yaeqpn8n

$$P(\mathbf{x}, \omega) = A(\omega)\ e^{\pm\sqrt{-k^2+k_x^2+k_z^2}\,y} e^{\imath(k_x x + k_z z)} \tag{5.13}$$

In this case the first exponential term is real, indicating an exponential decay in y-direction.[8] Both types of waves are illustrated in Fig. 5.1. Note that in this example the propagation direction of the propagating wave and the evanescent wave are the same.

For periodic functions the motion equation, Eq. 5.1, yields:

$$\nabla p(\mathbf{x}, t) = -\imath\omega\rho_0 \boldsymbol{v}(\mathbf{x}, t) \tag{5.14}$$

and in the frequency domain

$$\nabla P(\mathbf{x}, \omega) = -\imath k \rho_0 c \mathbf{V}(\mathbf{x}, \omega), \tag{5.15}$$

where $\mathbf{V}$ is the sound velocity in frequency domain.

5.1.5 Inhomogeneous Wave Equation

The homogeneous wave equation assumes a source free medium. But every sound field has at least one source which adds acoustic energy to the medium, propagating as a wave pursuant to the wave equation. To account for this, the eighths condi-

[8] Or an exponential increase which is ignored since it is non-physical, see Ahrens (2012), p. 23.

tion listed in Sect. 5.1.2 is dropped and a source term is added to the homogeneous wave equation. Then a solution $p(\mathbf{x}, t)$ is sought describing the temporal and spatial behavior of the source signal in the system.

$$\nabla^2 p(\mathbf{x}, t) - \frac{1}{c^2}\frac{\partial^2 p(\mathbf{x}, t)}{\partial t^2} = -4\pi\delta(\mathbf{x} - \mathbf{x_0}, t - t_0) \tag{5.16}$$

This wave equation is called *inhomogeneous wave equation.* $\delta(\mathbf{x}, t)$ is the Dirac delta function, an impulse. It is defined as being ∞ at point $\mathbf{x}_0$ at time t_0, otherwise it is 0. A transformation of the Dirac delta function into the spectral domain

$$\delta(\omega) = \int_{t=-\infty}^{\infty} \delta(t - t_0)\, e^{\imath\omega t} \mathrm{d}t = 1 \tag{5.17}$$

shows that its amplitude for every frequency is 1, i.e. all frequencies have an equal amplitude and are in phase. That means every arbitrary function $p(t)$ can be expressed by weighted and delayed Dirac delta functions $\delta(\mathbf{x}, t)$. Amplitude and phase of spectral components $P(\omega)$ of sound signals may be arbitrary so they can be expressed as multiplication of the spectra of the Dirac delta function by frequency-dependent complex amplitudes $A(\omega)$. That conforms to a convolution of a sound signal with the Dirac delta function in the time domain.

5.1.6 Point Sources

One solution for the inhomogeneous wave equation is the point source. A point source is a sound source with no volume. In the simplest case, its radiation is equal in each direction. This is referred to as a *monopole source* or *zero order radiator.* Amplitude and phase are dependent on frequency and distance but independent of direction. Therefore, a formulation in spherical coordinates is meaningful. For spherical coordinates the following holds:

$$\begin{aligned}
\mathbf{r} &= \begin{bmatrix} r \\ \varphi \\ \vartheta \end{bmatrix} \\
r &= \sqrt{x^2 + y^2 + z^2} \\
\varphi &= \arctan\left(\frac{y}{x}\right) \\
\vartheta &= \arccos\left(\frac{z}{r}\right) \\
\nabla_{\text{spherical}} &\equiv \frac{\partial}{\partial r} + \frac{1}{r}\frac{\partial}{\partial \vartheta} + \frac{1}{r\,\sin\vartheta}\frac{\partial}{\partial \varphi}
\end{aligned}$$

$$\mathbf{x} = \begin{bmatrix} x \\ y \\ z \end{bmatrix}$$
$$\begin{aligned} x &= r\ \cos\varphi\ \cos\vartheta \\ y &= r\ \sin\varphi\ \cos\vartheta \\ z &= r\ \sin\vartheta \\ \nabla_{\text{Cartesian}} &\equiv \frac{\partial}{\partial x} + \frac{\partial}{\partial y} + \frac{\partial}{\partial z} \end{aligned} \tag{5.18}$$

With radius r, azimuth angle φ and polar angle ϑ. Respectively, the position vector $\mathbf{x}$ is redefined to $\mathbf{r}$. In principle, this spherical coordinate system is consistent with the head related spherical coordinates used in Sect. 4.4 for describing directional hearing. But in this case the coordinate origin is not the listener's head but the source position. Figure 5.2 illustrates the relations of Cartesian and spherical coordinate systems.

Thus, the inhomogeneous wave equation (5.16) takes the form

$$\frac{1}{r}\frac{\partial\left(r^2\frac{\partial p}{\partial r}\right)}{\partial r} + \frac{1}{r^2\ \sin\vartheta}\frac{\partial\left(\sin\vartheta\frac{\partial p}{\partial\vartheta}\right)}{\partial\vartheta} + \frac{1}{r^2\ \sin^2\vartheta}\frac{\partial^2 p}{\partial\varphi^2} - \frac{1}{c^2}\frac{\partial^2 p}{\partial t^2} = -4\pi\delta\left(\mathbf{x}-\mathbf{x}_0, t-t_0\right). \tag{5.19}$$

Since radiation of a monopole is independent of φ and ϑ the wave equation simplifies to

$$\frac{\partial^2 p\,(r,t)}{\partial r^2} + \frac{2}{r}\frac{\partial p\,(r,t)}{\partial r} - \frac{1}{c^2}\frac{\partial^2 p\,(r,t)}{\partial t^2} = -4\pi\delta\left(r-r_0, t-t_0\right), \tag{5.20}$$

and the Helmholtz equation appropriately to

$$\frac{\partial^2 P\,(r,t)}{\partial r^2} + \frac{2}{r}\frac{\partial P\,(r,t)}{\partial r} - k^2 P\,(r,t) = -4\pi\delta\left(r-r_0, \omega\right). \tag{5.21}$$

The point source solution for this case is

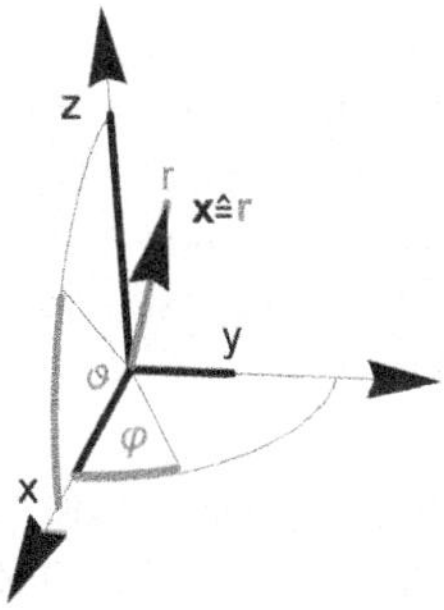

Fig. 5.2 Representation of the position vector $\mathbf{x}$ or, respectively $\mathbf{r}$ via Cartesian coordinates and spherical coordinates. After Ziemer (2018), p. 333

$$\begin{aligned} p(r,t) &= g(r,t) + \tilde{g}(r,t) \\ &= A(t)\frac{e^{-\imath(kr-\omega t)}}{r} + \tilde{g}(r,t)\,, \text{ or, respectively} \\ P(r,\omega) &= G(r,\omega) + \tilde{G}(r,\omega) \\ &= A(\omega)\frac{e^{-\imath kr}}{r} + \tilde{G}(r,\omega)\,. \end{aligned} \tag{5.22}$$

It is a *Green's function* comprised of a linear combination of a special solution—$g(r,t)$, or $G(r,\omega)$, respectively—and a general solution—$\tilde{g}(r,t)$, or $\tilde{G}(r,\omega)$—which are arbitrary solutions of the homogeneous wave equation, Eq. 5.4, and Helmholtz equation, Eq. 5.9. It is also called "impulse response" in the time domain and "complex transfer function" in the frequency domain.[9] Since the first term of the impulse response is already a complete solution of the inhomogeneous Helmholtz equation, the second term can be assumed to be zero. This case is called *free field Green's function* and describes the radiation of a monopole sound source. The exponential term describes the phase shift per distance of the propagating wave from the source. The fraction represents the amplitude decay per distance, the so-called *inverse distance law* or *1/r distance law*,[10] which is owed to the fact that the surface of the wave front increases with an increasing sphere radius, so the pressure distributes on a growing area.

The surface of a sphere S is given as

$$S = 4\pi r^2 \tag{5.23}$$

so the sound intensity I_0 in the origin of the point source at $r = 0$ spreads out to the surface with $I(r) = I_0 \frac{1}{4\pi r^2}$ and is thus directly proportional to $\frac{1}{r^2}$. Since I is proportional to p^2, $p(r)$ it is directly proportional to $\frac{1}{r}$[11]:

$$\begin{aligned} I(r) &\propto \frac{1}{r^2} \\ p(r) &\propto \frac{1}{r} \end{aligned} \tag{5.24}$$

The wave front of a propagating plane wave, in contrast, is assumed to be infinite and thus does not decay. In the far field—i.e. ignoring near field effects which show a complicated behavior close to the source—any stationary sound source can be simplified by considering it as point source.[12] These point sources, however, do not necessarily have to be monopoles. A dependence on direction $\Psi(\omega,\varphi,\vartheta)$ can be introduced ex post by reconsidering $A(\omega)$ as $A(\varphi,\vartheta,t)$ or, respectively, $\Psi(\omega,\varphi,\vartheta)\,A(\omega)$ for the far field:

[9] See e.g. Müller (2008), p. 65.

[10] See e.g. Vorländer (2008).

[11] See e.g. Roederer (2008), pp. 89f.

[12] See Ahrens (2012), p. 42.

$$
\begin{aligned}
p(\varphi, \vartheta, r, t) &= g(\varphi, \vartheta, r, t) + \tilde{g}(\varphi, \vartheta, r, t) \\
&= \Psi(\omega, \varphi, \vartheta)\, A(, t)\, \frac{e^{-\imath(kr-\omega t)}}{r} + \tilde{g}(r, t)\,, \text{ or, respectively} \\
P(\varphi, \vartheta, r, \omega) &= G(\varphi, \vartheta, r, \omega) + \tilde{G}(\varphi, \vartheta, r, \omega) \\
&= \Psi(\omega, \varphi, \vartheta)\, A(\omega)\, \frac{e^{-\imath kr}}{r} + \tilde{G}(r, \omega)\,.
\end{aligned}
\tag{5.25}
$$

Due to the complex factor $\Psi(\omega, \varphi, \vartheta)$, the amplitude $A(\omega)$ is modified for any direction. Note, that the Green's function with a direction-dependent radiation factor is not a solution to the inhomogeneous Helmholtz function as such.[13] It rather comprises the spherical harmonics, which are a solution to the angular dependencies of the Helmholtz equation in spherical coordinates over a sphere rather than a point. The radiation characteristic of point sources can be any arbitrary function of angles φ and ϑ, which can be composed by a linear combination of mono- and multipoles, as will be discussed in detail in Sect. 5.3.1.1. In the literature, point sources with a direction-dependent radiation factor are called "multipole point sources", "higher mode radiators" or "point multipoles", the directivity is called "far-field signature function".[14]

5.2 The Spatial Sound of Musical Instruments

The sound of musical instruments contains a lot of spectral, temporal and spatial features;

"[...] the main acoustic features of musical instruments include:

- musical scale,
- dynamics,
- timbre of sound,
- time envelope of the sound,
- sound radiation characteristics."[15]

The first four features are easily recordable via microphone and can be played back in a good approximation by any High-Fidelity (Hi-Fi) loudspeaker. Still, a listener is often able to distinguish simply recorded and played-back sound from the original instrumental sound. "Composers and musicians often complain about the way loudspeakers sound when aiming at reproducing or amplifying signals from musical instruments."[16] The reason for this is the so-called "Mauerlocheffekt".[17] It effects

---

[13] Cf. e.g. Ahrens (2012), p. 66.

[14] See e.g. Mechel (2013), p. 2, Magalhães and Tenenbaum (2004), p. 204, Ahrens (2012), p. 42.

[15] From Kostek (2005), p. 24.

[16] From Warusfel et al. (1997), p. 1.

[17] See Schanz (1966), p. 2.

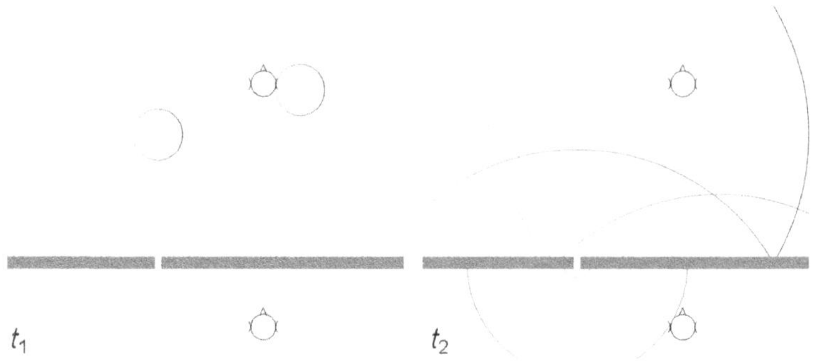

Fig. 5.3 Illustration of the Mauerlocheffekt. Wavefronts reach a small slit from all possible directions within a room. Behind the slit these wavefronts propagate like a monopole, originating at the slit location. A video can be found on https://tinyurl.com/y8ttnhf8

that a monophonic playback sounds like a single-slit diffraction, i.e., like hearing a concert through a keyhole.[18] Independent of directivity or incidence angle of sounds the further sound radiation will be the same for all wavelengths larger than the slit: From the slit on, as for monophonic sound, the feature "sound radiation characteristics" is missing and most information about the source location is lost. The Mauerlocheffekt is illustrated in Fig. 5.3 by two snap-shots of one auditory scene. The listener on top is situated in a room with two sound sources, the other listener is separated from this room by a wall with a small slit. The wave fronts are depicted by circles. The shading of the circles represents slightly different spectra at all positions on the wave front. That means the sound sources are no monopoles but have a direction-dependent radiation factor. The wave fronts will reach the listener inside the room from different angles. Slightly different spectra will reach both ears. Reflections from the wall will occur, also reaching this listener from different angles, having slightly different spectra at both ears. In contrast to that the listener outside the room will hear the two waves as coming from the slit. They will have the very spectrum which had reached the slid, so the spectra that reach the ears are the same. Radiation characteristic, source angle and distance are lost and only reflections from other walls will reach this listener. These will again arrive from the very slit location.

At time t_1.

The complicated radiation characteristics of musical instruments, especially for higher frequencies, create slightly different arrival times of wavefronts and relative amplitude and phase differences between each direction which lead to the impression of a natural, broad, spatial source for a listener due to ILDs and IPDs. It is an important characteristic of instrumental sound, sometimes referred to as "directional tonal color".[19] Even small instruments with simple geometries—such as the shakuhachi—create interaural sound differences for listeners which decrease as distance increases.

---

[18] Cf. Rossing (1990), p. 48.

[19] See Fletcher and Rossing (2008), p. 308.

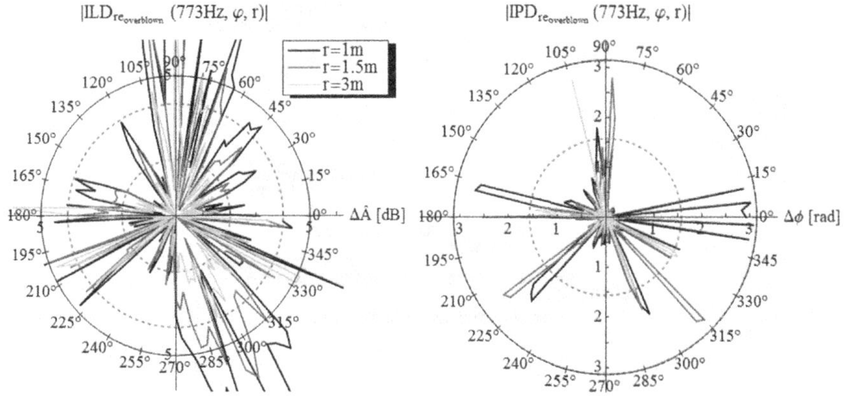

Fig. 5.4 Interaural level differences (left) and interaural phase differences (right) of one shakuhachu partial for listeners at different listening angles and distances. From Ziemer (2014), p. 553

ILD and IPD of a shakuhachi partial are illustrated in Fig. 5.4 for different listening positions.[20]

Without recreating this spatial aspect listeners will be able to distinguish between original instrumental sound and loudspeaker playback. "But only few systems incorporate the directivity characteristic of natural sources."[21] How this radiation characteristic occurs and how it is perceived by a listener is described in the following.

A body radiating with its whole surface in phase, as a so-called *breathing sphere*, radiates as a monopole. This approximately accounts for all wave lengths bigger than the radiating body. For musical instruments this is roughly the case for frequencies up to 500 Hz. Small acoustic sources compared to the radiated wave lengths can be considered as *point sources* whose wave front is a sphere increasing concentric around the source.

The amplitude on this wave front may be dependent on direction. This direction-dependency varies with frequency and is caused by interfering sound radiation from different areas on the body (modes), issues from different apertures (i.e. finger holes or f-holes),[22] directive radiation e.g. from the bell of a trumpet or from diffraction and acoustic shadow behind instrument and instrumentalist.[23] The directional characteristic of a frequency is typically independent of the dynamic but sometimes dependent on the played keynote, especially in the family of string- and woodwind instruments.[24]

[20] An examination of the relationship between features of direct sound and perceived source extent can be found e.g. in Ziemer (2014) and will be discussed in the context of room acoustics in more detail in Sect. 6.2.

[21] Albrecht et al. (2005), p. 1.

[22] Referred to as "structure- and air-borne sound", see e.g. Blauert and Xiang (2009), p. 177.

[23] See Hall (2008), pp. 290–294.

[24] See Meyer (2008), p. 156, Warusfel et al. (1997), p. 4, Pätynen and Lokki (2010) and Otondo and Rindel (2005).

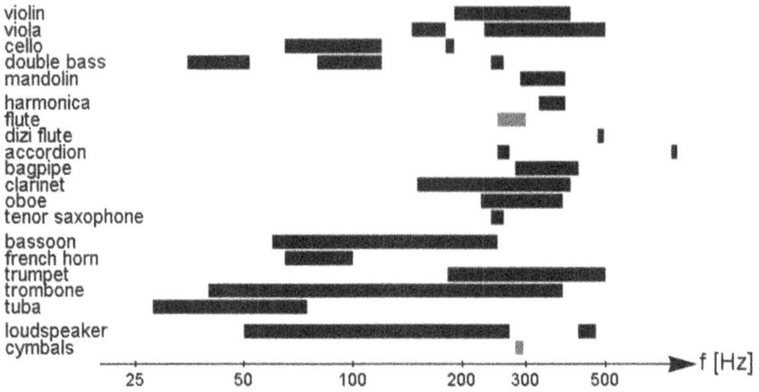

Fig. 5.5 Frequency regions with approximately monopole-shaped sound radiation (black) or dipole radiation (gray) of orchestral instruments. Data from Meyer (2009), p. 130, supplemented from measurements at the University of Hamburg

An overview of the sound radiation of musical instruments is given e.g. by Jürgen Meyers works, Pätynen and Lokki (2010) and Hohl (2009).[25]

Figure 5.5 illustrates frequency regions in which orchestral instruments show an omnidirectional radiation characteristic. Other frequency regions provide different spectra per direction. This means that the listening impression is dependent on the position of listeners and on movements of instrumentalists. This is especially the case in close proximity to the source. Thus, not only the audience but particularly the instrumentalist experiences spectral changes when moving relative to the instrument. A pianist for example is exposed to complicated interferences which may strongly vary with head movements. This is a natural experience which is typically not reproduced in electric pianos. This lack of spatial interference can make them sound static and boring.

The instrumental sound can be divided into phases with different spectral and temporal behavior, and possibly different sound radiation characteristics[26]:

The *transient phase* offers a dense, broad spectrum generated by the main mechanisms of sound production. For example the impulse of a hammer on a piano string, the irregular sticking and sliding of a bow and cello string or the wind burst of a trumpeter excite an amount of frequencies. The transient phase is additionally characterized by side noise like the grasping sound of a guitar, the clicking of trumpet valves or the quick inhaling of singers. For classical musical instruments the transients last for about 30 ms.[27] The duration depends on the instrument, pitch and playing technique. The transient sound plays an important role for the perception of timbre and the recognition of musical instruments as already discussed in Sect. 2.1.

[25]In Meyer (2009), pp. 129–177, Meyer (2008), pp. 123–180, Pätynen and Lokki (2010), and in Hohl (2009) and Hohl and Zotter (2010).

[26]See Meyer (2009), p. 24, Hammond and White (2008), pp. 4–7 and Hall (2008), pp. 124–125.

[27]See Bruhn (2002), p. 452.

The *quasi-stationary phase* is almost periodic. It contains the eigen frequencies of the instrument which established while other frequencies lost their energy e.g. by radiation, destructive interference of standing waves or energy transmission between modes. The long lasting steady sound of an organ or a viola are examples of a quasi-stationary phase which can also be damped as in case of a piano string.

5.3 Measurement of the Radiation Characteristics of Musical Instruments

As described previously in this chapter, the sound of musical instruments can radiate from their surface or containing air. The whole body can vibrate like a breathing sphere, or only parts of it, which may lead to complicated interferences and near field effects as well as complex radiation patterns in the far field. It is difficult to measure the vibrations of body and air without affecting the observed system. Therefore, methods exist to measure the propagated waves from which the magnitudes at their origin are concluded. This situation confronts us with the inverse problem, i.e. reconstructing the normal pressure on a source point or region given the pressure of propagated sound around the source.[28] Several methods supply different solutions. The results may strongly vary between the methods and even between different conditions within the same method. It depends on the individual case—the subject of interest and the research objective—which method delivers the most adequate and robust solution. The basic principles of three common methods—namely circular or spherical far field recordings, beamforming and nearfield acoustical holography—are introduced in the following subsections.[29] All of them quantify the spatial characteristics of sound sources and can be adapted to measure the instrumental sound radiation characteristic. This is done by measurements with microphone arrays in a free field room. Well established visualization methods for the instrumental sound radiation data gained by these three and other methods are subsequently presented and examined.

5.3.1 *Far Field Recordings*

The radiation characteristics of a musical instrument can be measured by simultaneous far field recordings. The far field is typically defined as $kr \gg 1$ or, respectively $r \gg \frac{\lambda}{2\pi}$.

From a distance greater than the dimensions of the musical instrument it is valid to consider the instrument a complex point source or, respectively, a spherical source

[28]See e.g. Williams (1999), p. 89 and p. 236.

[29]For an extensive revision of these and other methods, current research and prospects, see e.g. Magalhães and Tenenbaum (2004).

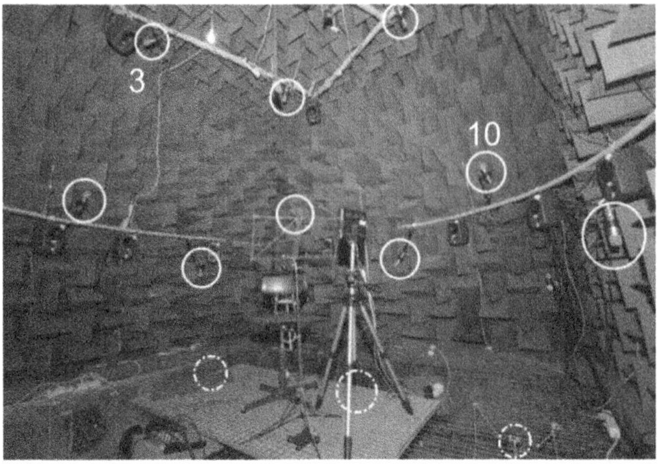

Fig. 5.6 Photo of a microphone array for far field recordings of musical instruments. Reproduced from Pätynen and Lokki (2010), p. 140, with the permission of Deutscher Apotheker Verlag

with infinitesimal volume. The measured wave field is assumed to originate in solely this point. This simplification only holds for big wave lengths compared to the dimensions of the source and is an oversimplification for small wavelengths. Furthermore, it is only valid in the far field of the source and does not inform about near field effects. Thus, the radiation characteristics of the point source can be calculated back from far field recordings. Choosing a meaningful position of the virtual point source in, on or very close to the actual body of the instrument is crucial for a reliable description. For circularly- or spherically-shaped instruments the position of choice might be the center, obviously. However, there is typically no single position which can be considered the "acoustical center" of the radiating sound.[30] There will hardly be a plausible argumentation to pick e.g. geometric centroid or center of gravity of the instrument's mass as acoustic center. One has to find a center position that fits the specific situation or intention. Then, microphones are arranged equidistantly around this center position, i.e. circularly or spherically. Figure 5.6 is a photo of a microphone array in a free field room for measuring the radiation characteristics of musical instruments. It is a spherical arrangement consisting of four groups of five circularly arranged microphones plus two additional microphones in front and above the investigated instrument.

Assuming the source to be a point rather than an area or a volume, the measured relative complex pressure at a microphone position represents not only the pressure at that very position but it can be regarded the pressure factor for that angle. Amplitude and phase per direction of one frequency of a shakuhachi tone are illustrated in Fig. 5.7. These may or may not be interpolated to approximate factors for the angles in

[30] See e.g. Pätynen and Lokki (2010), p. 139.

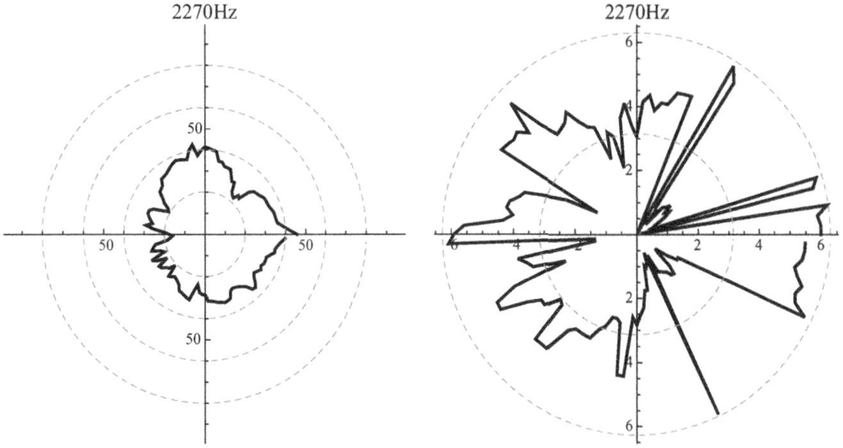

Fig. 5.7 Polar far field radiation pattern of amplitude (left) and phase (right) of one shakuhachi frequency measured at a distance of 1 m with 128 microphones, linearly interpolated. Note, that the phase is periodic, i.e. $\phi(2\pi) = \phi(0)$

between the measurement angles. Except a Fourier transform, Eq. 5.7, no calculation needs to be done.

The accuracy of this simple method can be increased by increasing the number of microphones. Complex factors for angles in between the measurement angles do not have to be approximated by interpolation but may as well be derived from spherical harmonic decomposition, as will be subsequently discussed. Often, the measured directional factors are not taken from single frequencies but are mean values of several frequencies within octave- or third octave bands.[31]

5.3.1.1 Spherical Harmonic Decomposition

As discussed previously in this chapter, the point source solution is a solution to the inhomogeneous wave equation. The zero order radiator, the monopole, is the simplest solution. It solves the simplified wave Eq. 5.20, assuming no dependency on direction. But to describe point sources which do have a directional dependency, a solution to the complete wave Eq. 5.19 must be found. Besides the monopole, such solutions are radiators of higher order, like dipoles, quadrupoles, octupoles etc. They are orthonormal eigen-solutions of the wave equation and referred to as *spherical harmonics*. An infinite series of spherical harmonics can compose any arbitrary wave field. Consequently, any measured wave field can be decomposed into a series of spherical harmonics. This principle is similar to the decomposition of wave forms to spectral components via Fourier transform as described in Sect. 5.1.3. Theory and

[31] See e.g. Otondo and Rindel (2004), p. 1179 or Otondo and Rindel (2005), p. 903, Pelzer et al. (2012), Pätynen and Lokki (2010) and Zotter et al. (2007).

application of spherical harmonic decomposition are illuminated next, followed by a brief discussion.[32]

To find a solution for the inhomogeneous wave equation, Eq. 5.19, it is split into a set of ordinary differential equations by separation of variables.[33] The solution $g(\omega, r, \varphi, \vartheta, t)$ becomes one function of radius $\Pi(\omega, r)$, one of azimuth angle $\Gamma(\omega, \varphi)$ and one of polar angle $\Theta(\omega, \vartheta)$ and one of time $T(\omega, t)$ for each frequency:

$$G(\omega, r, \varphi, \vartheta) = \Pi(\omega, r)\,\Gamma(\omega, \varphi)\,\Theta(\omega, \vartheta)\,T(\omega, t) \tag{5.26}$$

This yields the four following ordinary differential equations:

$$\begin{aligned}
\frac{\mathrm{d}^2\Gamma}{\mathrm{d}\varphi^2} + m^2\Gamma &= 0\\
\frac{1}{\sin\vartheta}\frac{\mathrm{d}\left(\sin\vartheta\frac{\mathrm{d}\Theta}{\mathrm{d}\vartheta}\right)}{\mathrm{d}\vartheta} + \left(n(n+1) - \frac{m^2}{\sin^2\vartheta}\right)\Theta &= 0\\
\frac{1}{r^2}\frac{\mathrm{d}\left(r^2\frac{\mathrm{d}\Pi}{\mathrm{d}r}\right)}{\mathrm{d}r} + k^2\Pi - \frac{n(n+1)}{r^2}\Pi &= -4\pi\delta(r - r_0)\\
\frac{1}{c^2}\frac{\mathrm{d}^2T}{\mathrm{d}t^2} + k^2T &= \delta(t - t_0)
\end{aligned} \tag{5.27}$$

The solution for the azimuth angle is $\Gamma(\varphi) = Ae^{\imath m\varphi} + Ae^{-\imath m\varphi}$. It is a complex exponential function referred to as "circular harmonics" or "cylindrical harmonics".[34] Here, m must be an integer to assure periodicity, i.e., a repetition every 360° and thus a unique function. The first five circular harmonics are plotted in Fig. 5.8. The polar plots show the absolute values of the real part per azimuth angle φ but the sign is indicated by the brightness. The dark gray lobes are positive values, the light gray lobes are negative values, i.e. lobes with different brightness are of opposite phase. The number of lobes is $2m$. Neighboring lobes have different signs, so for odd m lobes on opposed directions also have opposite signs, for even m they have equal signs. Naturally, lobes become narrower with increasing m.

The functions of the polar angle $\Theta(\omega, \vartheta) = P_n^m(\cos\vartheta)$ are *associated Legendre functions*, having two integer indices. Note that the polar angle ϑ lies in a range between 90° and −90° to the x-y-plane as defined in Sect. 5.1.6. Some exemplary associated Legendre functions are plotted in Fig. 5.9 in two different visualization methods for matters of clarification. Equal $|m|$ and n lead to a lobe which becomes narrower with increasing order. The number of lobes is $n + 1 - m$.

---

[32] Mainly based on Williams (1999), pp. 183–208, Teutsch (2007), pp. 41ff, Arfken (1985), pp. 111ff and pp. 573ff, Slavik and Weinzierl (2008) and Ahrens (2012), p. 24ff.

[33] See e.g. Williams (1999), p. 185.

[34] See e.g. Teutsch (2007), p. 44, Ahrens (2012), p. 31, Hulsebos (2004), pp. 16–19 and Zotter (2009), p. 35.

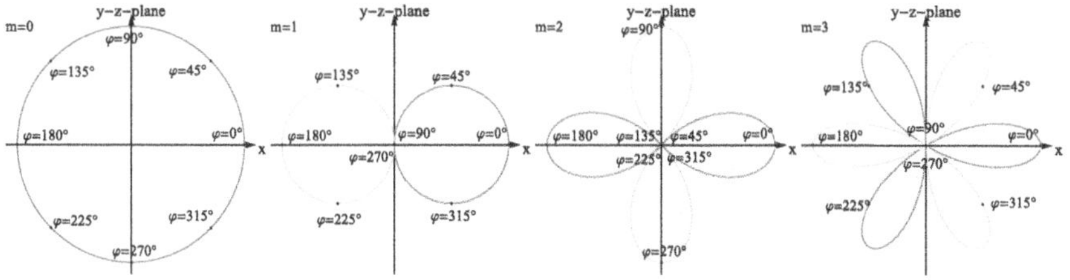

Fig. 5.8 Polar plots of the first five circular harmonics. The absolute values of the real part is plotted over azimuth angle φ. The different shadings illustrate inversely phased lobes, the points on the curve mark the values for the referred angles

Combining the circular harmonics and the associated Legendre functions yields the *spherical harmonics* $\Psi_n^m(\omega, \varphi, \vartheta) = \Gamma(\omega, \varphi)\,\Theta(\omega, \vartheta)$. These are orthogonal complex functions which describe the angular dependency over a complete sphere. Some of the lower order spherical harmonics are plotted in Fig. 5.10. Typically, the smallest sphere that tangents the actual source is taken as spherical source for this method. Another possibility is to chose a point in the center of the instrument as complex point source. This method is referred to as *point multipole method* or *complex point source model*.[35]

The solution for $\Pi_n^{(2)}(\omega, r) = J_n(r) + \imath I_n(r)$ is the *expansion coefficient*. It is a *spherical Hankel function* of the second kind and nth order which is defined as *spherical Bessel function* of first kind and nth order J_n and of second kind and nth order $I_n(r)$, also referred to as *spherical Neumann function*. This is analogous to taking $e^{\imath kr} = \cos(kr) + \imath \sin(kr)$.[36] Both, real and imaginary part of the spherical Hankel function, are plotted in Fig. 5.11. With $m = 0$, this function is equivalent to the free field Green's function as described earlier in this chapter, in Sect. 5.1.6.

5.3.2 Beamforming

Beamforming is a technique which is used not only in musical acoustics but also in the field of technical acoustics, sonar and many more.[37] It is used e.g. for the localization of sources, blind source separation and speech enhancement.[38] The theoretical basis to measure the pressure distribution on musical instruments is introduced in this subsection.[39]

---

[35] See e.g. Magalhães and Tenenbaum (2004), p. 204, Ziemer (2014, 2015, 2017), Ziemer and Bader (2015).

[36] See Arfken (1985), p. 604.

[37] See e.g. Kim (2007), p. 1079.

[38] See e.g. Gannot and Cohen (2008), p. 946.

[39] Mainly based on Hald (2008) and Michel and Möser (2010).

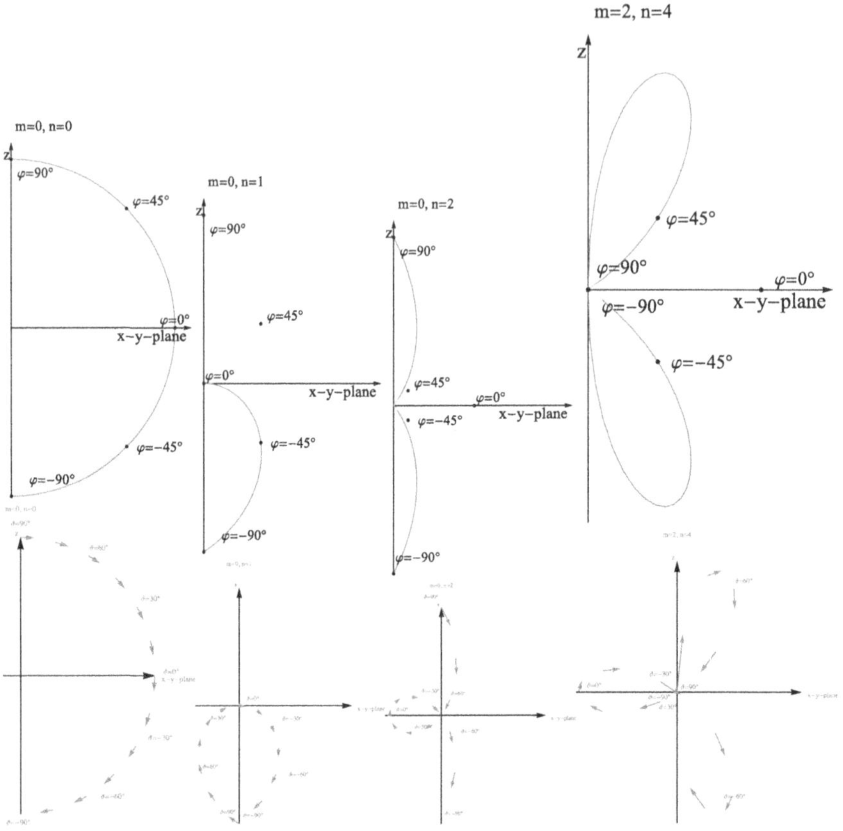

Fig. 5.9 Exemplary associated Legendre functions with different m and n. Upper row: Negative signs are gray. Lower row: Arrows and numbers indicate the course from 90 to $-90°$

The instrumental sound is simultaneously recorded with a regular or pseudo-random array of omnidirectional microphones—often referred to as "acoustical antenna"[40]—in the far field. As will be described in more detail in Sect. 6.1, traveling waves can be simplified as rays directly connecting the origin $\mathbf{Q}$ and a receiver point $\mathbf{X}$. The length of the ray $\mathbf{Q} - \mathbf{X}_m$ is proportional to the travel time of the wave $\Delta t_{\mathbf{Q}-\mathbf{X}_m}$. Here, $m = 1, 2, \ldots, M$ is the microphone number. For beamforming, one can choose an arbitrary position—e.g. on the surface of the instrument—and connect all microphone positions with this point via rays. Focusing on a certain angle instead of a point, the source direction of plane waves can be obtained. The microphone recordings are shifted back in time, by the exact duration derived from the ray length, and then added up and averaged:

[40]See e.g. Michel and Möser (2010).

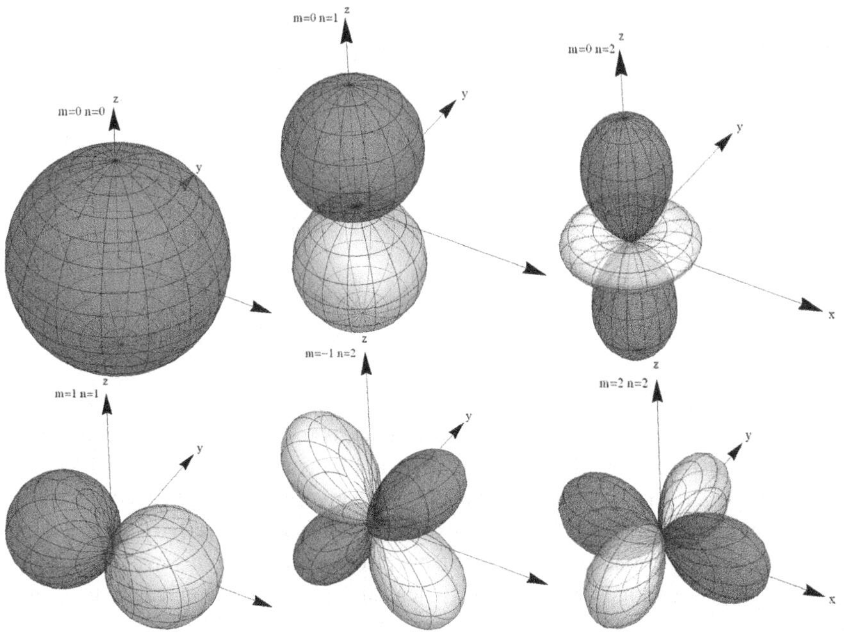

Fig. 5.10 Exemplary spherical harmonic functions with different m and n

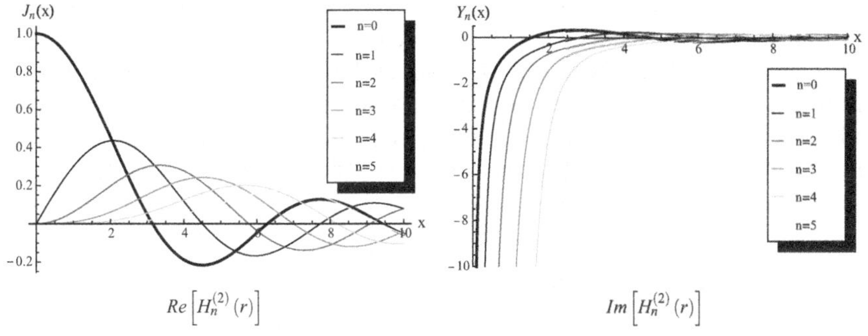

Fig. 5.11 Plot of real part (left) and imaginary part (right) of the spherical Hankel function of second kind and orders 0–5

$$p_{\mathbf{Q}}(t) = \frac{1}{M}\sum_{m=1}^{M} A_{\mathbf{Q}}\, p\left(t - \Delta t_{\mathbf{Q}-\mathbf{X}_m}\right) \tag{5.28}$$

This formulation is the basis of the so-called *delay-and-sum-beamformer*. The factor $A(\mathbf{Q})$ denotes the amplitude which can have two different forms, depending on the considered position. Focusing on a certain incidence angle, the measured wave is assumed to be a plane wave whose amplitude does not decrease with increasing

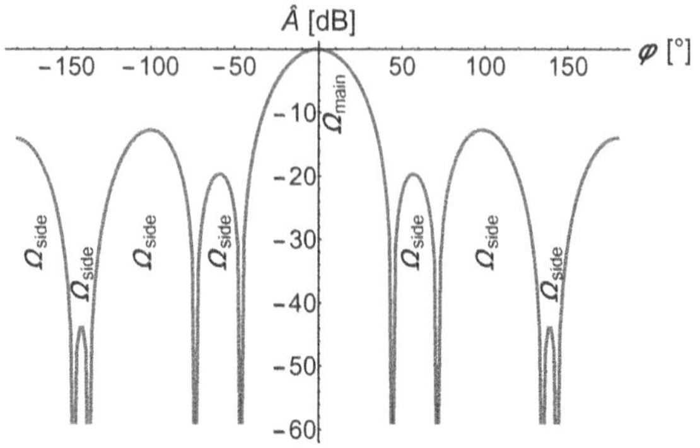

Fig. 5.12 Generic directional sensitivity of a beamformer including main lobe Ω_{main} and sidelobes Ω_{side}

distance, so A_{Q} can be chosen 1. If a source point is focused, one has to compensate the inverse distance law of point sources by implementing an exponential level increase. This can quickly lead to implausibly high amplitudes and errors due to noise, reflections and alike. Figure 5.12 is an example of a generic directivity pattern which illustrates the directional sensitivity of a beamformer for a given wavelength.

Regular microphone arrays cause typical aliasing errors which lead to *grating lobes* as in Fig. 5.12. Therefore, irregular arrays are preferably in use for several applications. These may decrease the height of the sidelobes but typically do not create such deep notches between the lobes. Low frequencies have a wide main lobe which can only be narrowed by increasing the number of microphones considerably. The relation between main lobe width and the number of microphones is $\Delta\Omega_{\text{main}} = \frac{2}{M}$. To increase the ratio of main lobe level to side lobe levels, an even higher increase of microphones is necessary. The relation between main lobe level and minimal side lobe level is $20\log M$. Several weighting and filtering methods and microphone distributions exist, to improve the performance quality of beamformers.[41]

5.3.3 *Nearfield Recordings*

Several nearfield recording techniques gain information about various acoustical quantities of a source surface, such as the sound pressure, the vector intensity field or the particle velocity field. This is accomplished by measuring the radiated sound pressures in the near field. Many methods are derived from optical holography, and are referred to as Nearfield Acoustical Holography (NAH). The complete derivation and formulation of NAH is given e.g. by Maynard et al. (1985). These and other methods can also be reconstructed from the derivation of the sound field synthesis theory, which will be illuminated from Sects. 8.2 to 8.3.2.2. For now, a general near field recording approach to measure how musical instruments radiate their sound is

[41]These are presented e.g. in Bader (2014), Michel and Möser (2010), and Gannot and Cohen (2008).

discussed based on Maynard et al. (1985), Hayek (2008) and Williams (1999), pp. 89–114.

The general relation between the sound pressure in a hologram plane $P(\omega, \mathbf{X})$ and the sound pressure distribution on a source plane $P(\omega, \mathbf{Y})$ can be described by the Rayleigh I integral. It is the basis of planar nearfield recording techniques:

$$P_{\mathbf{X}}(\omega) = -\frac{1}{2\pi} \iint_{S1} \left(G(\omega, \Delta \mathbf{r}) \frac{\partial P_{\mathbf{Y}}(\omega)}{\partial \mathbf{n}} \right) \mathrm{d}S. \tag{5.29}$$

$G(\omega, \Delta \mathbf{r})$ is a wave propagation function as introduced in Sect. 5.1.6. For planar nearfield recordings, an array of M microphones is placed at the hologram plane $\mathbf{X}$ in the nearfield of a musical instrument, parallel to a sound radiating surface at $\mathbf{Y}$, e.g., the sound board of a piano or the top plate of a guitar. Then all microphones simultaneously record the sound to receive the sound pressures in the sampled hologram plane. This sound pressure originated from the source surface. This source surface may be sampled to N point source positions. As stated in Eq. 5.22 in Sect. 5.1.6, a point source can have any spectrum $P_{\mathbf{Y}}(\omega)$. Its propagation describes amplitude- and phase modification of the spectrum per distance and may contain a directional modification $\Psi(\omega, \varphi, \vartheta)$ which describes the directional amplification of spectral components, e.g., creating a dipole-characteristic, as discussed in Sect. 5.1.6. This can be written as the following equation or linear equation system, respectively:

$$P_{\mathbf{X}_m}(\omega) = \sum_{n=0}^{N} G(\omega, \mathbf{Y}_n - \mathbf{X}_m) \frac{\partial P_{\mathbf{Y}}(\omega)}{\partial \mathbf{n}}$$

$$\begin{bmatrix} P_{\mathbf{X}_1} \\ \vdots \\ P_{\mathbf{X}_M} \end{bmatrix} = \begin{bmatrix} G_{1,1} & \cdots & G_{1,N} \\ \vdots & \ddots & \vdots \\ G_{M,1} & \cdots & G_{M,N} \end{bmatrix} \cdot \begin{bmatrix} \frac{\partial P_{\mathbf{Y}_1}}{\partial \mathbf{n}} \\ \vdots \\ \frac{\partial P_{\mathbf{Y}_N}}{\partial \mathbf{n}} \end{bmatrix} \tag{5.30}$$

Equation 5.30 states that the recorded complex pressures at any microphone position $P_{\mathbf{X}}(\omega)$ is the sum of all complex source pressures $P_{\mathbf{Y}}(\omega)$, each modified by their transfer function $G(\omega, \mathbf{Y}_n - \mathbf{X}_m)$. Because the source is sampled to a finite number of point sources whose sound propagation yield the measured sound field via superposition, this method is called *equivalent sources method*. The idea is similar to Huygens' principle. The equation system is only valid if the considered sources on the source plane are the only sources present. If the chosen number of point sources N equals the number of measurement microphones M, the equation has a unique solution. Equation 5.29 as well as Eq. 5.30 consider source and hologram plane to be infinite. Due to the sudden end of finite sources "truncation errors" and "wrap-around errors" occur.[42] Compensating methods for this problem are discussed in many publications.[43] Many alternatives to the equivalent sources method exist, such as the

[42] See e.g. Yang et al. (2008), p. 157.

[43] See e.g. Yang et al. (2008), Maynard et al. (1985), Hayek (2008), Kim (2007).

point multipole method mentioned above, where the directional factors of one complex point source just behind the considered source surface is calculated that would create exactly the measured complex amplitudes at the microphone positions. Another method is to decompose the measured sound field by two-dimensional spatial Fourier transform to plane waves and evanescent waves. This method is referred to as nearfield acoustical holography. An extensive overview of this and other methods is given in Magalhães and Tenenbaum (2004).

Since any radiation pattern can be composed of monopoles, as discussed in the preceding Sect. 5.3.1.1, it is meaningful to consider the source points on the source plane as zero order radiators with a constant directional factor, i.e., $G(\omega, \varphi, \vartheta) = 1\frac{e^{-\imath kr}}{r}$. Here, r is the distance between source and receiver point $||\mathbf{Y} - \mathbf{X}||_2$. Unfortunately, this can lead to an ill-conditioning of the propagation matrix if considered adjacent radiation points or receiver points are close to one another compared to the wave length. Then, the influence of adjacent source points on the wave field at one measurement point is very similar—so is the influence of one point source on two adjacent measurement points. This means the rows of the propagation matrix are not orthogonal. Small errors in the measurement can have massive effects on the calculated solution, often leading to implausibly high amplitudes. To make the approach more robust, a solution to this ill-conditioning problem must be found.

The ill-conditioning problem:

One approach to solve the ill-conditioning problem of the propagation matrix is to consider the point sources not as monopoles but as radiators with a directional dependence $\Psi_{\mathbf{Y}}(\omega, \varphi, \vartheta) \neq \text{const.}$ Then, as the angle changes, the contribution of closely adjacent point sources to the wave field at one microphone position is more differentiated. Of course, such a reconsideration must not be arbitrary. A solid argumentation for the choice of $\Psi_{\mathbf{Y}}(\omega, \varphi, \vartheta)$ is necessary since the reconstruction of the sound pressure distribution at the instrument surface is meant to represent the real nature of the physical conditions, which, unfortunately, cannot be verified due to the inverse problem.

A substantiated argumentation for a robust solution to the ill-conditioning problem is the Minimum Energy Method (MEM).[44] It considers $\Psi_{\mathbf{Y}}(\omega, \varphi, \vartheta)$ as having a lobe Ω, intermediate between a sphere at $\Omega = 0$ and a ray in normal direction at $\Omega = \infty$, formulated as

$$\Psi_{\mathbf{Y}}(\omega, \varphi, \vartheta) = 1 + \Omega \cdot (1 - \alpha) \tag{5.31}$$

with α being the angle between source position $\mathbf{Y}_n$ and microphone position $\mathbf{X}_m$, defined as inner product of both position vectors

$$\alpha_{m,n} = \left| \frac{\mathbf{Y}_n \mathbf{X}_m}{|\mathbf{X}_m|\,|\mathbf{Y}_n|} \mathbf{n} \right| . \tag{5.32}$$

---

[44] As proposed in Bader (2010) and discussed extensively in Bader (2014).

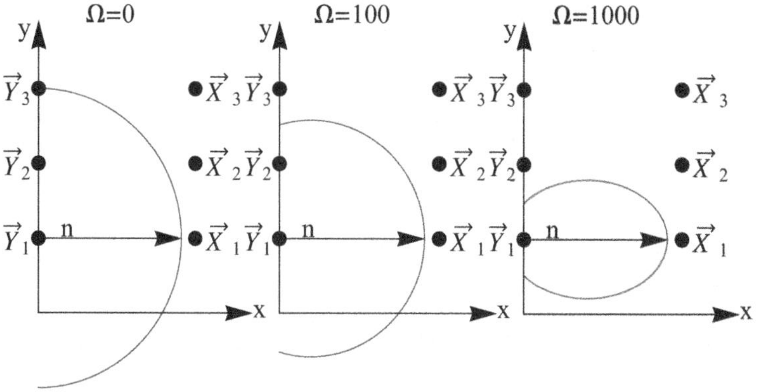

Fig. 5.13 Radiation patterns according to MEM with $\Omega = 0$, $\Omega = 100$ and $\Omega = 1000$. After Ziemer and Bader (2017), p. 485

Here, α is given by the distribution of source- and receiver positions and is 1 in normal direction **n** of the considered source position and 0 in the direction orthogonal to the normal direction. The correct value for Ω needs to be found to receive the correct function for $\Psi_{\mathbf{Y}}(\omega, \varphi, \vartheta)$. $\Psi_{\mathbf{Y}}$ cannot be calculated from the given linear equation system. Therefore, MEM defines that function the correct one which minimizes the reconstruction energy:

$$E \propto \sum_{n}^{N} \left| \frac{\partial P_{\mathbf{Y}_n}(\omega)}{\partial n} \right|^2 = \min \tag{5.33}$$

The energy E, which is proportional to the sum of the squared absolute pressure amplitudes on the considered structure, needs to be minimized to receive the correct function for $\Psi(\omega, \varphi, \vartheta)$ by iteratively finding the right value for Ω. Thus, MEM is a relaxation method, i.e. an iterative method to solve the given linear equation system. It also delivers an easily tunable parameter to adjust the reconstruction results and quickly receive plausible reconstructions. Figure 5.13 illustrates $\Psi_{\mathbf{Y}_n}(\omega, \varphi, \vartheta)$ for three different values of Ω. The minimum energy method has been applied to measure the vibration characteristics of numerous musical instruments, like grand piano, wind instruments, lounuet, Chinese ruan and yueqin.[45] In Chap. 9, especially Sect. 9.2.4, the ill-conditioning problem and the minimum energy method will be discussed more extensively, in the context of sound field synthesis.

A strength of near field recording techniques is that formulations for circular, cylindrical and spherical microphone arrays exist.[46] The sound can be extrapolated from the source structure itself towards all arbitrary points in space. As for the other methods discussed, a proper choice of microphone positions for any measured instru-

[45]See Bader et al. (2009, 2017), Richter et al. (2013), Münster et al. (2013), Bader (2011, 2012a, b), Pfeifle (2016), Takada and Bader (2012), and Plath et al. (2015).

[46]See e.g. Magalhães and Tenenbaum (2004), pp. 200ff.

ment is important to gain a correct distribution of sound pressures and velocities on the instruments surface. To measure more complicated geometries, nearfield recordings can be combined with the boundary element method (BEM) to include the surface of a source in the calculation.[47] Unfortunately, building a mesh of an instrument's surface is a complicated and time-consuming process itself and may suffer from sparse arrays and other issues. Although these methods define the sound source(s) in very different ways, leading to different solutions, each method can be reasonable and adequate for certain applications. One must find the most suitable and robust way for each individual case.

Far field recordings seem ideal to describe the radiation characteristics of musical instruments in two cases: First, if they can be considered as point sources, i.e. if they are small compared to the radiated wave lengths. This is the case for most musical instruments at low frequencies. Second, if the instrumental body can be considered as quasi-spherical. This is the case for many sorts of drums and one might find arguments to simplify even other geometries of musical instruments as being spherical. The calculated directivity pattern is only valid in the far field of the source and does not inform about near field effects. If the assumed position of the point source or spherical source is changed, this might lead to a significantly different solution. Choosing a meaningful position of the virtual point source in, on or very close to the actual body of the instrument is crucial for a meaningful solution that represents the actual physical conditions. If the measured radiation pattern is decomposed to a series by spherical harmonic decomposition, the measured wave field is approximated by a least mean square solution. The precision is limited by the number of receiver points. However, one can chose a solution intermediate between highest precision with highest computational costs and lowest precision with low computational costs. Because the calculated terms are already continuous, the results contain values for positions in between the receiver positions and no further interpolation is necessary. Whether these values meet the actual complex pressures is uncertain.

Near field recordings sample the radiated sound field near the radiating instrument surface. Equivalent sources methods sample this surface and calculate the contribution of each point to the sound radiation. This leads to a high number of virtual point sources on the instrument's surface. As for spherical harmonic decomposition, setups for nearfield recordings have been developed for specific geometries. The concept of planar NAH has been adapted to cylindrical and circular geometries. This makes NAH suitable for more different cases than spherical harmonic decomposition but results in between different NAH adoptions are not easily comparable. The amount of information gained by NAH is very high and especially in combination with BEM it is a superior method to describe the wave field properties of sound sources. But the calculation of resulting wave fields from a source whose properties are measured is very expensive. The wave field in space is the sum of all propagated point source signals. Recreating the wave field as created by a sound source whose properties are known from a NAH measurement e.g. by means of wave field synthesis demands

[47] See e.g. Ih (2008), Bai (1992), Veronesi and Maynard (1989).

very high computational costs since hundreds of virtual point sources would have to be rendered.

Beamforming is a method to localize sound sources. An advantage of beamforming is that it is able to handle any source geometry. The method is able to calculate the sound pressure at infinite positions in space and thus to sample a musical instrument in even higher precision than from many discrete nearfield recording techniques. But contrary to nearfield recording techniques, such as NAH, no information about near field effects can be gathered. Furthermore, the width of lobes and the unavoidable presence of sidelobes drastically restrict the effective measurement precision and reliability. Therefore, it seems more suitable for detection of source location than for analyzing the radiation characteristics of fine structures, such as musical instruments.

5.4 Visualization of the Radiation Characteristics of Musical Instruments

Many methods have been developed to visualize the spatial radiation characteristics of musical instruments. In the eighteenth century (Chladni 1787) visualized modes in a sounding structure by putting sand on it. Accelerations of the structure "throw" the sand away until it lands on areas where the maximum acceleration is smaller than gravitation constant and remains there. At high amplitudes these areas are the nodal lines. This visualization method not only reveals the excited mode of the structure, it also implies information about the sound accelerations in the structures. An example of a Chladni figure is given in Fig. 5.14.

Later, Chladni figures were created even from uneven structures, like the top plate of a violin. Sometimes, such Chladni figures are still obtained by putting sand on the structure. In other cases they are created from hologram interferometry[48] or laser

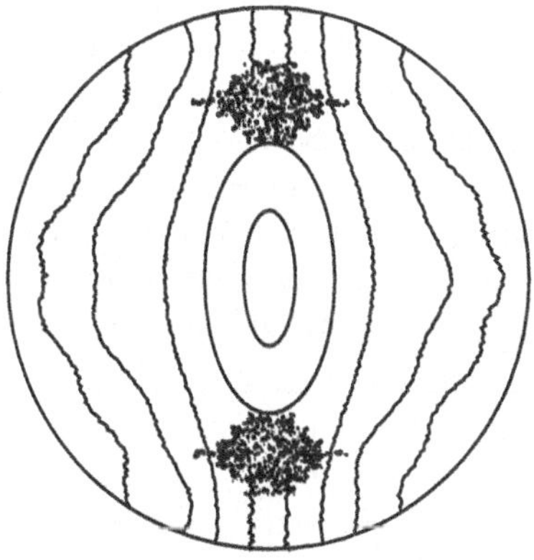

Fig. 5.14 Chladni figure showing nodes on a circular plate. After Chladni (1787), p. 89

[48]See e.g. Hutchins (1977, 1981) and Hutchins et al. (1971).

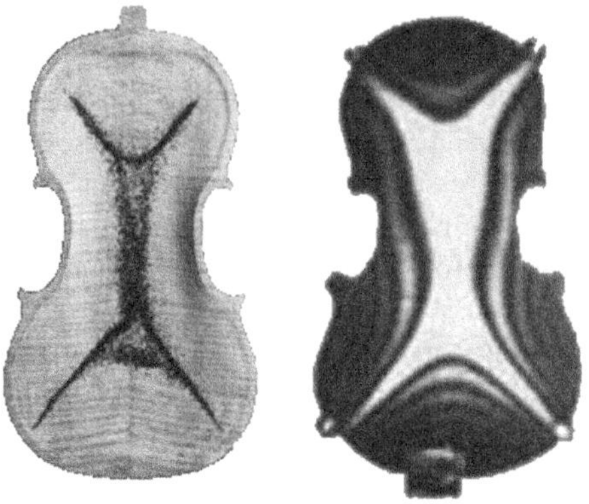

Fig. 5.15 Chladni figures of a violin back plate obtained by sand (left) and by hologram interferometry (right). From Hutchins (1981), p. 174 and 176

scanning vibrometers[49] as can be seen in Fig. 5.15. Eigenmodes of single components of musical instruments have been determined that way. Typically, several periods are necessary for a clear image. In real playing situations, however, components are attached to each other, like the top plate to the rim of a violin, leading to totally different spatial boundary conditions compared to single parts if the instrument. The coupling between those components introduces further temporal boundary conditions and effects such as forced vibrations and mode coupling.[50] Standing waves do not necessarily occur and thus do not have to contribute to the radiation characteristic in actual playing situations.

Electronic TV holography is a technique to measure instrumental body modes of components illuminating it with a strong laser and recording the interferences between sent and reflected light.[51] Saldner et al. (1997) used this method to measure vibrations of top- and backplate of a completely built violin with about 30 pictures per second. They found, however, that these vibration patterns create acoustical short circuits. Consequently they hardly contribute to the far field radiation of the instrument (Fig. 5.16).[52]

The sound radiation characteristics of musical instruments can be measured by several simultaneous recordings in the far field. Meyer (1995) did this for several symphonic instruments.[53] From such measurements directions of strongest sound radiation can be visualized by arrows as in Fig. 5.17 or by shading directions of maximum amplitude as in Fig. 5.18. The static directional factor Γ_{st} is the quotient of effective amplitude per direction and the average amplitude.

A more precise visualization method from far field recordings is the simple polar plot of the amplitude per angle as in 5.19. Pätynen and Lokki (2010) expand this

[49]See e.g. Fleischer (2000).

[50]See e.g. Bader (2013), p. 57 and p. 113.

[51]This and other optical measurement methods are explained e.g. in Molin (2007) and Molin and Zipser (2004).

[52]See e.g. Saldner et al. (1997).

[53]See e.g. Meyer (1995, 2008, 2009).

Fig. 5.16 Interferogram from a top plate of a guitar, created by the use of electronic TV holography. From Molin (2007), p. 1107

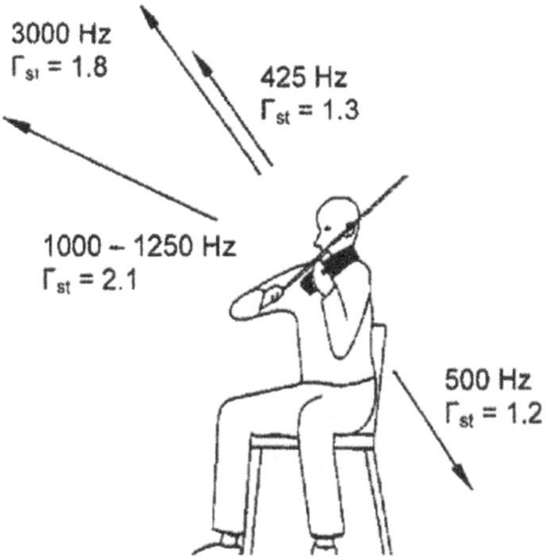

Fig. 5.17 Direction of strongest radiation of violin frequencies and their static directional factor Γ_{st}. Adapted from Meyer (2008), p. 158

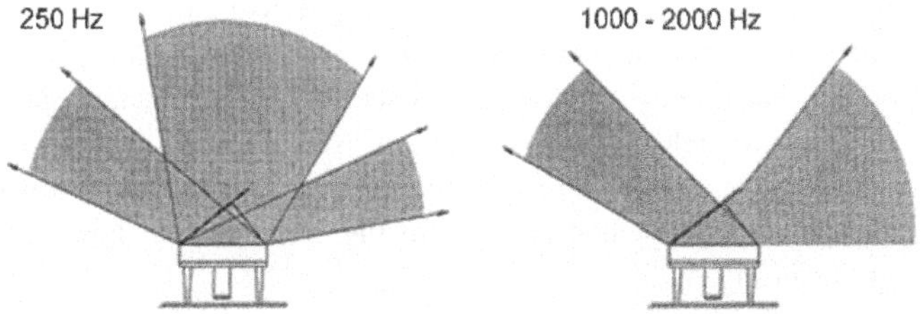

Fig. 5.18 Rough description of the far field radiation pattern from a grand piano for two different frequency regions. The gray areas show directions with an amplitude of 0 to −3 dB referred to the loudest measured amplitude. From Meyer (2008), p. 165

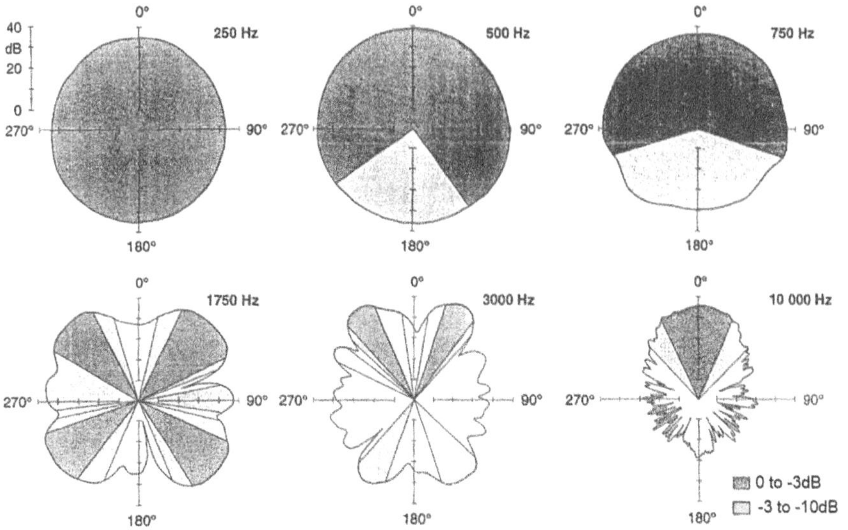

Fig. 5.19 Polar diagrams of an oboe for different frequencies. From Meyer (2009), p. 131

by plotting amplitudes as contour plots over frequency and azimuth angle for different polar angles as demonstrated in Fig. 5.20. These visualizations are gained from measurements discussed previously in Sect. 5.3.1 and illustrated in Fig. 5.6.

Both amplitude and phase per direction have already been illustrated in Fig. 5.7. They can be summarized in one plot that indicates amplitude by radius and phase by color for each angle along the horizontal plane as in Fig. 5.21. Polar patterns are the most common representations of sound radiation characteristics.[54] This can be expanded by the third dimension in so-called "balloon" diagrams[55] as illustrated in Fig. 5.22. Spherical harmonics have already been visualized in Sect. 5.3.1.1. Since they are orthogonal functions, they can simply be added to show a three-dimensional radiation pattern, gained from spherical far field recordings, as illustrated in Fig. 5.22. Decomposing measured complex amplitudes to spherical harmonics yields a complex amplitude not only for the measured direction but also for all angles in between because spherical harmonics are continuous. However, a truncated spherical harmonics decomposition yields a smooth diagram. Actual radiation patterns may look much more irregular, especially at high frequencies. This becomes obvious when comparing the rather complicated patterns in Figs. 5.19 and 5.21 with the smooth balloon diagrams in Fig. 5.22.

The same type of plot can be derived from nearfield measurements, as illustrated in Fig. 5.23. The pressure distribution on the surface is calculated and then forward propagated towards the surrounding air. The particle velocity field in a cross section of the air around a shakuhachi is illustrated in Fig. 5.24.

[54]See e.g. Meyer (2008), p. 156.

[55]See e.g. Vorländer (2008), p. 127.

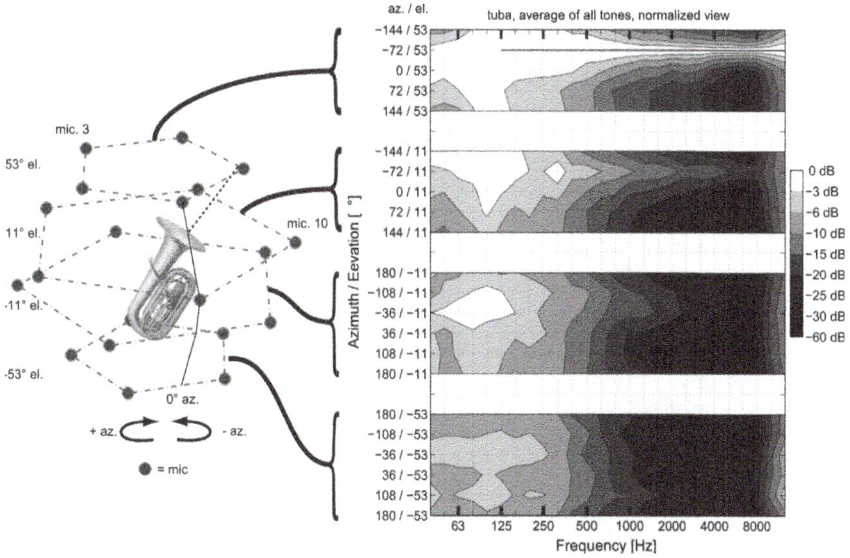

Fig. 5.20 Set of contour plots illustrating the radiation characteristic of a tuba for different angles and frequencies. Reproduced from Pätynen and Lokki (2010, p. 141), with the permission of Deutscher Apotheker Verlag

Fig. 5.21 Amplitude and phase of a single frequency from a played note as recorded at 128 angles around a violin. From Ziemer and Bader (2017), p. 484, with the permission of the Audio Engineering Society

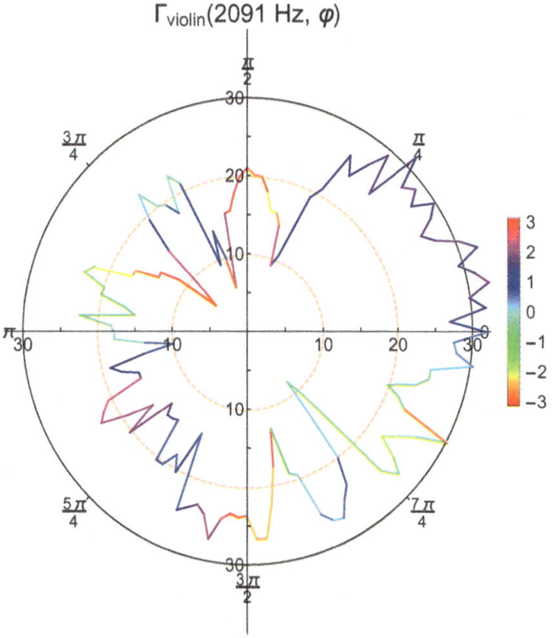

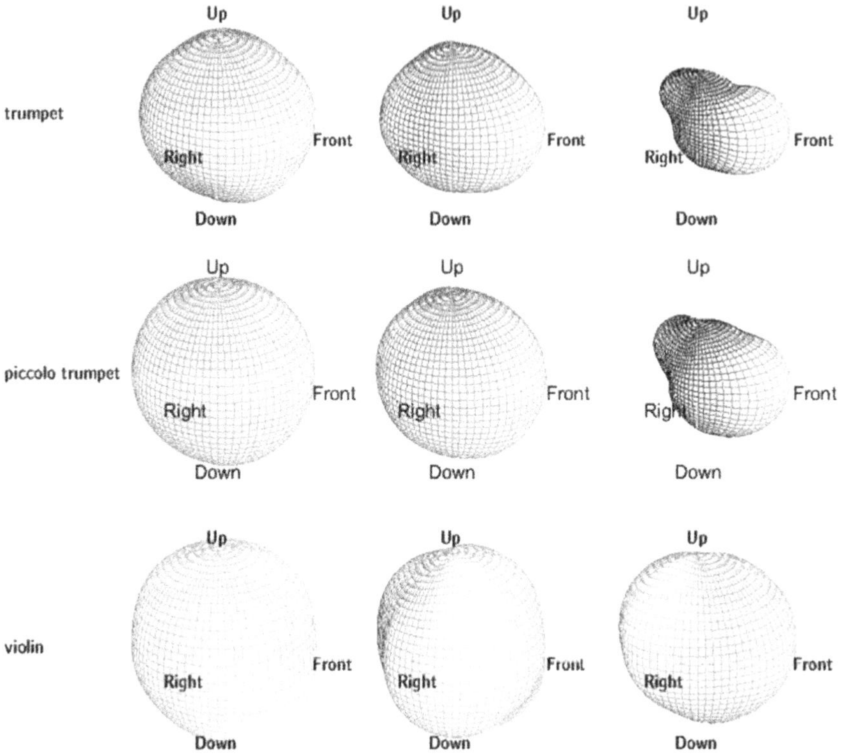

Fig. 5.22 Three dimensional polar plots of the radiation characteristics of different partials of musical instruments. From Vorländer (2008), p. 127

Fig. 5.23 Balloon diagram of a guitar radiation calculated from near field measurements. Reproduced from Richter et al. (2013), p. 7, with the permission of the Acoustical Society of America

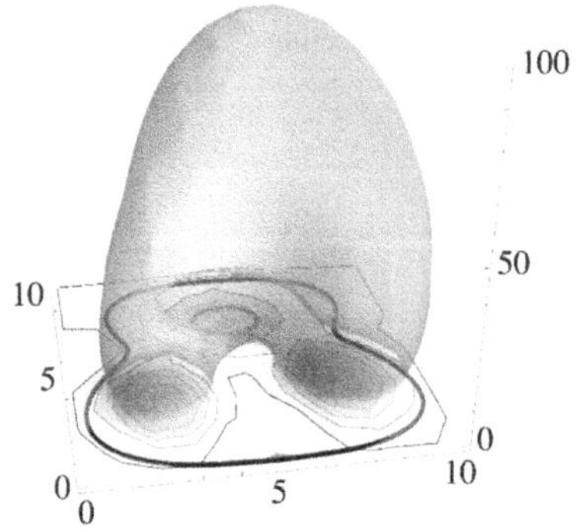

From all measurement techniques presented from Sects. 5.3.1 to 5.3.3, a lot of data is acquired. These can imply e.g. the position of pressure nodes, local complex sound pressures, particle velocities, accelerations and directions of motion.

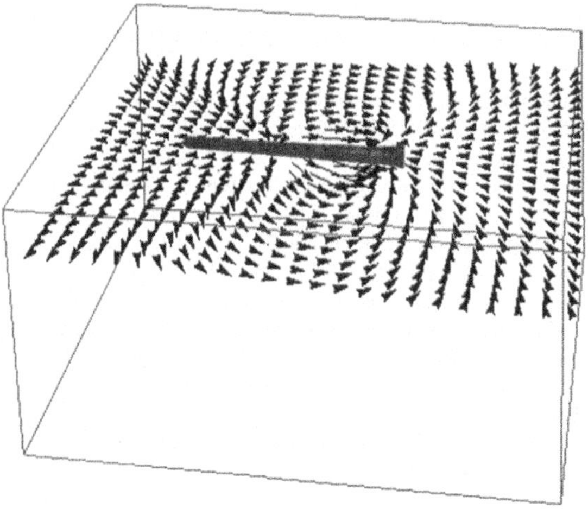

Fig. 5.24 Sound velocity in a cross section through a shakuhachi. The arrow length and direction indicate direction and velocity of particle motion

References

Ahrens J (2012) Analytic methods of sound field synthesis. Springer, Berlin, Heidelberg. https://doi.org/10.1007/978-3-642-25743-8

Albrecht B, de Vries D, Jacques R, Melchior F (2005) An approach for multichannel recording and reproduction of sound source directivity. In: Audio engineering society convention 119, Oct 2005

Arfken G (1985) Mathematical methods for physicists, 3rd edn. Dover

Baalman M (2008) On Wave Field Synthesis and electro-acoustic music, with a particular focus on the reproduction of arbitrarily shaped sound sources. VDM, Saarbrücken

Bader R (2010) Reconstruction of radiating sound fields using minimum energy method. J Acoust Soc Am 127(1):300–308. https://doi.org/10.1121/1.3271416

Bader R (2011) Characterizing classical guitars using top plate radiation patterns measured by a microphone array. Acta Acust United Acust 97(5):830–839. https://doi.org/10.3813/AAA.918463

Bader R (2012a) Radiation characteristics of multiple and single sound hole vihuelas and a classical guitar 131(1):819–828. https://doi.org/10.1121/1.3651096

Bader R (2012b) Outside-instrument coupling of resonance chambers in the New-Ireland friction instrument lounuet. In: Proceedings of meetings on acoustics, vol 15, no (1), p 035007. https://doi.org/10.1121/2.0000167, https://asa.scitation.org/doi/abs/10.1121/2.0000167

Bader R (2013) Nonlinearities and synchronization in musical acoustics and music psychology. Springer, Berlin. https://doi.org/10.1007/978-3-642-36098-5

Bader R (2014) Microphone array. In: Rossing TD (ed) Springer handbook of acoustics. Springer, Berlin, pp 1179–1207. https://doi.org/10.1007/978-1-4939-0755-7_29

Bader R, Münster M, Richter J, Timm H (2009) Measurements of drums and flutes. In: Bader R (ed) Musical acoustics, neurocognition and psychology of music. Peter Lang, Frankfurt am Main, pp 15–55

Bader R, Fischer JL, Abel M (2017) Minimum Energy Method (MEM) microphone array back-propagation for measuring musical wind instruments sound hole radiation. J Acoust Soc Am 141(5):3749–3750. https://doi.org/10.1121/1.4988269

Bai MR (1992) Application of BEM (boundary element method)-based acoustic holography to radiation analysis of sound sources with arbitrarily shaped geometries. J Acoust Soc Am 92:533–549. https://doi.org/10.1121/1.404263

Blauert J, Xiang N (2009) Acoustics for engineers. Troy lectures, 2nd edn. Springer, Berlin. https://doi.org/10.1007/978-3-642-03393-3

Bruhn H (2002) Wahrnehmung und Repräsentation musikalischer Strukturen. In: Bruhn H, Oerter R, Rösing H (eds) Musikpsychologie. Ein Handbuch, 4th edn. Rowohlt, Reinbek bei Hamburg, pp 452–459
Chladni EFF (1787). Entdeckungen über die Theorie des Klanges. Nabu, Leipzig
Fleischer H (2000). Schwingungen und Schall von Glocken. In: Fortschritte der Akustik—DAGA '00, Oldenburg
Fletcher NH, Rossing TD (2008) The physics of musical instruments, 2nd edn. Springer, New York
Gannot S, Cohen I (2008) Adaptive beamforming and postfiltering. In: Benesty J, Sondhi MM, Huang Y (eds) Springer handbook of speech processing, Chap. 47. Springer, Berlin, pp 945–978. https://doi.org/10.1007/978-3-540-49127-9_47
Hald J (2008) Beamforming and wavenumber processing. In: Havelock D, Kuwano S, Vorländer M (eds) Handbook of signal processing in acoustics, Chap. 9. Springer, New York, pp 131–144. https://doi.org/10.1007/978-0-387-30441-0_9
Hall DE (2008) Musikalische Akustik. Ein Handbuch. Schott, Mainz
Hammond J, White P (2008) Signals and systems. In: Havelock D, Kuwano S, Vorländer M (eds) Handbook of signal processing in acoustics, Chap. 1. Springer, New York, pp 3–16. https://doi.org/10.1007/978-0-387-30441-0_1
Hayek SI (2008) Nearfield acoustical holography. In: Havelock D, Kuwano S, Vorländer M (eds) Handbook of signal processing in acoustics, Chap. 59. Springer, New York, pp 1129–1139. https://doi.org/10.1007/978-0-387-30441-0_59
Hirschberg A, Gilbert J, Msallam R, Wijnands APJ (1996) Shock waves in trombones. J Acoust Soc Am l99(3):1754–1758. https://doi.org/10.1121/1.414698
Hohl F (2009) Kugelmikrofonarray zur Abstrahlungsvermessung von Musikinstrumenten. Master's thesis, University of Music and Performing Arts Graz, Technical University Graz
Hohl F, Zotter F (2010) Similarity of musical instrument radiation-patterns in pitch and partial. In: Fortschritte der Akustik—DAGA '10, Berlin
Hulsebos EM (2004) Auralization using wave field synthesis. PhD thesis, Delft University of Technology. http://www.tnw.tudelft.nl/fileadmin/Faculteit/TNW/Over_de_faculteit/Afdelingen/Imaging_Science_and_Technology/Research/Research_Groups/Acoustical_Imaging_and_Sound_Control/Publications/Ph.D._thesis/doc/Edo_Hulsebos_thesis.pdf
Hutchins CM, Stetson KA, Taylor PA (1971) Clarification of the 'free plate tap tones' by hologram interferometry. CAS Newsletter 16:15–23
Hutchins CM (1977) Acoustics for the violin maker. CAS Newsletter 28
Hutchins CM (1981) The acoustics of violin plates. Sci Am 285(4):170–180. https://doi.org/10.1038/scientificamerican1081-170
Ih J-G (2008) Inverse boundary element techniques for the holographic identification of vibro-acoustic source parameters. In: Marburg S, Nolte B (eds) Computational acoustics of noise propagation in fluids—finite and boundary element methods. Springer, Berlin, pp 547–572. https://doi.org/10.1007/978-3-540-77448-8_21
Kim Y-H (2007) Acoustic holography. In: Rossing TD (ed) Springer handbook of acoustics, Chap. 26. Springer, New York, pp 1077–1099. https://doi.org/10.1007/978-0-387-30425-0_26
Kostek B (2005) Perception-based data processing in acoustics. Springer, Berlin. https://doi.org/10.1007/b135397
Magalhães MBS, Tenenbaum RA (2004) Sound sources reconstruction techniques: a review of their evolution and new trends. Acta Acust United Acust 90:199–220. https://www.ingentaconnect.com/contentone/dav/aaua/2004/00000090/00000002/art00001
Maynard JD, Williams EG, Lee Y (1985) Nearfield acoustic holography: I. theory of generalized holography and the development of NAH. J Acoust Soc Am 78(4):1395–1413. https://doi.org/10.1121/1.392911
Mechel F (2013) Room acoustical fields. Springer, Berlin. https://doi.org/10.1007/978-3-642-22356-3
Mechel FP (2008) General linear fluid acoustics. In: Mechel FP (ed) Formulas of acoustics, 2nd edn, Chap. B. Springer, Berlin, pp 5–58. https://doi.org/10.1007/978-3-540-76833-3_2
Meyer J, Meyer P, Baird J (2001) Far-field loudspeaker interaction: accuracy in theory and practice. In: Audio Engineering Society Convention 110, May 2001

Meyer J (1995) Akustik und musikalische Aufführungspraxis. Ein Leitfaden für Akustiker, Tonmeister, Musiker, Instrumentenbauer und Architekten. PPV, Frankfurt am Main, 3. vollständig überarbeitete und erweiterte edition

Meyer J (2008) Musikalische Akustik. In: Weinzierl S (ed) Handbuch der Audiotechnik, Chap. 4. Springer, Berlin, pp 123–180. https://doi.org/10.1007/978-3-540-34301-1_4

Meyer J (2009) Acoustics and the performance of music. Manual for acousticians, audio engineers, musicians, architects and musical instrument makers, 5th edn. Springer, Bergkirchen. https://doi.org/10.1007/978-0-387-09517-2

Michel U, Möser M (2010) Akustische antennen. In: Möser M (ed) Messtechnik der Akustik, Chap. 6. Springer, Berlin, pp 365–425. https://doi.org/10.1007/978-3-540-68087-1_6

Müller S (2008) Measuring transfer-functions and impulse responses. In: Havelock D, Kuwano S, Vorländer M (eds) Handbook of signal processing in acoustics, Chap. 5. Springer, New York, pp 65–85. https://doi.org/10.1007/978-0-387-30441-0_5

Molin N-E (2007) Optical methods for acoustics and vibration measurements. In: Rossing TD (ed) Springer handbook of acoustics, Chap. 27. Springer, New York, pp 1101–1125. https://doi.org/10.1007/978-0-387-30425-0_27

Molin N-E, Zipser L (2004) Optical methods of today for visualizing sound fields in musical acoustics. Acta Acust United Acust 90(4):618–628. https://www.ingentaconnect.com/contentone/dav/aaua/2004/00000090/00000004/art00006

Morse PM, Uno Ingard K (1986) Theoretical acoustics. Princeton University Press, Princeton. https://doi.org/10.1063/1.3035602

Münster M, Bader R, Richter J (2013) Eigenvalue shapes compared to forced oscillation patterns of guitars. In: Proceedings of meetings on acoustics, vol 19, no (1), p 035001. https://doi.org/10.1121/1.4799103, https://asa.scitation.org/doi/abs/10.1121/1.4799103

Otondo F, Rindel JH (2004) The influence of the directivity of musical instrument in a room. Acta Acust United Acust 90:1178–1184. https://www.ingentaconnect.com/content/dav/aaua/2004/00000090/00000006/art00017

Otondo F, Rindel JH (2005) A new method for the radiation representation of musical instruments in auralization. Acta Acust United Acust 91:902–906. https://www.ingentaconnect.com/content/dav/aaua/2005/00000091/00000005/art00011

Pelzer S, Pollow M, Vorländer M (2012) Auralization of a virtual orchestra using directivities of measured symphonic instrument. In: Proceedings of the acoustics 2012 nantes conference, pp 2379–2384. http://www.conforg.fr/acoustics2012/cdrom/data/articles/000758.pdf

Pfeifle F (2016) Physical model real-time auralisation of musical instruments: analysis and synthesis. PhD thesis, University of Hamburg, Hamburg, 7. http://ediss.sub.uni-hamburg.de/volltexte/2016/7956/

Pierce AD (2007) Basic linear acoustics. In: Rossing TD (ed) Springer handbook of acoustics, Chap. 3. Springer, New York, pp 25–111. https://doi.org/10.1007/978-0-387-30425-0_3

Plath N, Pfeifle F, Koehn C, Bader R (2015) Microphone array measurements of the grand piano. In: Deutsche Gesellschaft für Akustik e.V., Mores R (eds) Seminar des Fachausschusses Musikalische Akustik (FAMA): "Musikalische Akustik zwischen Empirie und Theorie", Hamburg, pp 8–9. https://www.dega-akustik.de/fachausschuesse/ma/dokumente/tagungsband-seminar-fama-2015/

Pätynen J, Lokki T (2010) Directivities of symphony orchestra instruments. Acta Acust United Acust 96(1):138–167. https://doi.org/10.3813/aaa.918265

Rabenstein R, Spors S, Steffen P (2006) Wave field synthesis techniques for spatial sound reproduction. In: Hänsler E, Schmidt G (eds) Topics in acoustic echo and noise control. Selected methods for the cancellation of acoustical echoes, the reduction of background noise, and speech processing, Signals and communication technology, Chap. 13. Springer, Berlin, pp 517–545

Richter J, Münster M, Bader R (2013) Calculating guitar sound radiation by forward-propagating measured forced-oscillation patterns. Proc Mtgs Acoust 19(1):paper number 035002. https://doi.org/10.1121/1.4799461

Roederer JG (2008) The physics and psychophysics of music, 4th edn. Springer, New York. https://doi.org/10.1007/978-0-387-09474-8
Rossing TD (1990) The science of sound, 2nd edn. Addison-Wesley, Reading (Massachusetts)
Saldner HO, Molin N-E, Jansson EV (1997) Sound distribution from forced vibration modes of a violin measured by reciprocal and tv holography. CAS J 3:10–16
Schanz GW (1966) Stereo-Taschenbuch. Stereo-Technik für den Praktiker. Philips, Eindhoven
Slavik KM, Weinzierl S (2008) Wiedergabeverfahren. In: Weinzierl S (ed) Handbuch der Audiotechnik, Chap. 11. Springer, Berlin, pp 609–686. https://doi.org/10.1007/978-3-540-34301-1_11
Takada O, Bader R (2012) Body radiation patterns of singing voices. J Acoust Soc Am 131(4):3378. https://doi.org/10.1121/1.4708738, https://doi.org/10.1121/1.4708738
Teutsch H (2007) Modal array signal processing: principles and applications of acoustic wavefield decomposition. Springer, Berlin. https://doi.org/10.1007/978-3-540-40896-3
Veronesi WA, Maynard JD (1989) Digital holographic reconstruction of sources with arbitrarily shaped surfaces. J Acoust Soc Am 85:588–598
Vorländer M (2008) Auralization. Fundamentals of acoustics, modelling, simulation, algorithms and acoustic virtual reality. Springer, Berlin. https://doi.org/10.1007/978-3-540-48830-9
Warusfel O, Derogis P, Caussé R (1997) Radiation synthesis with digitally controlled loudspeakers. In: Audio engineering society convention 103, Sep 1997
Wöhe W (1984) Grundgleichungen des schallfeldes und elementare ausbreitungsvorgänge. In: Fasold W, Kraak W, Schirmer W (eds) Taschenbuch Akustik. Teil 1, Chap. 1.2. Verlag Technik, Berlin, pp 23–31
Williams EG (1999) Fourier acoustics. Sound radiation and nearfield acoustical holography. Academic Press, Cambridge
Yang C, Chen J, Xue WF, Li JQ (2008) Progress of the patch near field acoustical holography technique. Acta Acust United Acust 94(1):156–163. https://doi.org/10.3813/aaa.918018
Ziemer T (2011) Wave field synthesis. Theory and application. (magister thesis), University of Hamburg
Ziemer T (2014) Sound radiation characteristics of a shakuhachi with different playing techniques. In: Proceedings of the international symposium on musical acoustics (ISMA-14), Le Mans, pp 549–555. http://www.conforg.fr/isma2014/cdrom/data/articles/000121.pdf
Ziemer T (2015) Exploring physical parameters explaining the apparent source width of direct sound of musical instruments. In: Jahrestagung der Deutschen Gesellschaft für Musikpsychologie, Oldenburg, Sep 2015, pp 40–41. http://www.researchgate.net/publication/304496623_Exploring_Physical_Parameters_Explaining_the_Apparent_Source_Width_of_Direct_Sound_of_Musical_Instruments
Ziemer T (2017) Source width in music production. Methods in stereo, ambisonics, and wave field synthesis. In: Schneider A (ed) Studies in musical acoustics and psychoacoustics, vol 4. Current research in systematic musicoogy, Chap. 10. Springer, Cham, pp 299–340. https://doi.org/10.1007/978-3-319-47292-8_10
Ziemer T (2018) Wave field synthesis. In: Bader R (ed) Springer handbook of systematic musicology, Chap. 18, Berlin, Heidelberg, pp 175–193. https://doi.org/10.1007/978-3-662-55004-5_18
Ziemer T, Bader R (2015) Complex point source model to calculate the sound field radiated from musical instruments. In: Proceedings of meetings on acoustics, vol 25, Oct 2015. https://doi.org/10.1121/2.0000122
Ziemer T, Bader R (2017) Psychoacoustic sound field synthesis for musical instrument radiation characteristics. J Audio Eng Soc 65(6):482–496 https://doi.org/10.17743/jaes.2017.0014
Zotter F (2009) Analysis and synthesis of sound-radiation with spherical arrays. PhD thesis, University of Music and Performing Arts, Graz
Zotter F, Sontacchi A, Noisternig M, Höldrich R (2007) Capturing the radiation characteristics of the bonang barung. In: 3rd congress of the alps adria acoustics association, Graz

Chapter 6
Spatial Acoustics

The *direct sound* of an instrument is usually only one part of the sound reaching the listener. Only in case of a free field or a free field room—which simulates a free field by heavily damping the enclosures, see Fig. 4.15 in Sect. 4—diffracted and reflected sounds play a negligible role and we speak of a "free sound field".[1] Diffraction around obstacles and reflections from surfaces expand the direct sound by the spatial acoustics in interior and exterior areas. This indirect sound usually amounts to a much lager part of a sound heard by a listener. "Music lives and unfolds its effects with the room in which it resounds."[2] The spatial acoustics can be divided into two components:

Early reflections (ER) are the first reflections of a wave, usually single or double reflections. They appear in the first approximately 80 ms after the direct sound, are ascertainable individually by an impulse response measurement and distinguishable from direct sound and reverberation. Their direction of origin, delay relative to the direct sound, their intensity and spectral distribution characterize them.

ER fade to the *late reflections* (LR, also "reverberation" or "late reverb tail"[3]) consisting of considerably more—usually manifold—reflections. These appear so densely and chaotic that degree of diffusion, duration and sound coloration form its character. Altogether they yield the physical wave field at any position in space and amount to the psychoacoustically perceived sound characteristic of a listening room. The next section illustrates the fundamentals of room acoustics. After that, practical architectural considerations for a satisfying sound experience in concert halls are described by means of geometric room acoustics. What objective parameters have an impact on the subjective quality judgment of a listening room, how they can

[1] See Ahnert and Tennhardt (2008), p. 182.

[2] Loosely translated from David jr. (1988), p. 158. The influence of room acoustics on composition and performance practice is discussed in Sect. 2.2.

[3] See e.g. Berkhout et al. (1993), p. 2764 and Horbach et al. (1999), p. 6.

© Springer Nature Switzerland AG 2020

T. Ziemer, *Psychoacoustic Music Sound Field Synthesis*, Current Research in Systematic Musicology 7, https://doi.org/10.1007/978-3-030-23033-3_6

be measured and presumed and how they correlate with subjective parameters is discussed in the section about subjective room acoustics, Sect. 6.2.

6.1 Geometric and Architectural Room Acoustics

Numerous works address geometric and architectural room acoustics from its history over theory to practical application. This section is mainly based on Ahnert and Tennhardt (2008), Fuchs (2013), Knudsen (1988) and Blauert and Xiang (2009).[4]

For more than 2500 years, considerations regarding room acoustics exist, e.g., by Pythagoras. Also the ancient Romans had rules how to build amphitheaters with decent acoustics. In 1650 Athanasius Kirchner, a professor of Mathematics at the College of Rome, published "Musurgia Universalis" which deals with architectural acoustics, e.g., using a ray diagram to explain reflection and focusing from whispering galleries.[5] However, architecture has mostly been primarily a matter of vision concerning aesthetics, style, proportions etc. and concert halls have been modeled on existing halls with commendable acoustics. In the early nineteenth century the physicist W. C. Sabine founded the field of architectural acoustics with his methods for measuring and predicting room acoustical properties.

General requirements for proper acoustics in rooms for musical performance are[6]:

- Good sight lines
- Useful early reflections for performers and audience
- No protruding late reflections
- Homogeneous sound distribution

These criteria can be considered as a minimum claim and account for conventional musical performance. They do not necessarily hold for unconventional musical styles or rooms with electro-acoustic sound (re-)production systems. Multi-purpose rooms may furthermore require a variability of the acoustics, e.g., by means of adjustable reverberation chambers or mobile absorbers.[7]

A closer characterization of the minimum criteria as well as some simple rules to meet these are given in the following.

Good Sight Lines:

Good sight lines assure that direct sound arrives as first wave front and thus provide a correct localization of the sound source and a clear, distinct sound for the whole audience. This is mostly accomplished by an elevated stage and inclining audience seats or a constantly sloping floor. Furthermore, balconies can be used to provide

---

[4]See Ahnert and Tennhardt (2008), Fuchs (2013), Knudsen (1988) and Blauert and Xiang (2009).

[5]See Forsyth (1985), p. 235.

[6]According to Fuchs (2013), p. 221–223.

[7]As implemented e.g. in the Loyola concert hall and the Jupiter Hall Neuss, see Abdou and Guy (1996) and Blauert (1997) for detailed descriptions.

good sight lines while keeping distance between stage and audience short. This improves vision, but it especially increases the portion of direct sound compared to late reflections.

Useful Early Reflections for Performers and Audience:

Useful ER for performers and audience are especially lateral reflections from close walls or reflectors near the stage, close sidewalls or, in wider rooms, reflecting surfaces between the sidewalls.

No protruding Late Reflections:

Not useful but disturbing ER can be reflections from the rear wall. They appear to be early reflections for the listeners in the rear but the reflected wave front will arrive at the frontal listening positions and the stage with a high amplitude and a great delay and thus be a protruding LR. These protruding LR are heard as echoes and should be avoided. Therefore the rear wall can either steer the sound towards the rear seats or absorb the sound energy. High ceilings also produce protruding late reflections if no lower reflecting planes or sails provide earlier reflections and absorb some energy.

Homogeneous Sound Distribution:

Concave surfaces can focus sound and lead to a highly inhomogeneous distribution of sound energy. Also niches which resonate or which are damped too much should be avoided. For example the depth of balconies should not be larger than their height to avoid acoustic shadowing.[8] The shoebox shape is an established geometry for concert halls, like the Musikvereinssaal Wien, Boston Symphony Hall and the Konzerthaus Berlin. It promises an even distribution of reflections with a high contingent of lateral reflections while being easy to construct and calculate. Room modes f_{res} of rectangular rooms can be calculated as follows:

$$f_{\mathrm{res}} = \frac{c}{2}\sqrt{\frac{l^2}{\Delta x^2} + \frac{h^2}{\Delta y^2} + \frac{w^2}{\Delta z^2}}, \; l, h, w = 0, 1, 2, 3, \ldots \tag{6.1}$$

Resonance frequencies f_{res} are a function of sound velocity c and integers l, h and w for the length Δx, height Δy and width Δz of the room. Several relations were found to equally cover the whole frequency range, like *Volkmann's ratio* $2 : 3 : 5$ and *Boner's ratio* $1 : \sqrt[3]{2} : \sqrt[3]{4}$.[9]

To avoid flutter echoes, resonances and protruding late reflections while supporting diffusivity, parallel surfaces are slightly splayed or canted.[10] Large plane surfaces are subdivide into smaller structures using tilted, scattering surfaces, reliefs and absorbing materials with different absorption coefficients altogether covering the whole frequency range. According to Klepper (2008), tent-shaped architectures

[8]Cf. Everest and Pohlmann (2009), p. 389.

[9]See Everest and Pohlmann (2009), p. 230–250.

[10]By $> 5°$, see Blauert and Xiang (2009), p. 166.

might challenge the shoebox.[11] For larger halls, a leaf-shape delivers useful early reflections compared to fan-shape which lacks of lateral reflections.

6.1.1 Ray Tracing

To plan a room which fulfills all general requirements, it seems obvious to represent sound propagation by means of geometry which can be included in the preliminary blueprint, i.e. rays. Rays are straight connection lines between the origin of a sound—e.g., a point source—and an arbitrary point on the propagated wave front. The straight lines are consequence of the fact that wave fronts travel through air in straight lines as long as no obstacles or surfaces are in its way. Although geometrical, the time t can be derived from the length of a ray $\mathbf{x}$ and the sound velocity c:

$$c = \frac{\mathbf{x}}{t} \Leftrightarrow t = \frac{\mathbf{x}}{c} \tag{6.2}$$

Thus, propagation can be observed in a simplified practice and geometries of the room can be adjusted accordingly. However, phenomena, such as room modes and sound deflection, cannot be described by simple rays. Applying the law of reflection,[12] even sound reflections at surfaces can be modeled for small wave lengths compared to the surfaces as can be seen in Fig. 6.1. Instead of rays, it is possible to depict reflections using mirror sources. Mirror sources can be imagined as virtual sources. They play the original source signal simultaneously, but the wave front is mirrored and attenuated. This is also illustrated in Fig. 6.1 and in more detail in Fig. 6.2.

In general, ray-tracing can be sketched manually in blueprint drawings. However, computer models allow to calculate paths of much more rays with further quantities than only length and travel time. For example the inverse distance law as well as the high-frequency attenuation for long travel paths—expressed in Sect. 5.1.2 and 5.1.6—can be implemented in the ray. Mechel (2013) extensively discusses theory and application of mirror source models.[13] These can be used in computer simulations to extend the concept of geometrical ray-tracing. Original sources and mirror sources can be modeled as complex point sources, having directional amplitude- and phase-factors and an amplitude drop per distance, e.g., as formulated earlier in Sect. 5.1.6. Surfaces can obtain further properties as well, like an absorption- and scattering coefficient. Related methods, like *cone tracing* or *pyramid tracing* represent the increase of wave fronts with growing distance from the source.[14] Alternatively, the finite element method (FEM), finite difference method (FDM) or boundary element method (BEM) can be applied to simulate the acoustical properties of sources and

[11] See Klepper (2008).

[12] That is incidence angle ϑ equals the reflection angle ϑ'.

[13] See Mechel (2013).

[14] See e.g. Ahnert and Tennhardt (2008), pp. 244ff.

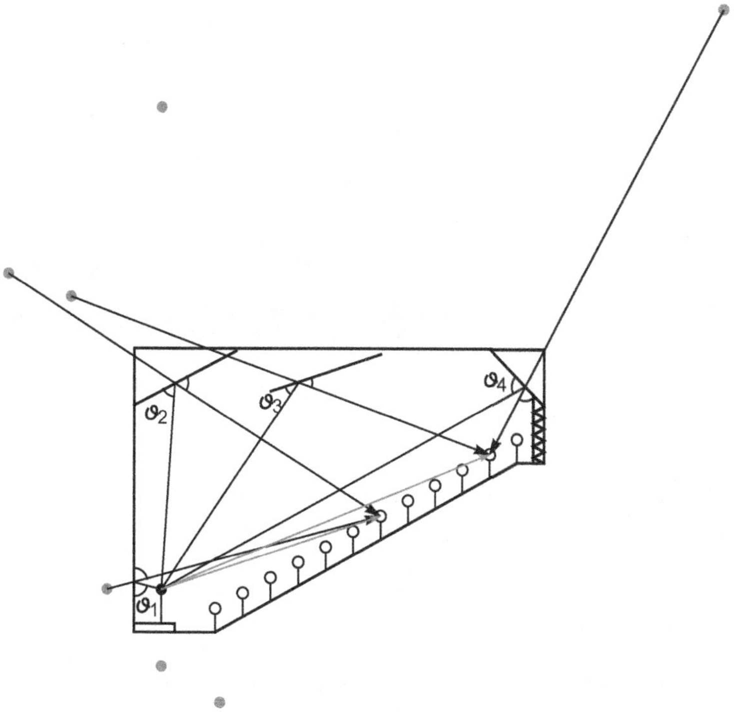

Fig. 6.1 A simple ray diagram of a concert hall including direct sound (gray arrows) some first-order reflections (black arrows) from mirror sources (gray dots). After Deutsches Institut für Normung (2004), p. 218

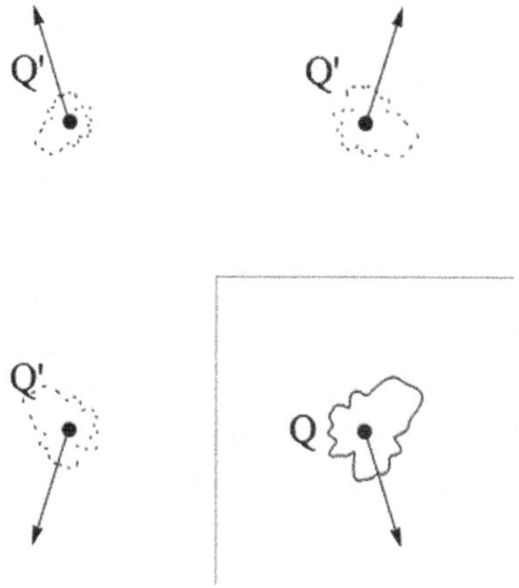

Fig. 6.2 Source Q and mirror sources Q' in a right-angled corner. Note that the contours represent the directional radiation factor of a complex point source, not the wavefront of the propagating wave which is assumed to be spherical. The arrows indicate the viewing direction of the instrument. The reduced contour size of the mirror sources is a result of sound absorption by the walls

Fig. 6.3 Model (left) in a scale of 1 : 20 and resulting hall (right) of the Konzerthaus Berlin. From Ahnert and Tennhardt (2008), p. 251

rooms. Computer aided design (CAD) models of multi purpose halls including ideas of FEM, FDM and BEM are widely used for investigations.[15] Escolano et al. (2005) introduced an *auralization* method via FDM and wave field synthesis. Auralization describes the recreation of rooms acoustics via headphones or loudspeakers from measurements, computations or both.[16] With CAD models one can even consider wave lengths and phase information to uncover resonance phenomenons and simulate diffraction. These virtual physical models can offer extremely high precision, unfortunately with high computational costs. Real life models in a scale of 1 : 10–1 : 20 can provide information about sound distribution, reflections, diffraction and resonances but they are less flexible concerning architectural changes. An example of a real life model and the resulting concert hall are illustrated in Fig. 6.3.

The latest methods and software for modeling rooms and calculating impulse responses exhibit good approximation to real rooms. The musicologists Bader and Schneider (2011) modeled the famous torn down *Star Club* in Hamburg using an auralization software which combines ray-tracing with the mirror source approach. Thereby, they reanimated its specific sound and conserved it. A convolution of dry recordings from original instruments and recording hardware of that era with the impulse response of the modeled room leads to a realistic sound, confirmed by contemporary witnesses.[17] Combining binaural room impulse responses from a ray-tracing software with stereoscopic visualizations the project "Virtual Electronic Poem" by Lombardo et al. (2005) even brings the multi-modal Gesamtkunstwerk "Poèmeme èlectronique"—as discussed in Sect. 2.2 and illustrated in Fig. 2.2—back to life.[18] It is illustrated in Fig. 6.4.

[15]See e.g. Vorländer (2008), pp. 175ff, Ahnert and Tennhardt (2008), pp. 242ff, Vassilantonopoulos and Mourjopoulos (2003), Choi and Fricke (2006), Vigeant and Wang (2008), Rindel et al. (2004).

[16]See e.g. Gade (2007), p. 316, Blauert (2005), pp. 14ff, Bleda et al. (2005), Wenzel et al. (2000) and many more.

[17]A complete description is given in Bader and Schneider (2011).

[18]See Weinzierl (2008), Lombardo et al. (2005, 2009).

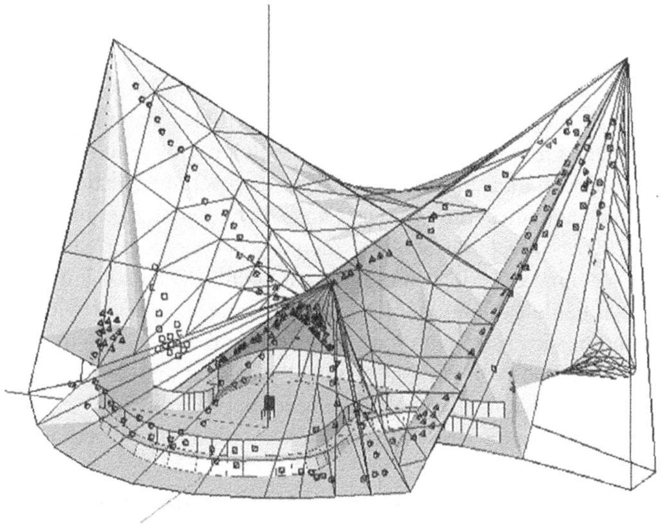

Fig. 6.4 Virtual reality implementation "Virtual Electronic Poem" reconstructing the "Poème Électronique" using stereoscopic visualization and binaural impulse responses gained from ray tracing software. Graphic by Stefan Weinzierl with friendly permission

Computer models more and more displace sculptural models of concert halls for matters of estimation of acoustical features. An overview about practical tools and methods for planning and simulating the acoustics of rooms from an architectural point of view is given, e.g., by Bergeron-Mirsky et al. (2010).[19] In room acoustical software sound sources are typically modeled as point sources with optional directivity function.[20] Acoustician Jürgen Meyer (1977) found that due to the radiation characteristics of orchestral instruments the ceiling plays a crucial role for the auditory brightness of sound, since high frequencies of the string section radiate mostly in that direction. In contrast, frequencies around 1 kHz of brass and string sections mostly radiate towards the sidewalls.[21] Findings of listening tests propose that implementing the spatial radiation characteristics of musical instruments even improves naturalness of auralizations.[22]

6.2 Subjective Room Acoustics

The direct sound of musical instruments, as described in Sect. 5.2, is rarely heard solely. Except from listening situations in the free field or in free field rooms, room acoustics enrich the pure direct sound of musical instruments and lead to several

[19]See Bergeron-Mirsky et al. (2010).

[20]See Pelzer et al. (2012), p. 2380.

[21]See Meyer (1977).

[22]See e.g. Vigeant and Wang (2008), Rindel et al. (2004) and Otondo and Rindel (2005).

(inter-)subjective impressions. For the audience these are especially spatial and spectral impressions from which judgments about the overall acoustical quality of the performance room are delivered. For musicians additional attributes are of importance concerning their playing solo and in an ensemble. Many investigations have been carried out to find physical parameters which correlate to subjective impressions concerning certain aspects of sound and the overall acoustical quality. Many of the objective parameters are standards and can be found, e.g., in DIN 18041 and ISO 3382-1:2009.[23] Since our acoustic memory is very short,[24] a direct comparison between listening experiences in different auditoria and concert halls is hardly possible. For reliable evaluations, listening tests are conducted with experts like conductors and music critics who have long-time experience with different concert halls. Another method is to present artificially created and systematically altered sound fields or even auralizations of existing rooms to listeners.[25]

This section is based on results from extensive research, carried out by Beranek (1996) and others.[26]

In this section the acquisition of objective acoustical data is explained. From these data, many parameters can be derived, which were found to be measures for subjective evaluations of musical performances. The objective parameters are described, followed by subjective parameters and their relations to the objective measures. Suggested values for conventional musical performances are outlined in the end of this section.

6.2.1 Objective Data

A room is usually not a creator of sound but a responder. Room acoustics are reflections of the sound which is produced by the original sources and reduced by absorptions. Therefore—assuming a linear room response—it can be considered as black box, filtering an input signal $A_{\text{in}}(\omega)$ with an unknown function $f(\omega)$, leading to an output signal $A_{\text{out}}(\omega)$:

$$A_{\text{in}}(\omega)\, f(\omega) = A_{\text{out}}(\omega) \tag{6.3}$$

To receive a quantity for the filter function $f(\omega)$, an easy and straightforward method is to use an input signal $A_{\text{in}}(\omega) = 1$. In this case the measured output equates to the

[23] See Deutsches Institut für Normung (2004, 2009).

[24] See Gade (2007), p. 304.

[25] As done, e.g., by Bradley et al. (2000) and Okano et al. (1998). Detailed information on auralization is given, e.g., in Vorländer (2008).

[26] Particularly Beranek (1996, 2004), Kuhl (1978), partly verified or revised by Winkler and Terhardt (1988), Barron and Lee (1988), Bradley et al. (2000), Okano et al. (1998), Okano (2002), Morimoto et al. (2007), Martellotta (2010) and Lokki et al. (2012) and summarized by Abdou and Guy (1996), Gade (2007), Meyer (2009), Ahnert and Tennhardt (2008), Vorländer and Mechel (2008), Kuttruff (2009) and Fuchs (2013).

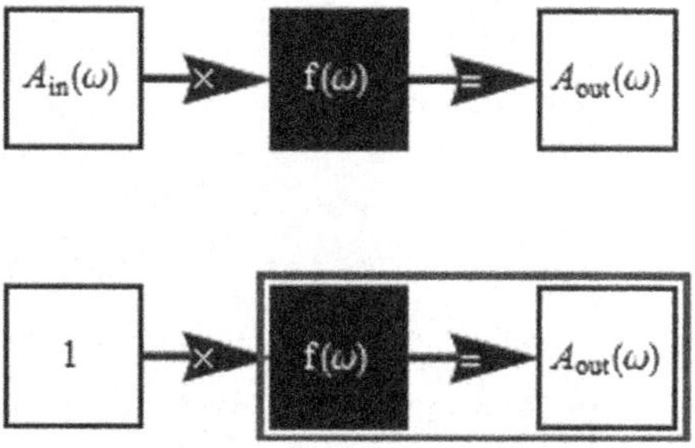

Fig. 6.5 Room acoustics represented as black box, filtering an input signal with an unknown filter function (top). When using an input signal $A_{in}(\omega) = 1$, i.e. a Dirac delta impulse, the output signal equals the filter function (bottom)

filter function as illustrated in Fig. 6.5. No further calculation is necessary. For sounds this means the response of a room to an omnidirectional Dirac impulse contains all its information for the specific source-receiver constellation. The impulse response can be considered equivalent to the filter function which characterizes the acoustics of a room.

Typically, a blank pistol is fired on the stage or at the position of a PA loudspeaker as source for an impulse response measurement in an empty music hall, as shown in the photograph 6.6. Alternatively, a popping balloon or an impulse, presented by an omnidirectional loudspeaker, is used. Microphones or microphone arrays measure the impulse and the room's response to it. The recordings are typically analyzed in six octave bands around center frequencies of 125, 250, 500, 1, 2 and 4 kHz to cover the temporal and spectral aspects of the room acoustics. The impulse response is recorded at average audience positions, with omnidirectional microphones. Furthermore, a dummy head, containing two microphones, and a pressure gradient microphone with a figure-of-eight characteristic pointing at the source are placed nearby. An omnidirectional microphone on stage records the impulse response in the musicians' area. Ideally, an additional unidirectional microphone with an opening angle of $\pm 40°$ is used. A recording of the pistol shot in a free field room in a distance of 10 m is used as calibration- and reference signal p_{ref}. The impulse response is recorded at several central, lateral and rear positions roughly covering the whole listening area. From these recordings the following objective parameters are gained which were found to correlate with subjective quality judgments.

Reverberation Time and Early Decay Time:

The reverberation time or decay time RT is a value introduced by W. C. Sabine. It is defined as the time the sound pressure needs to decrease by 60 dB_{SPL} after the switch-off of a continuous sound or after a loud impulse. Figure 6.7 shows the typical envelope of a room response to white noise after the switch-off. In a logarithmic scale the pressure decrease is approximately linear with some fluctuations. From such a recording the reverberation time or decay time RT can be calculated. Since at low sound pressures the impulse response fades to noise, it is not possible to just read it out from the time series. Therefore RT is extrapolated from the time span between a sound pressure level decrease of $-X_2$ to $-X_1$ dB:

Fig. 6.6 Shot of a blank pistol on the stage of the Docks Club in Hamburg as source signal for an impulse response measurement

$$\mathrm{RT} = 60\,\mathrm{dB}\frac{\left(t_{-X_2} - t_{-X_1}\right)}{|(-X_2\,\mathrm{dB}) - (-X_1\,\mathrm{dB})|} \tag{6.4}$$

t_{-X} is the time in which the sound pressure level decreases by X dB. X_1 needs to be a value considerably higher than the noise level to to not involve its disturbing influence. For X_2 two values exist, distinguishing two methods, leading to two slightly different outcomes. RT usually uses $X_2 = 5$ and $X_1 = 35$ or sometimes 25, indicated by a subscript $\mathrm{RT}_{X_1-X_2}$, i.e. $\mathrm{RT}_{35-5} = \mathrm{RT}_{30}$ and $\mathrm{RT}_{25-5} = \mathrm{RT}_{20}$. With $X_2 = 5$, direct sound and earliest reflections are not considered which means the method is quite robust and independent of the location in the room. The *early decay time* or *early reverberation time* EDT uses $X_2 = 0.1$ and $X_1 = 10.1$. This indicates that the EDT especially takes ER into account, which makes it more variable, depending on the location within the room. EDT is often shorter than RT.[27] Other methods with $X_2 = 0$ are the “Initial-Reverberation-Time” with $X_1 = 15$ and “Beginning-Reverberation-Time” with $X_1 = 20$.[28]

One can simply calculate RT, e.g., by doubling the time span the sound needs to decrease from -5 to -35 dB, and EDT by multiplying the time the sound needs to decrease by $10\mathrm{dB}_{\mathrm{SPL}}$ by the factor 6, as would result from Eq. 6.4. But this is prone to mistakes due to fluctuations in the impulse response. A more stable solution is to use the extrapolated least-square regression (LSR) of these regions, as done in Fig. 6.7. Averaging measurements from different receiver positions makes it even more robust.

[27] According to Kuttruff (2009), p. 237.

[28] See e.g. Meyer (2009), p. 189 or Fuchs (2013), pp. 155ff.

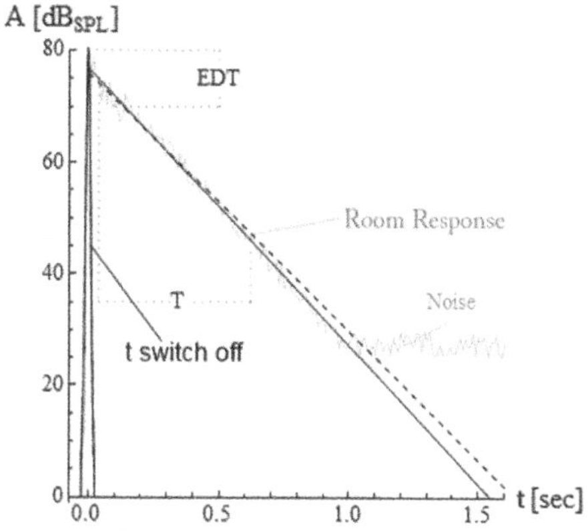

Fig. 6.7 Squared sound pressure level after the switch-off of long lasting white noise sound (gray). RT_{30} (solid black line) and EDT (dashed black line) are the least-square regression of the time span from a sound pressure level decrease from -5 to $-35\,dB_{SPL}$ and -0.1 to $-10.1\,dB_{SPL}$, as indicated by the dotted lines

RT_{30} is almost independent from the position in the room and can be predicted by

$$RT_{30,pre} = 0.163\frac{U}{\tilde{S}}\left[\frac{s}{m}\right], \tag{6.5}$$

with volume U and equivalent absorption area of the Surface $\tilde{S}$.[29] In general, higher frequencies tend to have shorter reverberation times. The JND for RT and EDT is about 5%.[30]

Initial Time-Delay Gap:

The delay between the arrival of the direct sound and the first reflection is called *initial time-delay gap* (ITDG) and can directly be read from the measured impulse response measurement, as shown in Fig. 6.8. It is dependent on source- and receiver-position and can be predicted or concluded, e.g., geometrically from the blueprint or footprint of a building. Of course, the value depends on the constellation of source and listener. Typically, the first reflections arrives from the floor, a close sidewall or a wall closely behind the source.

Hallmaß (Sound Proportion):

The *Hallmaß* (sound proportion) H compares the sound pressure of direct sound and early reflections with the sound pressure of the late reflections in the octave band around 1 kHz:

$$H = 20\lg\left(\frac{\int_{50\,ms}^{\infty} p^2(t)\,dt}{\int_0^{50\,ms} p^2(t)\,dt}\right) \tag{6.6}$$

Negative values denote a dominance of direct sound and ER compared to late reverberation.

[29]The equivalent absorption area is the sum of all areas times their individual absorption coefficient. $\tilde{S} = 0 \equiv 100\%$ absorption, $\tilde{S} = 1 \equiv 0\%$ absorption.

[30]See Gade (2007), p. 308 and Kuttruff (2009), p. 230.

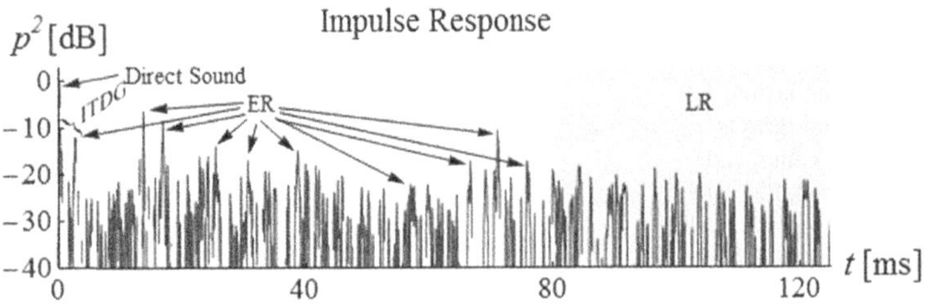

Fig. 6.8 Detail of a room impulse response. Direct sound, ER, LR and ITDG are marked. The increasing density of reflections and decreasing sound pressure over time can be observed

Clarity Factor:

The *clarity factor* or *early to late sound ratio* C_{80} is similar to H, except that 80 ms are chosen as limit for early reflections and the fraction in the logarithm is inverse. This leads to a positive value if the energy of the ER dominate and a negative value if LT contains more energy:

$$C_{80} = 10\lg\left(\frac{\int_0^{80\,\mathrm{ms}} p^2(t)\,\mathrm{d}t}{\int_{80\,\mathrm{ms}}^{\infty} p^2(t)\,\mathrm{d}t}\right) \tag{6.7}$$

Since a linear decay of sound pressure level in dB is expected for the reverberation, C_{80} can be estimated from a measured or predicted RT:

$$C_{80,\mathrm{pre}} = 10\lg\left(e^{\frac{1.104}{\mathrm{RT}}} - 1\right) \tag{6.8}$$

Center Time:

The *center time* t_s is the temporal center of gravity of the impulse response:

$$t_s = \frac{\int_0^{\infty}\left(\left|t\, p^2(t)\right|\right)\mathrm{d}t}{\int_0^{\infty}\left(\left|p^2(t)\right|\right)\mathrm{d}t} \tag{6.9}$$

In contrast, the ASC as already described in Sect. 2.5, is the spectral center of gravity. Due to a linear decrease of the reverberation on a dB scale, the value is predictable:

$$t_{s,\mathrm{pre}} = \frac{\mathrm{RT}}{13.8} \tag{6.10}$$

The JND for t_s is dependent on RT and lies around 8.5% of RT.

Binaural Quality Index:

The *binaural quality index* BQI can be calculated from the *interaural cross correlation coefficient* IACC, which is the maximum absolute value of the *inter aural cross correlation function* IACF:

$$\begin{aligned}\mathrm{IACF}_{t_1,t_2}(\tau) &= \frac{\int_{t_1}^{t_2} p_\mathrm{L}(t)\, p_\mathrm{R}(t+\tau)\,\mathrm{d}t}{\sqrt{\int_{t_1}^{t_2} p_L^2(t)\,\mathrm{d}t \int_{t_1}^{t_2} p_R^2(t)\,\mathrm{d}t}} \\ \mathrm{IACC}_{t_1,t_2} &= \max\left|\mathrm{IACF}_{t_1,t_2}(\tau)\right| \\ \mathrm{BQI}_{t_1,t_2} &= 1 - \mathrm{IACC}_{t_1,t_2}\end{aligned} \tag{6.11}$$

τ is the interval in which the interaural cross correlation is searched; $\tau \in (-1, 1)$ ms can be considered as standard, roughly covering the interaural time difference of a completely lateral sound, for easy comparability. In fact—as discussed in Sect. 4.4.2 and quantified in Eq. 4.8—a time window of ± 640 μs, which is the ITD of a wave with a completely lateral incidence, better fits the physical circumstances. Subscripts L and R refer to microphone at the left and right ear of the dummy head. t_1 and t_2 are chosen 0 and 80–100 ms for $\mathrm{BQI}_{\mathrm{early}}$ or 80–100 ms and 500–2000 ms for $\mathrm{BQI}_{\mathrm{late}}$ or, respectively 0 and 500–2000 ms for $\mathrm{BQI}_{\mathrm{all}}$. A second subscript indicates the octave bands, e.g., $\mathrm{BQI}_{\mathrm{early},500-2000\,\mathrm{Hz}}$. Note, that the left and right recording signal of the dummy head are cross correlated, not the squared sound pressures. The normalization by the denominator of the IACF leads to possible values between -1 and 1. Negative values denote out of phase relationships. The BQI is almost identical for empty and occupied rooms and is averaged over 8–20 seats.[31] The BQI should not be measured for octave bands below 500 Hz because large wave lengths always lead to a high correlation, since even for a completely lateral reflection the phase difference between both is is small. In the literature the BQI is often referred to as "1−IACC".[32] The BQI can have a value between 0 and 1 only since, in contrast to the IACF, it does not differentiate between in phase and out of phase relationships.

Lateral Energy Fraction:

The *lateral energy fraction* LEF—also referred to as "lateral fraction coefficient (LFC)"[33]—is the amount of lateral ER per absolute energy of direct sound and ER:

$$\mathrm{LEF} = \frac{\int_{5\,\mathrm{ms}}^{80\,\mathrm{ms}} p_8^2(t)\,\mathrm{d}t}{\int_0^{80\,\mathrm{ms}} p^2(t)\,\mathrm{d}t} \tag{6.12}$$

The ER as recorded by a figure-of-eight-microphone p_8 are compared with direct sound and ER of the omnidirectional microphone recording. The neutral line of the figure-of-eight-microphone points towards the source.

Since RT does not account for the direction of reflections, LEF cannot be predicted from a known or predicted RT. In a completely diffuse field it would have a value of

[31] See Beranek (2004), pp. 409f and p. 506.

[32] See e.g. Gade (2007), p. 310.

[33] See Ahnert and Tennhardt (2008), p. 204.

0.33.[34] Hence, it can be seen as an upper limit, hardly reached by early reflections. Ideal values lie between 0.2 and 0.3. The JND for LEF is about 5%.

Raumeindrucksmaß and Lateral Efficiency:

The *Raumeindrucksmaß* (spatial impression measure) R is the ratio of the sum of lateral reflections from 25 to 80 ms and all late reflections and the sum of frontal reflections from 25 to 80 ms and all reflections before 25 ms. "Lateral" in this case means from an angle outside of a ±40°-cone around the instrument. This is indeed quite complicated to measure. But bearing in mind that two reflections from directions symmetrical to the median plane are perceived as one frontal reflection, and considering a figure-of-eight characteristic an approximation to an exclusion of frontal and lateral signals within a cone of approx. ±40°, one can approximate R by:

$$R = 10\lg \frac{\int_{25\,\text{ms}}^{\infty} p^2(t)\,\mathrm{d}t - \int_{25\text{ms}}^{80\,\text{ms}} p_{40}^2(t)\,\mathrm{d}t}{\int_0^{25} p^2(t)\,\mathrm{d}t + \int_{25\,\text{ms}}^{80\,\text{ms}} p_{40}^2(t)\,\mathrm{d}t} \tag{6.13}$$

Here, $p(t)$ is the measurement with an omnidirectional microphone and $p_{40}(t)$ is measured with a directional microphone with an opening angle of ±40° facing the source. Easier to measure is the *lateral efficiency* LE, which is the ratio of the lateral sound pressure, recorded with a figure-of-eight microphone and ER from all directions:

$$\text{LE} = 10\lg \frac{\int_{25\,\text{ms}}^{80\,\text{ms}} p_8^2(t)}{\int_{0\,\text{ms}}^{80\,\text{ms}} p(t)} \tag{6.14}$$

Late Lateral Strength:

The *late lateral strength* or *late lateral sound level* LG relates the lateral reverberation with the sound pressure of the direct sound $p_{\text{ref}}(t)$, measured in free field or in the measured room at a distance of 10 m:

$$\text{LG} = \frac{\int_{t=80\,\text{ms}}^{\text{RT}} p_8^2(t)\,\mathrm{d}t}{\int_{t=0}^{t_{\text{dir}}} p_{\text{ref}}^2(t)\,\mathrm{d}t} \tag{6.15}$$

LG is often expressed in A–rated dB-values.[35] Again, as for LEF, the numerator is the squared sound pressure from a microphone with a figure-of-eight characteristic.

Sound Strength:

The *sound strength* G_X is the ratio of the sound pressure in the measured hall—including direct sound, early and late reflections—and the sound pressure of the impulse:

[34] According to Gade (2007), p. 309.

[35] Which basically means weighting lower frequencies considerably less than midrange frequencies, to resemble loudness perception of low-amplitude sound, see e.g. in Zwicker and Fastl (1999), pp. 203ff.

$$G_X = 10\lg \frac{\int_{t=0}^{\infty} p^2(t)\,\mathrm{d}t}{\int_{t=0}^{\mathrm{dir}} p_{\mathrm{ref}}^2(t)\,\mathrm{d}t} \tag{6.16}$$

G_X is usually measured in all frequency bands. If not, a subscript informs about the evaluated frequency band(s). G_{125}, also referred to as “bass strength”,[36] is the value of the 125 Hz-octave, G_{low} is the average of 125 Hz- and 250 Hz-band, G_{mid} is measured in 500 Hz and 1 kHz. Integrating only over 0–80 ms yields the early strength G_{early}. A completely damped room should have a value around 0 dB, depending on the distance of the receiver. However, in ordinary rooms, G_X should have an almost constant, positive value since the sound pressure of the reverberation is independent of the location. Therefore, with a given room volume U it is a predictable value:

$$G_{\mathrm{pre}} = 10\lg\left(\frac{\mathrm{RT}}{U}\right) + 45\,\mathrm{dB} \tag{6.17}$$

The JND of G_X lies around 0.25 dB, for G_{early} around 0.5 dB.

Bass Ratio and Treble Ratio:

Bass ratio BR_X and *treble ratio* TR_X are ratios of different frequency regions from objective parameters:

$$\begin{aligned} \mathrm{BR}_X &= \frac{(X_{125\,\mathrm{Hz}} + X_{250\,\mathrm{Hz}})}{(X_{500\,\mathrm{Hz}} + X_{1000\,\mathrm{Hz}})} \\ \mathrm{TR}_X &= \frac{(X_{2000\,\mathrm{Hz}} + X_{4000\,\mathrm{Hz}})}{(X_{500\,\mathrm{Hz}} + X_{1000\,\mathrm{Hz}})} \end{aligned} \tag{6.18}$$

Here, X can be one of the objective parameters mentioned above, typically RT, EDT or G. The ratios can be calculated from predicted RT and G if frequency-dependent absorption coefficient are known.

Early and Late Support:

The *support* ST is measured by an omnidirectional microphone on stage at a distance of 1 m from the source. It is the ratio of reflected to direct sound in dB:

$$\mathrm{ST} = 10\lg \frac{\left(\int_{t_1}^{t_2} p_{1\mathrm{m}}^2(t)\,\mathrm{d}t\right)}{\left(\int_0^{\mathrm{dir}} p_{1\mathrm{m}}^2(t)\,\mathrm{d}t\right)} \tag{6.19}$$

Choosing $t_1 = 20$ ms and $t_2 = 100$ ms yields the *early support* $\mathrm{ST}_{\mathrm{early}}$. $t_1 = 100$ ms and $t_2 = 1$ s yields the *late support* $\mathrm{ST}_{\mathrm{late}}$. Choosing $t_1 = 0$ ms and $t_2 = 80$ ms yields the *early ensemble level* (EEL). Both ST are typically measured in the octave bands from 250 Hz to 2 kHz. The lowest octave band is left out because it is difficult to isolate the direct sound from the reflections in a narrow band recording at such low frequencies.[37]

[36] See Beranek (2004), pp. 512f.

[37] See Gade (2007), p. 311.

Reverberation Ratio

The reverberation ratio

$$\mathrm{RR} = 10\lg\left(\frac{\int_{160\,\mathrm{ms}}^{320\,\mathrm{ms}} p(t)^2\,\mathrm{d}t}{\int_{0}^{160\mathrm{ms}} p(t)^2\,\mathrm{d}t}\right) \tag{6.20}$$

is measured at a distance of only 0.5 m to the source. This measure is proposed in Griesinger (1996). It is assumed to have a similar magnitude as the EDT at about 350 ms.

6.2.2 *Subjective Impressions*

Room acoustics is mainly the response of the room to sound created by a source inside it. It is obvious that many impressions which arise from the room acoustics are of spatial character. Finding standard terms for subjective impressions is not easy to accomplish, many terms have been used by different authors to describe these similar attributes. In the following, subjective impressions will be described which experienced broad agreement within the literature. Calling these impressions "personal preferences of listeners"[38] Beranek (1996) indicates the subjectiveness of impressions and preferences. Still, many listeners agree to a certain degree, indicating intersubjective validity, at least for the investigated present-day experts from the Western culture.

Of course, demands on the acoustics of a room for musical performance are not universal. For example they vary with different types of music, such as symphonic music, chamber music or popular music. They may also depend on different styles, such as a baroque, romantic or classical symphony. And of course they vary with the room itself. For larger halls, a longer reverberation is natural and desired. Furthermore, quality judgments for concert halls cannot simply be transferred to other locations for music performance and listening, such as living rooms, discotheques, cars, churches or sports stadiums. Furthermore, demands are culturally biased. Koreans for example were found to rate concert halls differently from Western subjects.[39] Therefore, in this work, preferences from present Western subjects towards symphonic music and opera music are summarized and ideal values are related to this sort of music. Subjects agree to a certain degree but (inter-)subjective impressions cannot be considered as ultimate everlasting truth free from influences such as mood, fashion or zeitgeist, of course. Results are commendable orders of magnitude rather than mandatory requirements.

Reverberance:

Outdoor, in an open-air musical performance, instrumental sounds are spatially, temporally and spectrally intelligible and distinct but sound dry. Such a performance

[38] See Beranek (1996), p. 285.

[39] See Everest and Pohlmann (2009), p. 385.

space has little reverberance. In a large, barely undamped room, like a cathedral, the sounds of instruments on different position fuse to an ensemble sound since the long lasting reverberation contains a mixture of all sounds. Successive notes bond or blend because there is no silence between them but reverberation sound. Frequency- and amplitude-modulations smudge since the reverberation contains all states of the modulation from the last several seconds and thereby averages it. Here, the high *reverberance* makes the musical performance sound full but less distinct. These are the two extremes of reverberance. A pleasant reverberance is a good compromise between distinctness and fullness.

The reverberation time RT was initially used as objective measure for the subjective impression of reverberance. The early decay time EDT especially considers the earlier and louder parts of the room acoustics. It is therefore related to the masking threshold of the reflections, which partly mask the direct sound, causing the indistinctness and unintelligibility. Furthermore, EDT can vary between different locations in the room, as can the impression of reverberance.[40] Hence, EDT shows better correlation with reverberance. However, there is a high correlation between both parameters as well. RT from 1.8 to 3 s or EDT in a range of 1.5–2.5 s is considered ideal for the performance of classical music, slightly more for baroque.

Clarity:

The term *clarity* has already been used in the description of reverberance. It is also called "definition" or "transparency"[41] and describes the degree to which details of performance can be perceived distinctly. This refers to simultaneous and successive sounds. As the name already implies, the clarity factor is one objective measure for clarity. Early reflections are integrated together with direct sound by the auditory system and therefore have an amplifying effect. These parts of the impulse response are compared to the reverberation, which can have a masking effect, reducing clarity. Another objective parameter is the center time $t_{s,1000\text{Hz}}$, in which low values indicate high clarity and vice versa. However, a higher value lowers clarity but increases the reverberance. Everest and Pohlmann (2009) consider a slightly adjusted treble ratio $\frac{\text{EDT}_{2000\text{Hz}}}{\text{EDT}_{500\text{Hz}}+\text{EDT}_{1000\text{Hz}}}$ as indicator for "clearness". As mentioned earlier, a good compromise between distinctness and fullness or, respectively, between reverberance and clarity is desirable. Dependent on the music, a C_{80} of -3.2 to 0.2 dB and a t_s of 70–150 ms is considered as ideal.

Spaciousness:

Sound coming from all possible directions, emanating from broad sources, is a pleasurable listening experience and can be described by several aspects, such as *liveness*, *spatial impression*, *intimacy*, *listener envelopment* (LEV) and a high *Apparent Source Width* (ASW). The degree of *spaciousness* is one of the most distinct indicator for the subjective judgment about the quality of a concert hall.[42] It is an unconscious but

[40]See Ahnert and Tennhardt (2008), p. 188.

[41]See Beranek (2004), p. 24 and Vorländer and Mechel (2008), p. 941.

[42]See Beranek (2004), p. 29.

pleasant experience.[43] BQI_{all} delivers an approximate value for the subjective impression of the spatial quality of a room, ideally having a value around 0.6. Furthermore, the presence of strong bass is desirable.

Liveness:

Liveness or "Halligkeit"[44] is the impression that there is more sound than just direct sound and repetitions of it. A "live" concert hall has a long reverberation, in contrast to a "dead" or "dry" hall.[45] It roughly corresponds to the RT in the frequency regions around 500 Hz and 1 kHz. An RT of 1.5–2.2 s can be measured in typical concert halls, slightly less in opera houses. A better measure is H, which compares the reverberation with direct sound and ER. Here, values between -2 and 4 dB are ideal.

Spatial impression:

The *spatial impression* or "Räumlichkeit" is the impression that the whole room itself is filled with sound, rather than only the area around the instruments. A spatial impression emerges when a listener experiences an amount of sound from many more or less distinct directions. Thus, it is influenced by the diffuse LR as well as lateral ER. Therefore, R is the measure of choice. Values from -10 to -5 dB are judged as little spatial, 1–7 dB as very spatial. Ideally, R lies in a range of -5–$+4$ dB.

Intimacy:

The term *intimacy* describes how close acoustic sources and surfaces seem to be, and thus, how small or big the room appears and how intimate musicians and audience are. It seems to be closely related with the ITDG.[46] An ITDG of less than 21 ms is measured in the best-rated concert halls, lower-rated halls show an ITDG of 35 ms, poor halls up to 60 ms. Lokki et al. (2012) found that perceived proximity of the musicians—a parameter related to intimacy—best fits the preference-rating of different concert halls, although they conducted listening tests with equal distances from the source through all simulated concert halls.[47] However, they were not able to find an objective parameter which could explain the subjective judgment.

Listener Envelopment:

Listener envelopment LEV is the feeling of being surrounded by sound. It is influenced by late lateral sounds. Therefore, LEF seems to be an adequate measure. LG is reported to show better correlation with LEV, yet its correlation has only been found in laboratory experiments with synthetic sound fields.[48] Another measure which

[43] See Kuhl (1978), p. 168.

[44] See e.g. Kuhl (1978), p. 168.

[45] See Beranek (2004), p. 29.

[46] See Okano (2002), pp. 217ff, Beranek (2004), p. 518 and Kuhl (1978), p. 168.

[47] See Lokki et al. (2012)

[48] See Okano et al. (1998).

asserted itself in listening tests in real concert halls is the late binaural quality index $\mathrm{BQI}_{\mathrm{late},500-2000\,\mathrm{Hz}}$. Additionally, LEV seems to depend on RT.[49] Negative LG and $\mathrm{BQI}_{\mathrm{late},500-2000\,\mathrm{Hz}}$ slightly over 0 are considered ideal.

Apparent Source Width:

As described in detail earlier in Chap. 5, the radiation characteristic of musical instruments has typically directional properties, leading to different amplitudes and phases at the listener's ears, determining the *apparent source width* ASW. This interaural difference can be highly increased by early reflection which are integrated with the direct sound in the auditory system and can have an amplifying effect. It results in a perceived widening of the source. The lateral efficiency LE is a plausible measure for ASW[50] as is the lateral energy fraction LEF, especially in the frequency region from 125 Hz to 1 kHz where the auditory system is most sensitive. Another measure is the early binaural quality index $\mathrm{BQI}_{\mathrm{early},500-2000\,\mathrm{Hz}}$. Especially in combination with G_{low} or, respectively, $G_{\mathrm{E,low}}$ it correlates to subjective ratings.[51] Some of these measures have been modified and applied to the pure direct sound of musical instruments and explained their physical extent fairly well.[52] The best explanation of physical source width could be achieved when combining one parameter that describes the incoherence of the ear signals with one that quantifies the level of low frequency content. Additional early reflections create the impression of an even wider source. However, late arriving reflections seem to be able to diminish ASW.[53]

It has been reported that for a LEF below 0.6, the relationship between LEF and IACC can be approximated by the formula

$$\mathrm{IACC} = 1 - \frac{\mathrm{LEF}}{1.5} \tag{6.21}$$

with a relative error of 5%.[54] Beranek (2004) also found a reasonable correlation between LEF and BQI.[55] In contrast to that, other authors report that LEF and BQI are not highly correlated.[56] Both measures consider different frequency regions important for the subjective impression. Furthermore, due to interferences, the BQI strongly varies even for small changes in listener location, which contradicts listening experiences.[57] That is why de Vries et al. (2001) doubt that the BQI is an adequate measure for the ASW. They propose modifications to the BQI measurement by means of temporal and spectral filtering as well as a combination with beamforming or wave

---

[49] See Morimoto et al. (2007).

[50] According to Ahnert and Tennhardt (2008), pp. 203f.

[51] As suggested by Okano (2002), Beranek (2004), p. 7 and Okano et al. (1998).

[52] See Ziemer (2011), Ziemer (2015).

[53] See Bradley et al. (2000).

[54] See Ouis (2003).

[55] See Beranek (2004), p. 528

[56] See e.g. Blau (2004) and Gade (2007), p. 310.

[57] See e.g. de Vries et al. (2001), Gade (2007), p. 310 and Kuttruff (2009), p. 241.

field decomposition. This way, they intend to reduce the effect of interference in the measurement, since the human auditory system does not seem to be affected much by them.[58] From several studies, Ando (2010) concluded that ASW does depend on the amplitude of the IACF, i.e. the BQI, but in combination with the width of this amplitude region W_{IACC}.[59] Apart from that, Abdou and Guy (1996) noted that a one-sided balance of ER can lead to an annoying source shift or the perception of a double-source,[60] which is not regarded in any of these measures, so there is a need for a more robust and comprehensive parameter.

Loudness:

Loudness is the perceived volume or force of sound. Reflections increase the loudness compared to pure direct sound. An objective measure for loudness is the sound strength G_X which describes the sound enhancement by the acoustics of the room. Dependent on the musical style, G_X between 1 and 4 is an ideal value.

Timbre/Tonal Color:

Timbre or *tonal color* is affected by the spectral balance of the sound, especially warmth and brilliance, which shall be described in terms of sound pressure ratios between frequency regions.

Warmth:

The *warmth* is a matter of audibility of bass frequencies. Therefore, the bass ratio BR_{RT} was suggested by Beranek (2004) as objective parameter.[61] Though commonly used and cited in the literature, Beranek (2004) himself found this measure to be inadequate to describe the categorical rating of concert halls from his own listening tests.[62] Rather, the strength of bass frequencies G_{125} correlates with warmth being 1.2 dB higher in empty halls than in occupied. Also the BR_G or BR_{EDT} are discussed as objective measures.[63]

Brilliance:

A brilliant, harsh or bright sound is experienced when high frequencies are present. The treble ratio of the reverberation time TR_{RT} is commonly used as a measure for *brilliance*. But as BR_{RT} it is criticized and therefore TR_{EDT} or TR_G are suggested as alternative measures.[64]

[58] See de Vries et al. (2001).

[59] See Ando (2010), pp. 127ff.

[60] See Abdou and Guy (1996), pp. 3217f.

[61] Commonly adopted, e.g. by Everest and Pohlmann (2009), p. 388.

[62] See Beranek (2004), pp. 512f.

[63] See Abdou and Guy (1996) and Gade (2007), p. 310.

[64] See e.g.Everest and Pohlmann (2009), p. 386, who only considers the 2000 Hz frequency band in the numerator, and Gade (2007), p. 310.

Table 6.1 Summary of subjective impressions, objective measures and ideal values of room acoustical parameters for symphonic music and operas

Subjective attribute	Objective measure	Ideal values
Reverberance	RT	1.5 to 2.4 s
	EDT	1.5 to 2.2 s
Clarity	C_{80}	−1 to 5 dB in empty rooms (≡ −4 to −1 dB in occupied rooms)
	$t_{s,1000}$	70 to 140 ms
Spaciousness	$\mathrm{BQI}_{\mathrm{all}}$	0.5 to 0.8
	$\mathrm{BR}_{\mathrm{RT}}$	>1
Liveness	$\mathrm{RT}_{500,1000}$	1.5 to 2.2 s
	H	−2 to 4 dB
Spatial impression	R	−5 to 7 dB
Intimacy	ITDG	≤21 ms
Listener envelopment	LG	−6 to −4 dB
	$\mathrm{BQI}_{\mathrm{late},500-2000\,\mathrm{Hz}}$	0.1 to 0.2
	LEF	0.2 to 0.3
Apparent source width	LE	$0.3 < 10\lg \mathrm{LE} < 0.8$
	$\mathrm{LEF}_{125-1000\,\mathrm{Hz}}$	0.2 to 0.3
	$\mathrm{BQI}_{\mathrm{early},500-2000\,\mathrm{Hz}}$	0.6 to 0.75
	$G_{\mathrm{early,low}}$	−1.5 dB
Loudness	G	3 to 6 dB
	$G_{500,1000\,\mathrm{Hz}}$	4 to 5.5 dB
Timbre/tonal color		
Warmth	G_{125}	
	$\mathrm{BR}_{\mathrm{RT}}$	1 to 1.3
	$\mathrm{BR}_{\mathrm{EDT}}$	1 to 1.25
Brilliance	$\mathrm{TR}_{\mathrm{RT}}$	1 to 1.3
	$\mathrm{TR}_{\mathrm{EDT}}$	0.7 to 0.8
Acoustical glare	SDI	1
Ease of ensemble	$\mathrm{ST}_{\mathrm{early},250-2000\,\mathrm{Hz}}$	−15 to −12 dB
	$\mathrm{EEL}_{500-1000\,\mathrm{Hz}}$	−15 to −10 dB
Support	$\mathrm{ST}_{\mathrm{late},250-2000\,\mathrm{Hz}}$	−15 to −12 dB

Acoustical Glare:

Hard, harsh reflections lead to the impression of *acoustical glare*, analogously to optical glare. How glary a sound is cannot be measured by a magnitude from the impulse response. Mellow, non-glary sounds result from diffusion of reflections, e.g. caused by irregularities or curvature of surfaces. The surface diffusivity index SDI is suggested as measure for acoustical glare. It considers the diffusivity of the surface of the room:

$$\mathrm{SDI} = \frac{\sum_{i=0}^{2} i \frac{S_i}{2}}{S}, i = 0, 1, 2 \tag{6.22}$$

S_0, S_1 and S_2 are the areas with low, medium, and high diffusivity. They are multiplied by a factor 0, 0.5 or 1 and divided by S, the area of ceiling and sidewalls. In the best-rated halls SDI is 1, in medium-rated halls 0.7 ± 0.1 and in the lowest-rated halls STI = 0.3.[65]

Texture:

Texture is a quality which describes the pattern of ER. It is nominal but has qualities like density, regularity and strength. A short ITDG in combination with homogeneously spaced ER lead to the impression of a good texture. The best-rated concert halls have more than 17 ER, medium-rated halls 10 to 16 and lower-rated 10.

Parameters for the musicians:

Musicians have other demands on the room acoustics than the audience, since their goal is to deliver a good performance, rather than to receive it. Therefore parameters for the musicians are evaluations of the conditions on stage.

Ease of Ensemble:

The *ease of ensemble* describes how easily musicians can play together, depending on how well they can hear each other and themselves. The early support $\mathrm{ST}_{\mathrm{early}}$ is an adequate measure for this subjective quality, ideally lying around −12 dB.

Support:

The *support* describes how much the room supports or carries the sound, or how much force is needed to fill the room with sound. A low support demands a hefty playing which is fatiguing and less delicate. A high late support $\mathrm{ST}_{\mathrm{late}}$ serves in this case. RR is supposed to quantify the self-support of an instrumentalist on the stage.

Table 6.1 summarizes subjective attributes with objective measures and their ideal values for symphonic music.[66] One must keep in mind that some of these measures are not independent of each other, which means their magnitudes must not be considered as individual measures of a quality. A high correlation can be found e.g. between RT and EDT and $C_{80,500-2000\,\mathrm{Hz}}$, reasonable correlation between $\mathrm{LEF}_{125-1000\mathrm{Hz}}$ and

[65] See Beranek (2004), pp. 521ff.

[66] Derived from tables in Abdou and Guy (1996), p. 3224-3225 and Gade (2007), p. 312 and from values in Everest and Pohlmann (2009), pp. 386ff and Blauert and Xiang (2009), p. 174 and in the literature named in the introduction of this section.

$BQI_{early,500-2000Hz}$. These measures are not independent but redundant in a way. Ando (2010) makes investigations on orthonormal factors to explain the subjective preferences for the acoustics in concert halls.[67]

A research trend is to include psychoacoustics, as discussed in Chap. 4, more thoroughly in room acoustical considerations.[68]

References

Abdou A, Guy RW (1996) Spatial information of sound fields for room-acoustics evaluation and diagnosis. J Acoust Soc Am 100(5):3215–3226. https://doi.org/10.1121/1.417205

Ahnert W, Tennhardt HP (2008) Raumakustik. In Weinzierl S (ed) Handbuch der Audiotechnik, Chap 5, pp 181–266. Springer, Berlin, Heidelberg. https://doi.org/10.1007/978-3-540-34301-1_5

Ando Y (2007) Concert hall acoustics based on subjective preference theory. In: Rossing TD (ed) Springer Handbook of Acoustics, Chap 10, pp 351–386. Springer, New York. https://doi.org/10.1007/978-0-387-30425-0_10

Ando Y (2010) Auditory and visual sensation. Springer, New York, Dordrecht, Heidelberg, London. https://doi.org/10.1007/b13253

Bader R, Schneider A (2011) Playing 'live' at the star club. Reconstructing the room acoustics of a famous music hall. In: Schneider A, von Ruschkowski A (eds) Systematic Musicology. Empirical and Theoretical Studies, pp. 185–209. Peter Lang, Frankfurt am Main. https://doi.org/10.3726/978-3-653-01290-3

Barron M, Lee L-J (1988) Energy relations in concert auditoriums. i. J Acoust Soc Am, 84(2):618–628. https://doi.org/10.1121/1.396840

Beranek LL (1996) Acoustics. American Institute of Physics, Woodbury (New York), reprint from 1954 edition

Beranek LL (2004) Concert halls and opera houses: music, acoustics, and architecture, 2nd edn. Springer, New York. https://doi.org/10.1007/978-0-387-21636-2

Bergeron-Mirsky W, Lim J, Gulliford J, Patel A (2010) Architectural acoustics for practitioners. In: Ceccato C, Hesselgren L, Pauly M, Pottmann H, Wallner J (eds) Advances in architectural geometry 2010, pp 129–136. Springer, Vienna. https://doi.org/10.1007/978-3-7091-0309-8_9

Berkhout AJ, de Vries D, Vogel P (1993) Acoustic control by wave field synthesis. J Acoust Soc Am 93(5):2764–2778. https://doi.org/10.1121/1.405852

Blau M (2004) Correlation of apparent source width with objective measures in synthetic sound fields. Acta Acust united Ac 90(4):720–730. https://www.ingentaconnect.com/content/dav/aaua/2004/00000090/00000004/art00015

Blauert J (1997) Hearing of music in three spatial dimensions. http://www.uni-koeln.de/phil-fak/muwi/fricke/103blauert.pdf. Last accessed 17 Feb 2013

Blauert J (2005) Analysis and synthesis of auditory scenes. In: Blauert J (ed) Communication acoustics, Chap 1, pp 1–25. Springer, Berlin, Heidelberg. https://doi.org/10.1007/3-540-27437-5_1

Blauert J, Xiang N (2009) Acoustics for engineers. Troy lectures, 2nd edn. Springer, Berlin, Heidelberg. https://doi.org/10.1007/978-3-642-03393-3

Bleda S, Escolano J, López JJ, Pueo B (2005) An approach to discrete-time modelling auralization for wave field synthesis applications. In: Audio Engineering Society Convention 118. http://www.aes.org/e-lib/browse.cfm?elib=13141

[67] See e.g. Ando (2010) and Ando (2007).

[68] See Vorländer (2018), p. 212.

Bradley JS, Reich RD, Norcross SG (2000) On the combined effects of early- and late-arriving sound on spatial impression in concert halls. J Acoust Soc Am 108(2):651–661. https://doi.org/10.1121/1.429597
Choi YJ, Fricke FR (2006) A comparison of subjective assessments of recorded music and computer simulated auralizations in two auditoria. Acta Acust united Ac 92:604–611. https://www.ingentaconnect.com/content/dav/aaua/2006/00000092/00000004/art00013
David jr EE (1988) Aufzeichnung und Wiedergabe von Klängen. In: Winkler K (ed) Die Physik der Musikinstrumente, pp 150–160. Spektrum der Wissenschaft, Heidelberg
de Vries Diemer, Hulsebos Edo M, Baan Jan (2001) Spatial fluctuations in measures for spaciousness. J Acoust Soc Am 110:947–954. https://doi.org/10.1121/1.1377634
Deutsches Institut fr Normung (2004) Hörsamkeit in kleinen bis mittelgroßen Räumen
Deutsches Institut für Normung (2009) Akustik—Messung von Parametern der Raumakustik—Teil 1. Aufführungsräume (ISO 3382-1:2009); Deutsche Fassung EN ISO 3382-1:2009
Escolano J, Pueo B, Bleda S, Lépez JJ (2005) An approach to discrete-time modelling auralization for wave field synthesis applications. In: Audio Engineering Society Convention 118, Barcelona. http://www.aes.org/e-lib/browse.cfm?elib=13141
Everest FA, Pohlmann KC (2009) Master handbook of acoustics, 5th edn. Mcgraw-Hill, New York
Forsyth M (1985) Buildings for music. The architect, the musician, and the listener from the seventeenth century to the prenent day. MIT Press, Cambridge. https://doi.org/10.2307/3105495
Fuchs H (2013) Applied acoustics. Concepts, absorbers, and silencers for acoustical comfort and noise control. Alternative solutions-innovative tools-practical examples. Springer, Heidelberg. https://doi.org/10.1007/978-3-642-29367-2
Gade AC (2007) Acoustics in halls for speech and music. In Thomas D. Rossing, editor, *Springer Handbook of Acoustics*, chapter 9, pages 301–350. Springer, Berlin, Heidelberg. https://doi.org/10.1007/978-0-387-30425-0_9
Griesinger D (1996) Spaciousness and envelopment in musical acoustics. In: Audio Engineering Society Convention 101. http://www.aes.org/e-lib/browse.cfm?elib=7378
Horbach U, Karamustafaoglu A, Rabenstein R, Runze G, Steffen P (1999) Numerical simulation of wave fields created by loudspeaker arrays. In: Audio Engineering Society Convention 107. http://www.aes.org/e-lib/browse.cfm?elib=8159
Klepper DL (2008) Tent-shaped concert halls, existing and future. J Acoust Soc Am 124(1):15–18. https://doi.org/10.1121/1.2932342
Knudsen VO (1998) Raumakustik. In: Winkler K (ed) Die Physik der Musikinstrumente, pp 136–149. Spektrum der Wissenschaft, Heidelberg
Kuhl W (1978) Rāumlichkeit als Komponente des Raumeindrucks. Acustica 40:167–181. https://www.ingentaconnect.com/contentone/dav/aaua/1978/00000040/00000003/art00006
Kuttruff H (2009) Room acoustics. Taylor & Francis, Oxon, 5th edition. https://doi.org/10.1201/9781315372150
Lokki Tapio, Pätynen Jukka, Kuusinen Antti, Tervo Sakari (2012) Disentangling preference ratings of concert hall acoustics using subjective sensory profiles. J Acoust Soc Am 132(5):3148–3161. https://doi.org/10.1121/1.4756826
Lombardo V, Fizch J, Weinzierl S, Starosolski R (2005) The virtual electronic poem (VEP) project. In: International Computer Music Conference Proceedings. http://hdl.handle.net/2027/spo.bbp2372.2005.153
Lombardo V, Valle A, Fitch J, Tazelaar K, Weinzierl S, Borczyk W (2009) A virtual-reality reconstruction of poème Èlectronique based on philological research. Comput Music J 33(2). https://doi.org/10.1162/comj.2009.33.2.24
Martellotta F (2010) The just noticeable difference of center time and clarity index in large reverberant spaces. J Acoust Soc Am 128(2):654–663. https://doi.org/10.1121/1.3455837
Mechel F (2013) Room acoustical fields. Springer, Berlin, Heidelberg. https://doi.org/10.1007/978-3-642-22356-3

Meyer Jürgen (1977) Der Einfluß der richtungsabhängigen Schallabstrahlung der Musikinstrumente auf die Wirksamkeit von Reflexions- und Absorptionsflächen in der Nähe des Orchesters. Acustica 36:147–161

Meyer J (2009) Acoustics and the performance of music. Manual for acousticians, audio engineers, musicians, architects and musical instrument makers, 5th edn. Springer, Bergkirchen. https://doi.org/10.1007/978-0-387-09517-2

Morimoto Masayuki, Jinya Munehiro, Nakagawa Koichi (2007) Effects of frequency characteristics of reverberation time on listener envelopment. J Acoust Soc Am 122(3):1611–1615. https://doi.org/10.1121/1.2756164

Okano Toshiyuki (2002) Judgments of noticeable differences in sound fields of concert halls caused by intensity variations in early reflections. J Acoust Soc Am 111(1):217–229. https://doi.org/10.1121/1.1426374

Okano Toshiyuki, Beranek Leo L, Hidaka Takayuki (1998) Relations among interaural cross-correlation coefficient ($iacc_e$), lateral fraction (lf_e), and apparent source width (ASW) in concert halls. J Acoust Soc Am 104(1):255–265. https://doi.org/10.1121/1.423955

Otondo F, Rindel JH (2005) A new method for the radiation representation of musical instruments in auralization. Acta Acust united Ac 91:902–906. https://www.ingentaconnect.com/content/dav/aaua/2005/00000091/00000005/art00011

Ouis D (2003) Study on the relationship between some room acoustical descriptors. J Audio Eng Soc 51(6):518–533. http://www.aes.org/e-lib/browse.cfm?elib=12220

Pelzer S, Pollow M, Vorländer M (2012) Auralization of a virtual orchestra using directivities of measured symphonic instrument. In: Proceedings of the Acustics 2012 Nantes Conference, pp 2379–2384. http://www.conforg.fr/acoustics2012/cdrom/data/articles/000758.pdf

Rindel JH (2004) Felipe Otondo, and Claus Lynge Christensen. Design and Science, Hyogo, April, Sound source representation for auralization. In: International Symposium on Room Acoustics

Vassilantonopoulos SL, Mourjopoulos JN (2003) A study of ancient greek and roman theater acoustics. Acta Acust united Ac 89:123–136. https://www.ingentaconnect.com/content/dav/aaua/2003/00000089/00000001/art00015

Vigeant Michelle C, Wang Lily M (2008) Investigations of orchestra auralizations using the multi-channel multi-source auralization technique. Acta Acust United Ac 94:866–882. https://doi.org/10.3813/aaa.918105

Vorländer M (2008) Auralization. fundamentals of acoustics, modelling, simulation, algorithms and acoustic virtual reality. Springer, Berlin, Heidelberg. https://doi.org/10.1007/978-3-540-48830-9

Vorländer M (2018) Room acoustics–fundamentals and computer simulation, pp 197–215. Springer, Berlin, Heidelberg. https://doi.org/10.1007/978-3-662-55004-5_11

Vorländer M, Mechel FP (2008) Room acoustics. In: Mechel FP (ed) Formulas of acoustics, 2nd edn, pp 378–944. Springer, Berlin, Heidelberg, New York. https://doi.org/10.1007/978-3-540-76833-3_13

Weinzierl S (2008) Virtuelle Akustik und Klangkunst. In: Fortschritte der Akustik—DAGA '08, pp 37–38. Dresden. http://pub.dega-akustik.de/DAGA_1999-2008/data/articles/003709.pdf

Wenzel EM, Miller JD, Abel JS (2000) Sound lab: a real-time, software-based system for the study of spatial hearing. In: Audio Engineering Society Convention 108, Paris. http://www.aes.org/e-lib/browse.cfm?elib=9198

Winkler H, Terhardt HT (1988) Die Semperoper Dresden, das neue Gewandhaus Leipzig und das Schauspielhaus Berlin und ihre Akustik. In: Fortschritte der Akustik—DAGA '88, pp 43–56. Bad Honnef. https://www.dega-akustik.de/publikationen/online-proceedings/

Ziemer T (2011) Psychoacoustic effects in wave field synthesis applications. In: Schneider A, von Ruschkowski A (eds) Systematic musicology. Empirical and theoretical studies, pp 153–162. Peter Lang, Frankfurt am Main. https://doi.org/10.3726/978-3-653-01290-3

Ziemer T (2015) Exploring physical parameters explaining the apparent source width of direct sound of musical instruments. In: Jahrestagung der Deutschen Gesellschaft für Musikpsychologie, pp 40–41. Oldenburg. http://www.researchgate.net/publication/304496623_Exploring_Physical_Parameters_Explaining_the_Apparent_Source_Width_of_Direct_Sound_of_Musical_Instruments

Zwicker E, Fastl H (1999) Psychoacoustics: facts and models, 2nd edn. Springer, Berlin, Heidelberg. https://doi.org/10.1007/978-3-662-09562-1

Chapter 7
Conventional Stereophonic Sound

In this chapter the demands on stereophonic sound systems are listed and it is discussed how well established audio systems meet these demands. Their strengths and weaknesses led to the idea of sound field synthesis as technique to overcome several constraints while preferably keeping the opportunities and benefits of conventional stereophonic sound system.

7.1 Technical Demands

Today, a variety of microphones exists which are able to record the sound pressure with relatively linear frequency and phase response in the audible range with an accurate dynamic and temporal linearity. These recordings are usually presented to one or more listeners via playback on a stereophonic loudspeaker system. General requirements in loudspeakers are—at least ever since the introduction of High Fidelity (Hi-Fi) in 1960[1]—a great bandwidth, uniformly, omnidirectional sound radiation,[2] a minimal distortion factor and a flat, smooth frequency- and phase-response.[3] Demands on an audio system consisting of such loudspeakers are especially spatial

[1]See Schubert (2002), p. 15.

[2]Suggested as a standard by the NWDR in the 1950s, see Hiebler (1999), pp. 728f. Even today this is a prerequisite for use in many broadcasting companies, cinemas and recording studios (Goertz 2018, p. 072)—but dependent on the application area other radiation characteristics may be preferred, see e.g. Goertz (2008), p. 483.

[3]Cf. Mäkivirta (2008), p. 649.

© Springer Nature Switzerland AG 2020

T. Ziemer, *Psychoacoustic Music Sound Field Synthesis*, Current Research in Systematic Musicology 7, https://doi.org/10.1007/978-3-030-23033-3_7

Table 7.1 Demands on a stereophonic sound system

	Demand on loudspeaker systems	Description
1.	Correct localizability of sound sources	Direction, distance and size/width/expansion of the sound source, especially determined by ITD, ILD and direction-dependent filtering (Head Related Transfer Function HRTF) and by time- and level-difference between direct sound and early reflections (ER) and interaural coherence
2.	Correct spaciousness	Determined especially by the relation of direct sound to number, delay time, intensity, filtering and direction of ER and the interaural degree of coherence
3.	Correct reverberation	Duration, degree of diffusion and sound coloration of the reverberation

Table 7.2 Supplement of demands on stereophonic sound systems

	Demand	Description
4.	Creation of unnatural sounds	Manipulability of all parameters

ones,[4] which require several parameters[5] to be perceived correctly. These demands are listed in Table 7.1.

These criteria are geared towards a natural sound reproduction which used to be the initial aim of Hi-Fi.[6] But the option to create an unnatural sound or spatial acoustic can be intended by composers or sound engineers and is therefore an additional desirable criterion (Table 7.2).

These demands should ideally apply for every listener in the listening room. Since audio systems are usually made for human beings, the word "correct" in Table 7.1 is meant perceptually and not compulsively physically.

7.2 Audio Systems

Since Thomas Edison first recorded and played back sound on December 6th 1877, multitudes of proceedings were developed to conserve sound as true to life as possible. The most widespread proceedings and systems are historically allocated[7] and

[4]After Verheijen (1997), p. 9, Pulkki (2008), p. 747, Schanz (1966), pp. 8–18, Berkhout et al. (1993), p. 2764, Faller (2009), p. 641 and Spors et al. (2013).

[5]1. See Berkhout et al. (1992), p. 2, Baalman (2008), p. 17 and Faller (2009), pp. 638–641, 2. See Rossing (1990), p. 503, Berkhout et al. (1992), p. 2, Blauert (1997), p. 353, Schanz (1966), pp. 13f, Verheijen (1997), p. 8, Gade (2007), p. 309, 3. See Schanz (1966), pp. 8 ff, Gade (2007), pp. 307 and 333f, Rossing (1990), p. 503, Blauert (1997), p. 351 and Favrot and Buchholz (2010).

[6]Cf. Toole (2008), p. 14.

[7]Mainly after Schubert (2002), Schanz (1966), Hiebler (1999), Owsinski (2014) and Slavik and Weinzierl (2008).

Table 7.3 Overview over time of origin and number of channels of diverse loudspeaker systems. An additional subwoofer is indicated by "0.1"

Audio system	Begin of dissemination	Channels front/back+overhead
Mono	1877	1/0
Stereo	≈1950	2/0
Quadraphonic sound	≈1970	1/3, 2/2
Dolby surround	1975	2/1, 3/1, 3/2, 3.1/2
Discrete surround systems	≈1990	3.1/1, 3.1/2, 3.1/3, 3.1/4, 5.1/2
Immersive audio	≈2006	3.1/2+2, 3.1/2+4, 3.1/2+5, 3.1/2+6, 3.1/4+5
Head related stereophonic sound	≈1960	e.g., 2/0, 2/2

explained in their functionalities and options.[8] It can be observed that advancements over the years of development are especially related to a wider panning panorama and an increased immersion. Table 7.3 lists loudspeaker systems, their time of origin and their number of speakers, divided in speakers in the front and the back of the listening position plus elevated loudspeakers. Note that for a better readability the nomenclature used in this book deviates from the nomenclature widely used in the literature.

7.2.1 Mono

Until the 1930s a pure monophonic sound recording and playback was common. Semantic information of a speaking person is rarely based on spatial attributes. Also, compositional information in scores is typically to a minor degree of spatial nature but rather contains information on instrumentation, played notes and chords in time, dynamics and articulation. Therefore, it is comprehensible that the initial focus of sound recording and playback did not lie on spatial accurateness. Sound playback via one single loudspeaker offers few possibilities to present spatial sound.

One step further than pure mono is the "pseudo-stereo-effect".[9] Here, one channel is not only played through a frontal loudspeaker but also through one or more additional loudspeakers. These are, e.g., placed in room corners, often facing away from the listener.

[8]After Davis (2007), Damaske (2008), Webers (2003), Henle (2001), Huber (2002), Verheijen (1997), Pulkki (2008), Faller (2009), Schanz (1966), Slavik and Weinzierl (2008), Dickreiter (1987), and Mores (2018).

[9]See Schanz (1966), p. 2 and p. 19.

Today mono is mostly used for inexpensive transmission of pure information rather than an enjoyable music listening experience.[10]

Localizability of Sound Sources:

The auditory event direction of a played sound on a monophonic loudspeaker system is usually the speaker position itself. Thus, the sound of a complete orchestra sounds as if originating from a single position. This "Mauerlocheffekt" has been described already in Sect. 5.2. Solely by exploiting monaural localization parameters the sound event direction can be shifted. In theory, front-back localization and elevation manipulation by spectral filtering are possible, as described in Sect. 4.4 and demonstrated in Fig. 4.23. The perceived source distance could be manipulated to a certain degree by the playback gain and the gain ratio between direct sound and ER. But these theoretic possibilities are barely used in practice because these monaural cues are somewhat weak and may require information about the individual listener's HRTF for a systematic use. Thus, an alteration of the perceived source size or location is hardly viable.

Spaciousness:

Monophonic sound seems suitable for solo-instrumental sound. But since the radiation pattern of a loudspeaker usually does not match the radiation characteristics of the instrument, even the playback of dry solo instrumental signals is clearly distinguishable from the original performance, especially concerning extent of the source and liveness of the performance. Since it is almost impossible to deflect the auditory event from the loudspeaker position by conventional monophonic sound reproduction, the essential lateral reflections for the perception of spaciousness in terms of source width and listener envelopment are missing. Reflections of the listening room itself create a spatial sound but it is not influenceable by the audio system. The creation of an arbitrary spatial acoustic is impossible via mono. However, the number and distribution of additional loudspeakers in pseudo-stereo setups affect the perception of spaciousness.

Reverberation:

In mono, duration and sound coloration of the reverberation can be altered by adding a colored reverberation to the dry signal. But a spatially distributed reverberation is not viable via one single speaker. Therefore, a "pseudo-stereo-effect" was created by playing the same signal through several loudspeakers distributed in the room.[11] These are especially placed in corners to create many early reflections. Tweeters are turned away from the listener so their direct sound hardly reaches him or her. Pseudo-stereo can be considered an intermediate step between mono and stereo. Today mono is especially used for inexpensive transmission of pure information.[12]

[10]Cf. Henle (2001), p. 111.

[11]See Schanz (1966), p. 2 and p. 19.

[12]See footnote 10.

7.2.2 Stereo

On December 14th 1931 Alan Dower Blumlein patented the first stereophonic recording method via two microphones with dipole characteristic, shifted 90° to each other. Two years later he patented the stereo groove (45/45 system) which was initially used by radio stations and globally established in private households after the first open market releases in 1958. Stereo became standard for radio, TV, audio-CD, audio cassette and further audio media.

Two channel stereophonic sound systems offer many more possible applications than monophonic ones. Two identical speakers with two meters distance are set two to three meters in front of a listener so that they form an equilateral triangle. Consequently, the loudspeakers are located at ±30° from the viewing direction of the listener. The optimal listening position—the so-called *sweet spot*—is at a distance between $s = 1.70$ m and $s = 2.60$ m from the middle of the connection line between the speakers, the *loudspeaker base*. According to DIN 15995 a deviant distance of the listener is acceptable as long as the angle between listener and loudspeakers ranges between 45° and 80°, i.e. between ±22.5° and ±40° relative to the listener's viewing direction.[13] The recommended stereo setup is illustrated in Fig. 7.1. Phantom sources can be distributed along the black circumference segment by panning. The dashed lines show the stereo triangle and the viewing direction of the listener. The loudspeaker setup is symmetric along the viewing direction of the listener.

Relative amplitude- and time-shifts between the loudspeaker signals are called *amplitude panning* and *time based panning*. They create *phantom sources* on an angle between the speakers and a little bit further. Both methods work for all positions on a line from the loudspeaker base through the sweet spot. Psychoacoustic phenomenons—especially summing localization, ILD and ITD, as described in Sects. 4.4.1 and 4.4.5—are used for this. The distance between the speakers is a compromise: A small distance creates a stable horizontal auditory event direction

Fig. 7.1 Stereo setup. Robust phantom sources can be distributed between ±30°

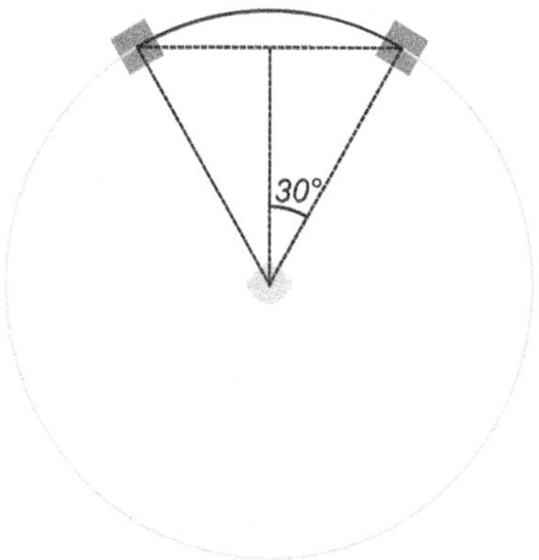

[13]See Deutsches Institut für Normung (1996).

but the panorama range is small. A larger distance enables a wider panorama but the auditory event direction becomes unstable already with slight head movements and undesirable elevation effects may appear.[14]

Amplitude based panning offers several options for manipulating the auditory event direction which are based on similar mathematical basic deliberations. The formula

$$\sin\varphi_Q = \frac{\hat{A}_\mathrm{L} - \hat{A}_\mathrm{R}}{\hat{A}_\mathrm{L} + \hat{A}_\mathrm{R}} \sin\varphi_0 = \frac{1 - \frac{\hat{A}_\mathrm{R}}{\hat{A}_\mathrm{L}}}{1 + \frac{\hat{A}_\mathrm{R}}{\hat{A}_\mathrm{L}}} \sin\varphi_0 \tag{7.1}$$

describes the *sine law of stereophony*[15] by the angle of the phantom source to the sweet spot φ_Q, the amplification factors of the left and right loudspeaker signal $\hat{A}_\mathrm{L}$ and $\hat{A}_\mathrm{R}$ and the angle between the speakers and the listener φ_0 which is usually $\pm 30°$. A simpler form of the sine law is

$$\frac{\sin\varphi_Q}{\sin\varphi_0} = \frac{\hat{A}_\mathrm{L} - \hat{A}_\mathrm{R}}{\hat{A}_\mathrm{L} + \hat{A}_\mathrm{R}}. \tag{7.2}$$

With this formula and Fig. 7.2 it becomes clear that the sine law derives the gain ratio $\Delta\hat{A}$ from the leg ratios of two triangles. The first triangle is one half of the stereo triangle, so a right triangle between the listener, the loudspeaker and the center of the loudspeaker base. In the other triangle the loudspeaker is replaced by the phantom source position. Naturally, the sine law considers the ratio of the opposite leg and the hypotenuse of both triangles.

The equation

$$\tan\varphi_Q = \frac{\hat{A}_\mathrm{L} - \hat{A}_\mathrm{R}}{\hat{A}_\mathrm{L} + \hat{A}_\mathrm{R}} \tan\varphi_0 = \frac{1 - \frac{\hat{A}_\mathrm{L}}{\hat{A}_\mathrm{R}}}{1 + \frac{\hat{A}_\mathrm{L}}{\hat{A}_\mathrm{R}}} \tan\varphi_0 \tag{7.3}$$

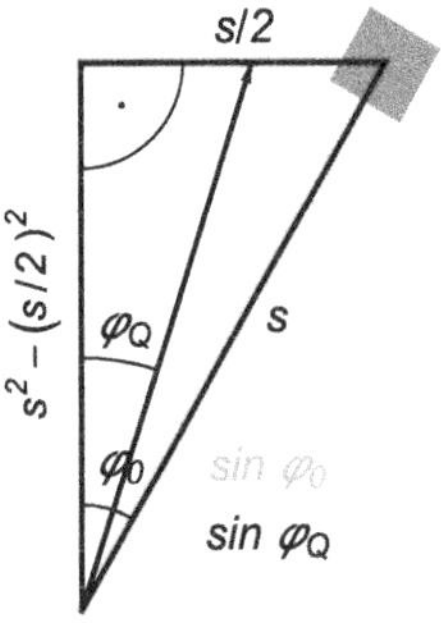

Fig. 7.2 The sine panning law considers the ratio of the opposite leg and the hypotenuse of two triangles

[14] See Damaske (2008), pp. 8 f.

[15] Derived in Bauer (1961).

is the *tangent law* which slightly differs from the sine law but is reported to be more stable in the case of head movements.[16] Again, with a modified form

$$\frac{\tan \varphi_Q}{\tan \varphi_0} = \frac{\hat{A}_\mathrm{L} - \hat{A}_\mathrm{R}}{\hat{A}_\mathrm{L} + \hat{A}_\mathrm{R}} \tag{7.4}$$

and Fig. 7.3, the main difference between the tangent panning law and the sine panning law becomes obvious. They consider different leg ratios of the same set of triangles.

For both the sine and the tangent law, two gains $\hat{A}_\mathrm{L}$ and $\hat{A}_\mathrm{R}$ are searched but only one equation is given. That means these problems are under-determined. To find a valid solution one can choose one of these gains to be fixed and solve the equation to find the other gain to create a desired phantom source angle φ_Q. Alternatively, one can add a second equation like

$$\sqrt[n]{\hat{A}_\mathrm{L}^n + \hat{A}_\mathrm{R}^n} = \text{const.} \tag{7.5}$$

Choosing $n = 1$, the cumulated pressure amplitude stays constant, no matter which phantom source position is chosen. If $n = 2$ is chosen, the cumulated sound energy stays constant. The first roughly creates the impression of constant loudness under anechoic conditions, the latter is preferred in rooms with some reverberation.[17] A constant loudness is particularly important for moving sources. In this case the gain ratio changes gradually while the notes are playing. So if loudness would change as a result of panning, an undesired tremolo effect would occur.

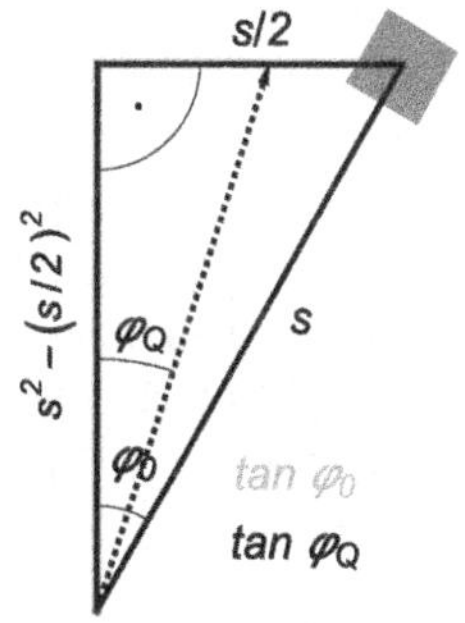

Fig. 7.3 The tangent panning law considers the ratio of the opposite leg and the adjacent leg of two triangles

[16]Introduced in Bernfeld (1973), revisited e.g. in Pulkki (1997), p. 457 and Pulkki (2001).

[17]See, e.g., Pulkki (2001), pp. 12f.

A third panning law

$$\hat{A}_L = \sqrt{\frac{\varphi_m - \varphi_Q}{\varphi_m - \varphi_n}}$$
$$\hat{A}_R = \sqrt{\frac{\varphi_n - \varphi_Q}{\varphi_n - \varphi_m}} \tag{7.6}$$

is proposed by Chowning (1971).[18] Here, φ_m and φ_n are the position angles of two loudspeakers relative to the viewing direction of a listener. As in the stereo setup, they have the same distance to the listener. The gain ratios $\Delta\hat{A}$ over phantom source angle are plotted in Fig. 7.4 for all three panning laws. In a range up to ±20° the phantom source position moves approximately linearly 2.1° to 2.5° per dB. At a level difference of approximately 30 dB the signal seems to radiate from the louder speaker only.[19] In the literature divergent values occur,[20] which seem to arise because from 12 to 15 dB level difference the angle of the phantom source is already so lateral that the perceived angle does barely differ from the speaker position.[21] Only at angles between about 8° and 28° the panning laws exhibit a considerable difference. Figure 7.5 zooms in the graphs to emphasize their differences in this region. A level difference of 10 dB yields a phantom source angle of about 15° according to the sine law, almost 17° according to the tangent law and almost 18° according to Chowning's panning law. The deviation from the linear coherence between angle and amplitude—as described in Sect. 4.4.2—is caused by crosstalk. Since the radiation of one speaker reaches both ears—in contrast to a dichotic pair of headphones—the so-called "double-arrival problem"[22] occurs. This implies that one speaker signal leads to an ITD of about 250 ms which causes slight smear-ups of sharp transient sounds and a comb filter effect with the first notch around 2 kHz.[23] The dependence of auditory event angle on frequency—mentioned in Sect. 4.4.2 and shown in Fig. 4.19—is not considered in the three panning methods.

Especially the sine and the tangent law have been manifested in many listening attempts and exploit the psychoacoustic property that people evaluate ILDs for localization as discussed in Sect. 4.4.1.

Occasionally, the room acoustics of the listening room can have a huge influence on the perceived sound. Event position and width can be influenced particularly by early reflections.[24] For example it has been observed that phantom sources were

[18] See Chowning (1971).

[19] See Verheijen (1997), p. 12 and Fig. 7.4.

[20] 16 dB according to Webers (2003), p. 184, "12–15 dB" according to Damaske (2008), p. 6, only 10 dB according to David jr. (1988), p. 159.

[21] See Dickreiter (1987), p. 127.

[22] Davis (2007), p. 776.

[23] See e.g. Theile (1980), pp. 10ff.

[24] Discussed in detail in Chap. 6, especially Sect. 6.2.

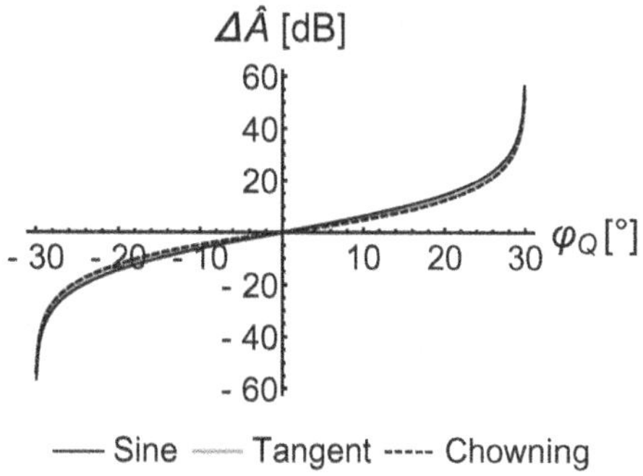

Fig. 7.4 Angle of a phantom source φ_Q by utilization of the sine law (black), the tangent law (gray) and Chowning's panning law (dashed)

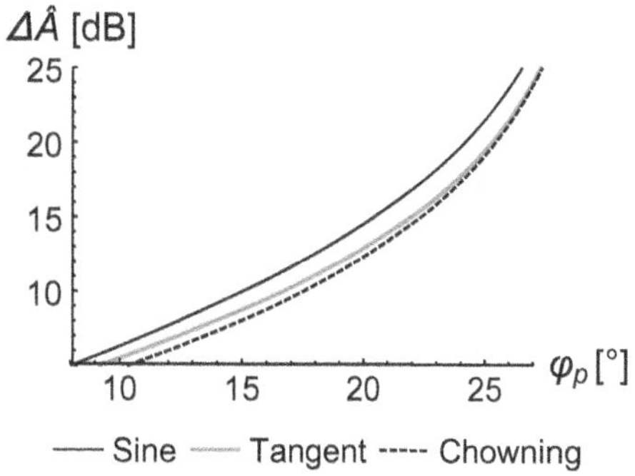

Fig. 7.5 Gain ration $\Delta\hat{A}$ over phantom source angle φ_Q according to the sine law (black), the tangent law (gray) and Chowning's panning law (dashed)

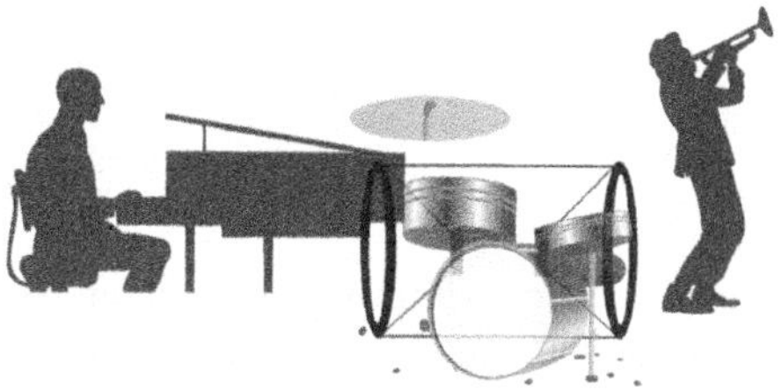

Fig. 7.6 Stereo speakers with a shared cabinet can create the impression of phantom sources beyond the loudspeaker base

localized outside the stereo base.[25] This can be achieved by stereo loudspeakers with a shared cabinet as illustrated in Fig. 7.6. It is quite possible that the perceived source positions not only result from the interplay of loudspeakers and room reflections. Since the typical arrangement of a jazz ensemble is well-known, they may also be affected by imagination, i.e., top-down processes as described in Sects. 4.4.1 and 4.5.

Time based panning manipulates the auditory event angle by ITD resulting from inter-channel time differences. As for amplitude panning, deviations from Blauert's results, as illustrated in Fig. 4.20 in Sect. 4.4.2, are based on crosstalk. Since signals of both loudspeakers reach both ears, an extended time difference is needed to cause a lateral shift of the perceived source position. Some authors, like Dickreiter (1978) and Friesecke (2007), give plots or tables to specify the relationship between inter-channel

[25] Reported, e.g., in Schanz (1966) for jazz recordings.

Table 7.4 Phantom source deflection at different ICTDs according to Friesecke (2007), p. 146

ICTD in ms	Deflection in %
0	0
0.04	5
0.08	10
0.13	15
0.18	20
0.23	25
0.28	30
0.33	35
0.38	40
0.43	45
0.48	50
0.53	55
0.59	60
0.66	65
0.73	70
0.81	75
0.91	80
1	85
1.13	90
1.31	95
1.5	100

time differences (ICTDs) and phantom source deflection.[26] An example is given in Table 7.4. Other authors report that the exact time difference for a certain phantom source angle is dependent on properties of the signal, especially on frequency,[27] and is very subjective and susceptible to extreme localization alteration when leaving the sweet spot.

For inter-channel time differences of up to 5 ms, a creation of point-shaped phantom sources is possible. For larger time differences, between 3 and 30 ms, initially the law of the first wave front steers the auditory event position towards the speaker with the earlier signal. Then, at time differences between 20 and 65 ms, the source seems expanded. For for even larger time differences at about 50 ms or more, two sources are localized, as described in Sect. 4.4.5 and illustrated in Fig. 7.7. The A-B recording technique creates phantom sources by ITD.[28]

[26]See Dickreiter (1978), p. 82 and Friesecke (2007) pp. 138–146.

[27]See Verheijen (1997), p. 16.

[28]See Dickreiter (1987), p. 129 and David jr. (1988), p. 159.

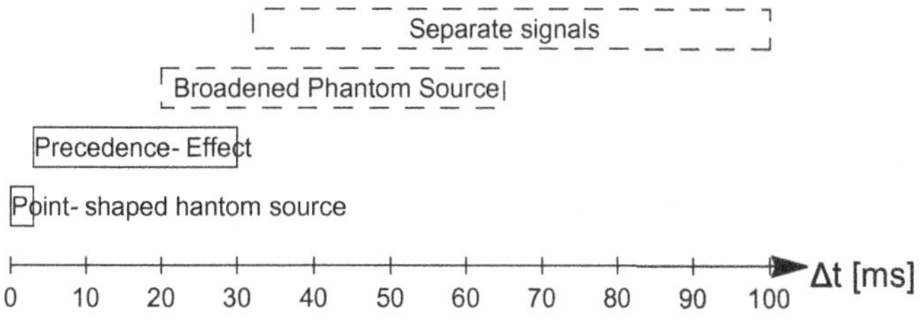

Fig. 7.7 Phenomenons appearing with the play back of equal signals time shifted between two loudspeakers. After Dickreiter (1987), p. 129

A combination of both panning techniques is difficult, since an ITD leading to the same auditory event angle as an ILD is heavily dependent on frequency. It is roughly realizable by ORTF microphoning technique.[29] According to Blauert (2008), a combined stereophony enlarges the sweet spot and improves spectral naturalness but leads to an increased localization blur and mixing is difficult or even impossible.[30] Even amplitude based panning does not always work. The low- or broadband loudspeaker membrane and the tweeter of one loudspeaker cabinet may exhibit a phase difference. For example crossover networks can create slight phase shifts between both membranes. However, as both membranes of a loudspeaker are located at the same angle in a stereo triangle, this phase effect should not lead to unexpected interaural phase differences. Still, with this setup, the two stereo loudspeaker positions may be localized in addition to the desired phantom source. It is not exactly clear why phantom sources work improperly when the loudspeaker membranes exhibit this phase difference in the crossover frequency region.[31]

Localizability of Sound Sources:

Stereo uses the psychoacoustic phenomenon of summing localization to create a source origin direction and expansion via phantom sources. Especially the localization of phantom sources is very precise. However, the direction is limited to the area between the speakers and a little further and panning techniques only work on the sweet spot and slightly in front and behind it. In case of phantom sources the signals at the ears are not identical to the signals of a real source at the same position but they hold as equivalent and are perceived as quite natural although an audible comb filter effect arises.[32] Contradictions between amplitude and phase in the dry signal can create the impression of an expanded source. Nevertheless, the radiation characteristic—though this feature was already known when stereo was invented[33]—and the distance of the source cannot be reproduced properly. With roughly 5 cm the localization accuracy is best at a distance of 3 m. The average localization blur for

[29] Details, see Blauert and Braasch (2008), pp. 110ff.

[30] See Blauert (2008), p. 25.

[31] See Friesecke (2007), p. 137.

[32] See Dickreiter (1987), p. 135.

[33] Cf. Scheminzky (1943), p. 38.

phantom sources in stereo setups is ±5° which is only slightly larger than the localization blur for real sources in the frontal region as already discussed earlier in Sect. 4.4.2. Furthermore, moving objects are localized better than still ones.[34]

Spaciousness:

In stereo, the early reflections can be manipulated in the same manner as the direct sound. Their delay, filtering and direction can be emulated from a natural environment but the angle is still restricted to roughly ±35°. Early reflections from the room in which the stereo system is set up typically deviate a lot from reflections of a real source at the phantom source position

Reverberation:

As for mono, duration and sound color of the reverberation can be regulated by a stereo system. The restriction of source directions to the frontal area can be suppressed a bit for the reverberation by the possibility to create diffuse sound images by inter-channel time-differences and filtering. Stereo recordings are usually construed for a listening room with some reverb, so the interference of reverb in the signal and reverberation of the room has usually been accounted for when generating the signals.

7.2.3 *Quadraphonic Sound*

Quadraphonic sound was especially applied in the 1960s and 1970s for film sound and electro acoustic music but could never accomplish extensive commercial success. Merely its matrix system was applied in later systems. It stores the information of four channels in two channels via overlay, capable to regain four signals. This approach is called 4 : 2 : 4 matrixing. The equation

$$\begin{bmatrix} L_T \\ R_T \end{bmatrix} = \begin{bmatrix} a_{11} & a_{12} & a_{13} & a_{14} \\ a_{21} & a_{22} & a_{23} & a_{24} \end{bmatrix} \cdot \begin{bmatrix} L_F \\ R_F \\ L_B \\ R_B \end{bmatrix} \tag{7.7}$$

describes the encoder with the left and right transfer channels L_T and R_T (also "Transmission"/"Total"/"Track"[35] T), front channel F and back channel B. The encoding factors a are complex numbers which manipulate amplitude and phase of the signals. The decoder

---

[34]See Strube (1985), p. 69 and Schanz (1966), p. 54.

[35]Verheijen (1997), pp. 23f/Dolby Laboratories Inc. (1998), p. 8-1 and Henle (2001), p. 115/Webers (2003), p. 220.

$$\begin{bmatrix} L'_F \\ R'_F \\ L'_B \\ R'_B \end{bmatrix} = \begin{bmatrix} b_{11} & b_{12} \\ b_{21} & b_{22} \\ b_{31} & b_{32} \\ b_{41} & b_{42} \end{bmatrix} \cdot \begin{bmatrix} L_T \\ R_T \end{bmatrix} \tag{7.8}$$

extracts the four speaker channels from the two transfer channels. The decoding factors b are the reciprocal values of a. The four decoded signals deviate from the original ones due to crosstalk between adjacent channels (See Fig. 7.8). Quadraphonic sound systems consist of an arrangement of four loudspeakers forming equally spaced along a circle that surrounds the listener. Two quadraphonic sound setups are common, namely the *Scheiber array* and the *Dynaquad array*. The Scheiber array is illustrated in Fig. 7.9. The loudspeakers are placed at ±45° and ±135°. The Dynaquad array is illustrated in Fig. 7.10. It is identical to the Scheiber array if the listener rotates by 45°.

From stereo it is already known that amplitude based panning does not work well if the opening angle between the loudspeakers is wider than 80°. Consequently, the localization of phantom sources is ambiguous in quadraphonic audio setups.[36] Furthermore, it could be proven that amplitude based panning works worse, if the loudspeaker pair is not placed symmetric to the facing direction of the listener. Theile and Plenge (1976) rotated the stereo triangle around the listeners and tested several phantom source angles. He found that the larger the rotation the more vague the localization gets and the more inconsistent the reported phantom source locations become.[37] This principle is depicted in Fig. 7.8. So amplitude based panning should somewhat work in the frontal and the rear region of the Scheiber array. But panning between a front and a rear speaker or between any pair of the Dynaquad array is very indistinct.

Because the stereo record was widely spread since 1958, it was usually chosen as data carrier. Although companies such as Toshiba, Sansui, Sony and Kenwood developed quadraphonic sound systems or, respectively, matrix based sound systems, the crosstalk problem could not be overcome. No more than 3 dB channel separation was achieved. Furthermore, no standardization took place. The company JVC

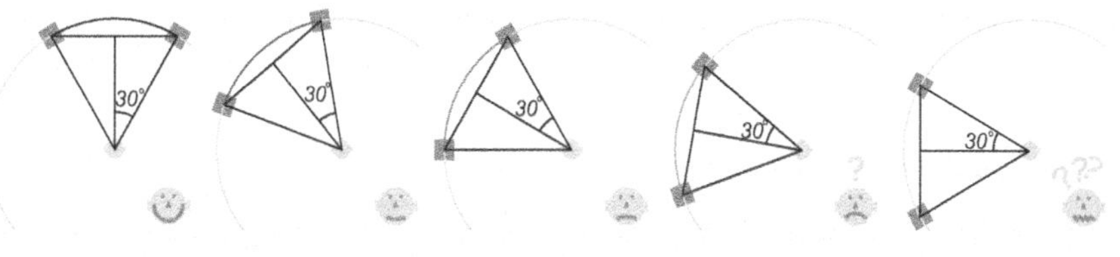

Fig. 7.8 Amplitude based panning between pairs of loudspeakers. The more the loudspeakers are rotated away from the viewing direction, the more ambiguous the phantom source position becomes (indicated here by the lightness of the loudspeaker base and the facial expression)

[36] Cf. e.g. Toole (2008), p. 278.

[37] The listening test can be found in Theile and Plenge (1976).

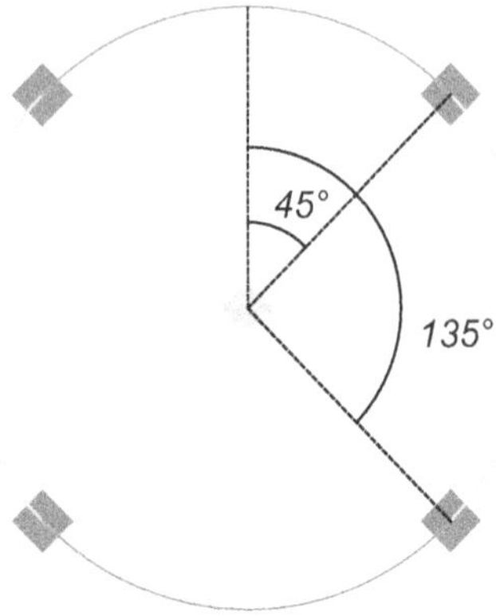

Fig. 7.9 Scheiber setup. Phantom sources can be distributed in the front and the rear (gray). But localization precision is weak

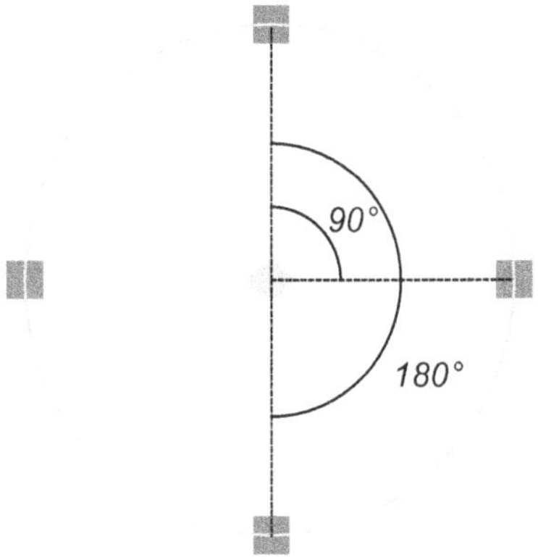

Fig. 7.10 Dynaquad setup. Panning does not create stable phantom source positions

developed a proceeding to add two additional channels to a record via frequency modulation technique with a 30 kHz carrier frequency. Thereby four channels could be stored on a record but demanded a more subtle stylus and a more homogeneous medium which was not compatible with common record players. Later, when digital technology became accessible to a wide audience, discrete quadraphonic sound material established. This led to much more control because the problem of crosstalk was solved. But probably the unsatisfying panning stood in the way of commercial success.

Localizability of Sound Sources:

In quadraphonic sound, the spatial distribution of the speakers offer a wide panorama for virtual source positions in the front and additionally in the rear. As in stereophonic sound, source angle can be constructed but not the distance. The wide angles between listening position and speakers cause the phantom sources to be not particularly stable. The matrixing can encode and decode single channels unambiguously, whereas all channels together create massive crosstalk effects. The missing channel separation corrupts the construction of parameters, so the source position may move due to influences of other channels. Furthermore, increasing the perceived source extent by inter-channel time differences leads to unwanted sounds from the rear speakers after decoding.

Spaciousness:

The advantages and disadvantages in terms of localizability of sound sources also count for early reflections. Due to the wide angles between listener and adjacent speakers and due to the rear speakers, the panorama width in the front is wider and early reflections can originate from many more directions. But the crosstalk problem persists.

Reverberation:

The reverberation can arrive from many directions and be played back in different sound colorations and more or less diffuse. But the degree of diffuseness can not be controlled purposefully. Because of the crosstalk different phases always lead to different phantom source directions since phase-shifted signals are decoded to the rear speakers. Thus, a reverberant sound may create permanent jumps of phantom sources.

7.2.4 Dolby Surround

Based on the 4 : 2 : 4 matrix of quadraphonic sound, Dolby Stereo—an analogue optical sound format—was developed in 1975, initially for cinemas. It contains four channels with additional noise suppression—Dolby A or Dolby SR—but had the same dissatisfying channel separation of 3 dB, just like quadraphonic sound. For home use without noise suppression and independent of sound carrier medium Dolby Stereo established under the name *Dolby Surround*. The four channels are front left L, front right R, center C and a rear/back/surround channel B, especially used as effects channel.[38] Dolby Surround is compatible with stereo and offers a *downmix* function—which allows to set a mixing ratio of the channels—for compatibility with mono (Fig. 7.11).

The encoder

$$\begin{bmatrix} L_T \\ R_T \end{bmatrix} = \begin{bmatrix} 1 & \frac{1}{\sqrt{2}} & 0 & \frac{-j}{\sqrt{2}} \\ 0 & \frac{1}{\sqrt{2}} & 1 & \frac{j}{\sqrt{2}} \end{bmatrix} \cdot \begin{bmatrix} L_F \\ C \\ R_F \\ B \end{bmatrix} \tag{7.9}$$

codes the channels L and R unmodified to the transfer channels L_T and R_T. Channel C is coded to both transfer channels reduced by 3 dB ($\frac{1}{\sqrt{2}} \hat{=} -3\,\text{dB}$) and the surround channel is also coded to both transfer channels reduced by 3 dB and phase-shifted

[38] See Webers (2003), p. 219.

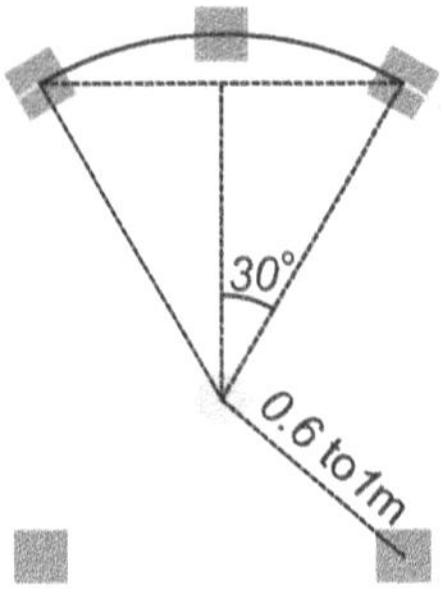

Fig. 7.11 Loudspeaker array for dolby surround sound systems. The frontal speakers are positioned on a circle line around the sweet spot facing the center. The surround loudspeakers are placed between 0.6 and 1 m both behind and above the listening position, not facing the sweet spot

by ±90°. In case of only one active channel the en- and decoding is unique and loss-free. Only multiple active channels cause crosstalk. The center speaker has the aim to create a signal which is localized in the center of the stereo base, even for positions beyond the sweet spot. So the whole cinema audience perceives the dialogs as coming from the screen to avoid an "eye/ear conflict".[39] It transduces all mono information, this means equally loud signals, in phase. The surround channel has a bandwidth limited to a range from 100 Hz to 7 kHz. It radiates signal contingents with equal amplitude but phase inversion. The L-signal is the decoded L_T signal without B and C contingent. Built-in signal processing promise some adjustment of the sound to the demands of listener and listening room, like "wide". "This corresponds to an acoustic widening of the basis. The listener gets the impression that the loudspeakers are farther apart".[40] The sound of the rear speakers can be delayed by a value between 20 and 150 ms and manipulated in intensity.[41] According to Dolby especially the delay serves to tune the arrival time of loudspeaker signals based on the distance ratio between the listening position and a front speaker a surround speaker.[42] The adjustable delay has an additional advantage which can be explained by a short example: Direct sound, like dialog, tends to com from the front center speaker. A large portion of the reverberation is played by the rear speakers because both recorded and artificial reverberation tend to contain many out-of-phase components. If many listeners are distributed over an area of say 10 × 40 m, as in many cinemas, the arrival time of the front and the rear loudspeaker signals is dependent on the seat of each listener. The signal from the center speaker may take 100 ms to the last tiers, the signal of the rear speakers only 10 ms. In this case the delay of the rear speakers is increased by up to 90 ms so that even for listeners in the rear the frontal sound arrives before the rear speaker signals. Of course, this adjustment is a compromise.

[39] Dolby Laboratories Inc. (1998), p. 2–8.

[40] From Schneider Rundfunkwerke-AG (1995), p. 29.

[41] See e.g. Dolby Laboratories Inc. (1998), p. 3–14, Schneider Rundfunkwerke-AG (1995), pp. 28–29.

[42] For details see e.g. Dolby Laboratories Inc. (2000), p. 5.

A 90 ms-delay is satisfying for listeners on the rear seats but the delay for listeners in the front is a bit too large in this case, so a delay of 60 ms may be chosen as a compromise. This delay is too long for listeners in the front tier, works well for listeners in the center and is too short for listeners in the rear.

Since the surround signal arises from phase differences between L_T and R_T even decoding of pure stereo signals can create a surround sound. Inversely phased signals may occur by A-B recording, by electronic reverb and phase effects—such as phaser, flanger, chorus, reverb, delay etc.—and by synthetic sounds and the many pseudo-stereo effects as discussed in Sect. 2.3. These will be played by the rear speakers, whereas centered signals, typically bass drum or dry singing, will sound from the center speaker. The surround sound decoded from a pure stereo signal is called "magic surround".[43]

In 1987 the active, adaptive, standardized decoder named *Pro Logic* was released to improve the channel separation. The more stable surround panorama was realized by a "steering"[44] function which permanently calculates the dominant source origin direction and amplifies the appropriate channels. Its successors *Pro Logic II* and *Pro Logic IIx* achieved further channel separation advancements. This is achieved by a better technique, more phase-stable storage and transfer media and voltage controlled amplifiers VCAs. A perceived channel separation of up to 40 dB, $5:2:5$ and $6:2:6$ matrixing—to code 3/2 and 3.1/2 channel sound with two surround channels limited to a range of 100 Hz to 20 kHz—are realized.

Since Dolby Surround is compatible with mono and stereo and can be transferred via two channels, as standardly used by TV, CD, radio, stereo record, video (VHS) and audio cassette, it established itself as standard, especially for analogue film and TV.[45] Dolby Surround has especially been used as affordable solution for DVD-players and gaming consoles but is displaced recently by newer formats such as Dolby Digital, Dolby True HD, SDDS and DTS-HD Master Audio. For example Sony Playstation 2 and Nintendo Wii are compatible to Dolby Surround whereas the follow-up models Playstation 3 and Wii-U have Dolby Digital compatibility implemented.[46]

Localizability of Sound Sources:

In Dolby Surround, frontal sound events can be localized correctly even beyond the sweet spot, due to the center loudspeaker. Since the advantaged channel separation by Pro Logic II in the year 2000, a stable positioning of phantom sources via amplitude panning is possible but time based panning creates phase differences which will be interpreted as rear signals by the decoder. Because the two rear channels in the initial Dolby Surround setup play the same sound, a rear positioning of phantom sources is almost impossible. This shortcoming is slightly improved with the introduction of a second rear channel in Pro Logic II and the steering function.

---

[43]See Dolby Laboratories Inc. (1998), pp. 5-2 to 5-3 and Slavik and Weinzierl (2008), p. 624.

[44]See Henle (2001), p. 115.

[45]Henle (2001), p. 117.

[46]See Dolby Laboratories Inc. (2010), Games->Dolby Pro Logic II and Nintendo Co. (2013).

Spaciousness:

The ER can be created by amplitude panning in the same manner as in stereo. Additionally, rear reflections can be created. However, due to crosstalk and steering, there is little control over the direction of early reflections.

Reverberation:

Additionally to frontal reverb, rear reverberation can be created. Direction and degree of diffusion can be varied by choice since the introduction of Pro Logic II. But the typically random phase relations in reverberations lead to unwanted jumps between front and rear channels, reinforced by the steering function. The signal for the surround speakers is often delayed and low-pass filtered to simulate a realistic reverberation.[47]

7.2.5 Discrete Surround Sound

The International Telecommunication Union (ITU) describes several multi-channel loudspeaker layouts. The 5.1 setup is certainly the most-widespread. It is illustrated in Fig. 7.12. The line-up of the five speakers is similar to Dolby Surround. But in this case the surround speakers have individual channels, no limited bandwidth and they face the sweet spot. A subwoofer, also referred to as *Low Frequency Effect* (LFE) or simply as *boom*, is placed beside the center speaker on the loudspeaker base. It has a limited bandwidth between 3 and 120 Hz to add audible and haptically sensible vibrations. The 5.1-setup contains the stereo triangle, one center loudspeaker and two additional rear speakers at ± 100 to $120°$.

Driving signals for the loudspeakers can be embedded in any format that is compatible to 5 audio channels, like a multi-channel wave file, MP3 Surround, DVD-Audio and Super Audio CD (SACD).

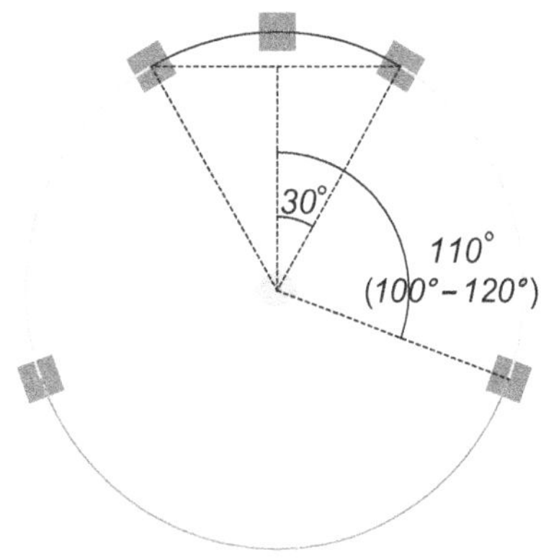

Fig. 7.12 5.1 loudspeaker arrangement after ITU-R BS.775

[47] Faller (2009), p. 635.

Table 7.5 Overview of advanced dolby digital formats

Format	Innovations
Dolby E	Extension to 3.1/3 channels and distribution- and production-bitstreams for the transmission of programs via single channels
Dolby digital live	For live concerts and interactive sound imaging e.g. for video games. Encodes 5.1 sound in real-time
Dolby digital plus	Extension to up to 13.1 channels between 30 kbit and 6 Mbit per second via HDMI connection
Dolby true HD	Lossless coding of the audio data via *Meridian Lossless Packing* MLP, for DVD audio. Samplerates up to 96 kHz are possible, data rates up to 18 MBit per second allow up to 20 channels
Dolby mobile	For mobile phones, allows (pre-)settings of users, e.g. concerning spaciousness, gain, spectrum etc.
Dolby virtual speaker	Simulates 5.1 loudspeaker sound via headphones

One audio format for the 5.1 layout is *Dolby Digital*. It was invented by Dolby Laboratories Inc. and released in 1991 as film sound format. It comprises the six discrete channels (5.1) under the term "program". 1995 the first consumer products with Dolby Digital compatibility entered the market. Since Dolby Digital was able to handle input signals with sample depths up to 24 bit at a sample rate of up to 48 kHz the multichannel audio coding system "Adaptive Transform Coder no. 3"[48] AC-3 was deployed. This psychoacoustic audio compression format allows for data rates between 32 and 640 kbit per second. It makes use of perceived loudness, dynamic range, and auditory masking to compress the amount of data.[49] As data carrier every bit transparent medium can be utilized. Therefore, Dolby Digital is widely spread all over the world and became part of standards and recommendations, like those of the American and European digital broadcasts ("Advanced Television System Committee" ATSC and "Digital Video Broadcast" DVB) and the DVD standard. 5.1 audio systems almost displaced stereo in the mid and upper Hi-Fi segment.[50]

Beyond the pure audio data the format contains meta data which inform about volume, dynamic and downmix information for mono, stereo and Dolby Surround compatibility. An error detection ("Cyclic Redundancy Check" CRC) accompanies the data transfer. The reproduction latency lies between 179 and 450 ms.[51] To be applicable in realtime, *Dolby Digital live* was introduced e.g. for powerful gaming consoles. In the year 2010 the first mobile phone with Dolby Digital sound was introduced.[52] Further advancements of Dolby Digital are listed in Table 7.5.

[48]Slavik and Weinzierl (2008), p. 627.

[49]Masking is treated extensively in Sect. 4.3.

[50]See Goertz (2008), p. 423.

[51]Slavik and Weinzierl (2008), p. 629.

[52]See Dolby Laboratories Inc. (2010) ->GAMES and -> MOBILE.

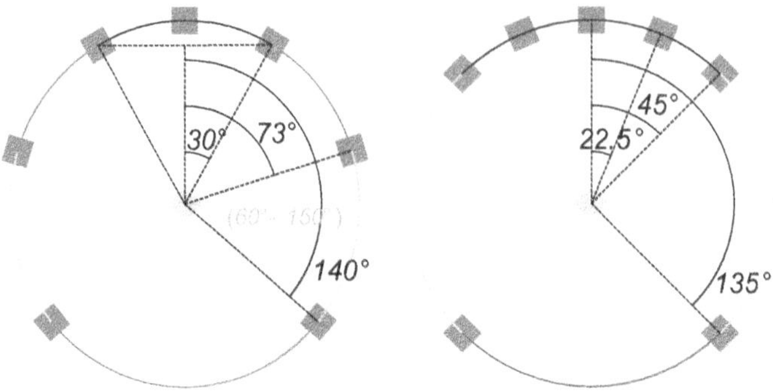

Fig. 7.13 7.1 loudspeaker arrangements recommended by ITU (left) and for SDDS (right)

Another widely used 5.1 audio format is DTS Digital Surround which debuted in 1993 in the movies *Dr. Giggles* and *Jurassic Park*. Here, the audio content was delivered on a number of synced CD-ROMs.[53] DTS typically uses lossless audio compression and is part of the Laser Disk (LD), DVD and Blu-Ray specifications. Many professional hardware and gaming consoles like Sony's Playstation 3 and 4, Microsoft's Xbox 360 and Xbox One support the DTS Digital Surround and newer lossless and lossy DTS codecs.[54]

In some setups, a third back channel is added to enable for a more robust panning in the rear. It is especially used for "cinema-effect-sound-localization".[55] Of course, three equidistant back loudspeakers improve the stability of phantom sources. The ITU also recommends other discrete surround systems with more than 5 channels. Examples are the two 7.1 loudspeaker layouts illustrated in Fig. 7.13. The left one is a 3.1/4 system which adds lateral loudspeakers to the 5.1 setup. This setup is often recommended by Dolby and DTS. Controlled signals from these side-loudspeakers can increase the perceived source width or envelopment by adding early or late lateral reflections to the sound scene. Formats for 7.1 audio are e.g. DTS Neo:X, DTS-HD Master Audio, Dolby Digital Plus and Dolby True HD. The layout on the right hand side is a 5.1/2 system. It is used for Sony's Dynamic Digital Sound format (SDDS) and adds both a higher precision and a wider panorama for panning in the front. SDDS was introduced 1993 in the movie *The Last Action Hero* and uses Sony's psychoacoustic ATRAC codec to compress the audio material.[56] ATRAC is well-known for the audio compression in mini discs.

As no matrixing or alike is necessary, every device with five or more output channels can play back sound for discrete surround setups like 5.1. Formats like

[53]See Davis (2003), p. 565.

[54]For more information on DTS, refer to Davis (2003) and DTS, Inc. (2016), with articles like DTS, Inc. (2015a).

[55]See Owsinski (2014), p. 55.

[56]Details on 7.1 can be found e.g. in Apple (2009), pp. 1161f.

Dolby Digital and DTS:Digital Surround mainly compress the audio content and/or add compatibility to mono, stereo and other loudspeaker arrangements.

Localizability of Sound Sources:

Discrete surround setups are mainly used for film sound, so their focus lies on dramaturgical aspects rather than on realistic spatial sound. Dialog comes from the center speaker, atmospheric sound from the surround loudspeakers.[57] The principle of phantom sources via amplitude panning is usually kept.[58] Due to the center loudspeaker, localizability of sources in the frontal region is improved compared to stereo or quadraphonic sound. With discrete channels no crosstalk occurs and the localizability in the frontal and rear area is rather good even in the case of several simultaneous sources. An additional rear speaker, as used in the 6.1 audio setup, improves the localizability of rear sources over 5.1 systems. But as demonstrated in Fig. 7.8 lateral auditory events are not realizable via amplitude panning. Thus, the 5.1/2 SDDS setup certainly offers a wider panorama for stable panning in the front compared to the stereo triangle or 3.1/4 discrete surround.

Spaciousness:

The spaciousness of discrete surround systems can increase with an increasing number of loudspeakers. Due to discrete channel signals, the sound distribution can be adjusted much better than in Dolby Surround since gain and phase for each loudspeaker can be controlled individually. Also, the limited bandwidth is omitted. In the 3.1/4 setup lateral and rear speakers can be used to create a homogeneous distribution of ER. Especially lateral reflections affect the impression of source extent and envelopment.

Reverberation:

Five ore more discrete channels offer control over the degree of diffusion of the reverberation. In addition to the front, reverb can arrive from the sides and the back. Without phase and frequency restrictions the degree of diffusion and sound color of the reverberation can be controlled more accurately than in quadraphonic audio systems or Dolby Surround setups. Reverberation still cannot completely surround the listener but a large part of the 360° in the horizontal plane is covered.

7.2.6 Immersive Audio Systems

In recent years, discrete surround systems have been expanded by additional channels for elevated loudspeakers. These tend to add a second, elevated surround setup—a so-called *height layer*—rather than expanding the discrete surround setup to an actual three-dimensional system. Often, such systems are called "advanced multichannel

[57] See Henle (2001), p. 116.

[58] See Reisinger (2002), p. 49.

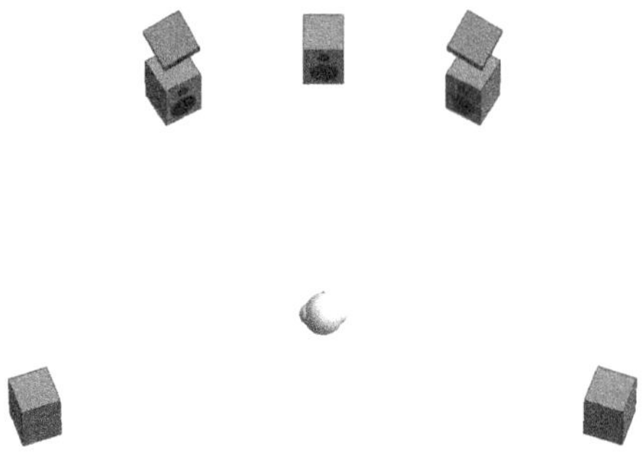

Fig. 7.14 Immersive 7.1 loudspeaker arrangement (3.1/2+2)

sound systems" or "immersive" audio.[59] Again, several companies provide formats for different loudspeaker configurations.

The classical 7.1 codecs can be used to feed a 3.1/2+2 loudspeaker array as illustrated in Fig. 7.14. Here, a 3.1/2 setup is extended by two additional front loudspeakers which are elevated by 30 to 55°. This is the simplest Dolby Atmos setup.[60] With nine channels, Dolby Atmos can expand a 5.1 or a 7.1 setup as illustrated in Fig. 7.15. In the figure another specialty of Dolby Atmos can be observed: As a space-saving solution for home theater, the "elevated" speakers are not really elevated. Instead, highly directive speakers are integrated in the cabinet of the front and rear speakers in the bottom layer. They mainly radiate towards the ceiling, so the reflection shall create the elevation effect. In the illustration these directive speakers are depicted by circles on the top of the loudspeaker cabinets. Dolby Atmos supports up to 34 channels for loudspeaker configurations like 5.1/19+10.[61] Although Dolby Atmos does not come along with many technological advancements over earlier multi-channel approaches, some practicists like Owsinski (2014) judge it as "totally revolutionary multispeaker" system.[62]

The Auro-3D setup looks similar to Dolby Atmos. It does, however, provide an optional *voice of god* (VOG) loudspeaker, also referred to as *zenith*, at the top. In cinemas, it can be used for a voice from the off, like the heavenly voice of god. Or it is used as effects channel for objects in the sky. It can even increase immersion by playing some reverberation from above. For the Auro-3D loudspeaker setups the audio material can be delivered in two ways. In the channel-based method, the driving signals for the elevated layer and the optional VOG are encoded in the 5 to 7 channels of the bottom layer. Without a decoder they are inaudible and the system is compatible to conventional discrete surround sound. The Auro-3D decoder extracts the loudspeaker signals. Alternatively, source signals are stored together

[59] See e.g. International Telecommunication Union (2016), pp. 5, 12 and 38 or Dolby Laboratories Inc. (2015), p. 3.

[60] See Dolby Laboratories Inc. (2015), pp. 16f.

[61] See Dolby Laboratories Inc. (2015), pp. 3f and 28.

[62] See Owsinski (2014), p. 53.

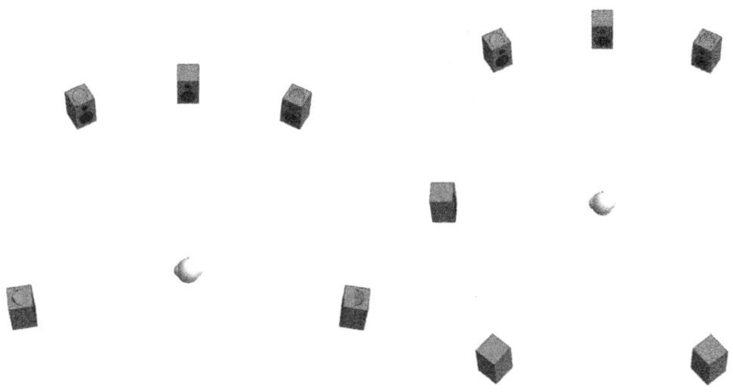

Fig. 7.15 Dolby atmos setups 3.1/2+4 (left) and 3.1/4+2 (right)

with metadata which describe their desired starting time, location and trajectory. The combination of Auro-3D with object-based audio is named AUROMAX®. The location of each loudspeaker is fed to the decoder system before use. The decoder then applies the appropriate amplitude panning to create the phantom source locations and trajectories which are stored in the metadata.[63]

Another format for immersive audio is DTS:X, the successor to DTS-HD Master Audio. The format launched in 2015. It is object-based, so audio tracks are stored together with metadata that may define, e.g., location and trajectories.[64] The advantage of object-based coding is that it is not restricted to a certain loudspeaker layout. Here, the decoder has to take care of the audio rendering. Source locations could be realized by means of amplitude based panning or by approaches that synthesize the according sound field, such as ambisonics and the approaches of wave field synthesis discussed in Chaps. 8 and 9. In the simplest case, the encoded objects are just 6 static point sources, each associated with one loudspeaker. This way it the audio scene is simply played back by a 5.1 audio system.

MPEG-H is a novel audio format.[65] Its core is a powerful compression algorithm which leverages both psychoacoustics and signal theory. Audio material can either be stored in the ambisonics B-format, channel-based, or object-based. Again, the final audio rendering has to be done for the individual loudspeaker arrangement. However, downmix options for conventional loudspeaker systems, such as two-channel stereo and 5.1 setups, are implemented.

Most immersive audio systems bring nothing new in terms of source panning. The amplitude based panning approach, which is used since two-channel stereophony, is typically applied here as well. Panning between two different height layers is not intended. As the precedence effect is very effective in the median plane, amplitude based panning between loudspeakers of different height is only possible if all involved

[63]See Auro Technologies (2015) for details and much more information on Auro-3D.

[64]See e.g. DTS, Inc. (2015a, b, 2016) for details.

[65]For further information, see e.g. Herre et al. (2014, 2015) and ISO/IEC JTC 1/SC 29: Coding of audio, picture, multimedia and hypermedia information (2016).

loudspeakers have the same distance to the listener. In this case, to achieve elevation panning, the tangent law is re-formulated to three dimensions. Here, phantom sources are panned between a triplet of loudspeakers. These are placed on a spherical surface with the sweet spot in its center. The locations of $N = 3$ loudspeakers $\mathbf{Y}_1$ to $\mathbf{Y}_3$ are described by a coordinate system, i.e.

$$\mathbf{Y}_n = [x_n \; y_n \; z_n] . \tag{7.10}$$

A loudspeaker vector $\mathbf{y}$ contains the positions of all loudspeakers

$$\mathbf{y} = \begin{bmatrix} \mathbf{Y}_1 \\ \mathbf{Y}_2 \\ \mathbf{Y}_3 \end{bmatrix} = \begin{bmatrix} x_1 \; y_1 \; z_1 \\ x_2 \; y_2 \; z_2 \\ x_3 \; y_3 \; z_3 \end{bmatrix} , \tag{7.11}$$

the desired phantom source position is

$$\mathbf{Q}_p = [x \; y \; z] \tag{7.12}$$

and the three loudspeaker amplitude factors necessary to create the chosen phantom source position are

$$\hat{\mathbf{A}} = \left[\hat{A}_1 \; \hat{A}_2 \; \hat{A}_3 \right] . \tag{7.13}$$

The amplitude factors or gains are found by solving a linear equation system

$$\mathbf{Q}_p = \hat{\mathbf{A}}\mathbf{y}. \tag{7.14}$$

This equation is referred to as *vector base amplitude panning* (VBAP).[66] A simple selection criterion for the three active loudspeakers is that the loudspeakers and the listener span the smallest possible three-dimensional space that contains the phantom source. This implies that neither two nor all three loudspeakers should be in-line with the listener. The loudspeaker triplet is the smallest possible, so there are no overlapping triplets. The closer the three loudspeakers are placed to each other, the more robust the created phantom source position becomes. This is especially true for listeners besides the sweet spot. As for the sine and tangent panning law, vector base amplitude panning is under-determined until either one amplitude factor is chosen or an assumption is made, like the constant energy assumption, Eq. 7.5 with $n = 2$.

Note that Eq. 7.14 is nothing but a matrix formulation of the tangent panning law, Eq. 7.3. When choosing $N = 2$, they deliver the exact same amplitude factors. And in fact, if the phantom source lies exactly on the connection line between two loudspeakers, the gain of the third loudspeaker will be 0. If the phantom source position coincides with a loudspeaker position, this loudspeaker will be the only active loudspeaker. If a phantom source moves beyond the loudspeaker triplet, another triplet is

[66] For details, see Pulkki (2001).

Fig. 7.16 Active loudspeakers when applying vector base amplitude panning in three cases. Left: The phantom source position coincides with a loudspeaker position. Middle: The phantom source lies on the boundary of a loudspeaker triplet. Right: The phantom source lies within a loudspeaker triplet. The gray arrow points at the phantom source, the black arrows at the active loudspeakers

active, according to the selection criterion. Some examples are given in Fig. 7.16. The black arrows point at the active loudspeakers which create the phantom source whose location is indicated by the gray arrow. Vector base amplitude panning reformulates the tangent panning law to an N-dimensional matrix formulation which makes sense for two loudspeakers in a one-dimensional speaker arrangement and three loudspeakers in a two-dimensional speaker arrangement. VBAP assumes the listener to face the phantom source. Still, many users apply VBAP for phantom sources at all arbitrary locations. In a way, this can be meaningful. The localization capability of people is relatively poor for lateral sources, especially if they are elevated. So if a phantom source is panned to surround a listener, its position accurately recognized in the front, becomes vague towards the sides and a bit clearer in the rear. The same is true for a real source that surrounds a listener. Furthermore, there may be applications in which a listener will always look at a panned source and even follow its movements. This is especially likely in audiovisual systems. However, it is a common misbelief that VBAP enables robust three-dimensional source panning of objects that surround listeners.

Already in two-dimensional setups, it is criticized that the perceived width of phantom sources is not stable. It rather changes with the phantom source location. When a phantom source position coincides with a loudspeaker position, the source sounds very point-like. The further the phantom source is deflected away from the loudspeakers, i.e. the more central the phantom source, the larger its directional spread. In principle, a wider sounding source may be desired. But especially for moving sources, this variation in perceived source extent may be unwanted. To oppose this variation, additional loudspeakers can be activated whenever the phantom source comes close to a loudspeaker location. *Multiple Direction Amplitude Panning* (MDAP) creates a homogeneous spatial spread. This is achieved by replacing a phantom source by several phantom sources which are distributed around the phantom source. In a two-dimensional setup, i.e. with $N = 2$, two phantom sources are placed with equal distance to the left and the right of the desired phantom source position. The spacing between these replacement phantom sources is at least the length of the loudspeaker base. This way, a central phantom source is the only case in which only two loudspeakers are active. For all other phantom source positions, three loudspeakers are active. An example is illustrated in Fig. 7.17. Note that the methods works best when the listener faces the phantom source. As mentioned above, in Sect. 7.2.3, this stereo panning method works best when the listener faces the phantom source. It

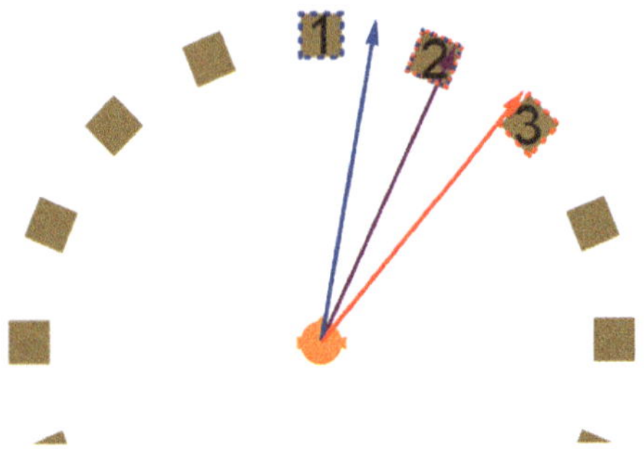

Fig. 7.17 Example of multiple direction amplitude panning. Panning between loudspeakers 1 and 2 creates the blue phantom source. Panning between loudspeakers 2 and 3 creates the red phantom source. Together, they create the violet phantom source with an increased spatial spread

does not work well for lateral phantom sources, especially when the spacing between speakers is too large.

Localizability of Sound Sources:

In principle, VBAP offers the possibility for vertical panning in addition to horizontal panning. But loudspeaker setups are usually not conceptualized as hemispherical setup but as two to three layers. So phantom sources can now move along the azimuth of two height layers. This offers new creative possibilities not only for film sound but also for music. On the one hand, the additional layer facilitates a natural source placement, e.g. in terms of an elevated organ or an opera singer above the orchestra pit. On the other hand, music can be mixed creatively in one additional dimension. However, sources do not tend have vertical trajectories. Vertical panning is only possible if all loudspeakers that contribute to the phantom source panning have the same distance to the listener. For listeners beyond the sweet spot, amplitude based panning is not very robust as the precedence effect takes action. The localization error becomes smaller, the more densely the loudspeakers are distributed besides each other. So a higher number of loudspeakers within one layer comes along with a stabilized phantom source position. MDAP may decrease the source localizability in exchange for a more stable phantom source width, especially for moving sources.

Spaciousness:

Multiple direction amplitude panning increases the control over the perceived source extent, when enough loudspeakers are present. This mainly affects the direct sound. The more loudspeakers are involved, the more potential directions for early reflections exist. This adds some fidelity for spatial sound reproduction. Elevated reflections can certainly add some vertical source spread in addition to width. This way, sources may appear more bodily present in terms of a three-dimensional extent.

Reverberation:

With elevated loudspeakers, immersive audio setups can create a much higher degree of diffusion. Due to the elevated speakers, reflections from above can be added, which is much more natural than the restriction of reflections coming from the height of the head. The addition of elevated reflections creates a high listener envelopment, which is the main strength of immersive audio systems.

7.2.7 *Head Related Stereophonic Sound*

From binaural stereophonic sound, which often plays back dummy head recordings via dichotic headphones, diverse loudspeaker systems have been developed. In binaural tetraphonic sound, a dummy head recording is played via two frontal speakers supported by two rear speakers which are delayed and manipulated in phase- and frequency response. Thereby, the sweet spot is broadened a little compared to stereo but accuracy in distance and direction localization is substantially worse than binaural stereophonic sound.[67]

For other binaural speaker systems, the speakers act as "virtual headphones".[68] Their signals are filtered in a way that the signals at the listener's ears are the same as from a real source at the posed position. Because a HRTF is very individual, it often has to be measured for each listener beforehand. Of course, it is possible to use the HRTF of a dummy head or derive it from models like the ITD models Eqs. 4.8 and 4.9 discussed in Sect. 4.4.2.[69] But since the direction-dependent filtering of a dummy head or a model may be different from the individual HRTF of the listener, playback of sounds pre-filtered with these non-individualized HRTF can still lead to an increased localization blur, front-back confusion and localization inside the head. Therefore, these non-individualized HRTFs must be adapted to the individual or a good fitting HRTF must be chosen from a database.[70]

"Cross talk cancellation"[71] diminishes crosstalk between two speakers by a value of 20 to 25 dB. A speaker plays the sound for one single ear plus another sound to cancel crosstalk of the second speaker on that ear. Since this canceling sound also reaches the other ear, it has to be diminished by a canceling signal from the speaker for the other ear. This vicious circle is not infinite since the amplitude of the needed canceling signal reduces with every step. But highly absorbing walls and a still listener are necessary.[72] This principle can further be controlled dynamically by adaptive filters so that the sweet spot follows the listener.[73] Of course, the double-arrival problem remains. Cross talk cancellation works best if the active loudspeakers are frontal and very close to each other. This minimizes the delay between the targeted

[67] Webers (2003), pp. 224f.

[68] Vorländer (2008), p. 293.

[69] As proposed, e.g. in Sodnik et al. (2006) and Busson et al. (2004).

[70] Individualization is proposed, e.g. by Seeber and Fastl (2003). Kan et al. (2009) propose a method to create near field HRTF from far field HRTF measurements. Several HRTF databases exist, like "AUDIS" and "CIPIC", see Blauert et al. (1998) and Algazi et al. (2001).

[71] Vorländer (2008), p. 295.

[72] See Blauert (1997), p. 360.

[73] Vorländer (2008), p. 297.

ear and the other ear as well as the influence of complicated deflections around the head.

Localizability of Sound Sources:

Except from completely frontal sounds a good direction- and distance-localization is possible if the listener stands still, especially when the HRTF of the listener himself is used. Combined with a head tracking system head related stereophonic sound gets good results.

Spaciousness:

For the playback of lateral reflections head related stereo is very suitable. But reflections of the listening room disturb the natural impression even worse than in case of stereophonic sound since the canceling signal will be reflected, too. It is dependent on the recording method and the signal processing whether the signals at the listener's ears are natural and thus provide a natural spaciousness.

Reverberation:

In principle, a natural reverberation can be created for the listening position. The target sound field at the listeners ears can be created by a binaural recording or a convolution of the direct sound with a binaural room impulse response. However, if the listening room is not heavily damped the reflections of the canceling signal will affect the reverberation as well.

7.3 Discussion of Audio Systems

Stereo introduced the quality of spatial sound to audio playback which was missing in mono. Direction and expansion of sources and reflections can be transmitted to a certain degree from the recording environment to a listener in a listening room by diverse recording techniques or signal processing via phantom sources. The spatial aspect of instrument and the room in which it is played are transferred in good psychoacoustic resemblance. But a restriction arriving with conventional stereophonic audio systems is the drastically reduced ideal listening area, the sweet spot. Also, the direction of phantom sources is restricted to the loudspeaker base. Nevertheless, stereo became a worldwide standard for stereo record, audio cassette, audio CD, radio and TV and other forms of broadcast. The more recent audio formats are often stereo compatible and work with the same methods, especially amplitude based panning. After the widely spread Dolby Surround format, Dolby Digital followed up as quasi standard, especially for film and multi media applications. Dolby Digital works with a 5.1 audio setup which widens the possible auditory source direction area. But due to the origin of Dolby Digital in film sound, the speakers are not used equivalently and the aim is usually not to recreate realistic spatial sound. The frontal speakers dominate the sound whereas the rear speakers are used for atmosphere and environmental sound. Phantom sources can be distributed in the frontal region and

Table 7.6 Advantages and disadvantages of conventional stereophonic sound systems, especially stereo and 5.1 surround

Advantage	Disadvantage
Natural spectrum and dynamics	Restriction to one listening spot
Good direction-localizability	Insufficient depth-localizability
Standardized, widely spread sound format, media and carrier	No stable surround sound from all directions possible
Standardized number of channels	Barely no presentation of the radiation characteristic of musical instruments
Real-time processing possible	Room acoustics in the signal and real room acoustic superimpose
Compatible between each other	No representation of height

somewhat vaguely in the rear. The standardized loudspeaker arrangement on a circle line around a listener in a room with some reverberation creates similar listening conditions for producers and listeners. But the loudspeaker arrangement is impractical for many listening rooms, e.g. living rooms. The radiation characteristics of sound sources is reproduced in neither of these audio systems.[74]

Table 7.6 illustrates the advantages and disadvantages of the most widespread audio systems, i.e., stereo and 5.1.

"WFS is considered as *the* solution for providing large listening areas".[75] Furthermore, it is considered as an approach which may fill the lateral gaps to create a truly surrounding sound and circumvent the disadvantages of conventional stereophonic sound.[76] Wave field synthesis is discussed in the next chapter.

References

Ahrens J, Geier M, Spors S (2008) The soundscape renderer: a unified spatial audio reproduction framework for arbitrary rendering methods. In: 124th audio engineering society convention

Algazi VR, Duda RO, Thompson DM, Avendano C (2001) The CIPIC HRTF database. In: IEEE workshop on applications of signal processing to audio and acoustics. New York, pp 99–102. https://doi.org/10.1109/aspaa.2001.969552

Apple (2009) Logic pro 9. User manual. https://documentation.apple.com/en/logicpro/usermanual/Logic%20Pro%209%20User%20Manual%20(en).pdf

Auro Technologies (2015) Auro-3d®home theater setup, rev. 6. http://www.auro-3d.com/wp-content/uploads/documents/Auro-3D-Home-Theater-Setup-Guidelines_lores.pdf

Baalman M (2008) On wave field synthesis and electro-acoustic music, with a particular focus on the reproduction of arbitrarily shaped sound sources. VDM, Saarbrücken

[74]See Warusfel and Misdariis (2004), p. 4.

[75]Daniel et al. (2003), p. 1.

[76]See Ahrens et al. (2008), p. 3.

Bauer BB (1961) Phasor analysis of some stereophonic phenomena. J Acoust Soc Am 33(11):1536–1539. https://doi.org/10.1121/1.1908492
Berkhout AJ, de Vries D, Vogel P (1992) Wave front synthesis: a new direction in electroacoustics, vol 10. In: Audio engineering society convention 93. https://doi.org/10.1121/1.404755
Berkhout AJ, de Vries D, Vogel P (1993) Acoustic control by wave field synthesis. J Acoust Soc Am 93(5): 2764–2778. https://doi.org/10.1121/1.405852
Bernfeld B (1973) Attempts for better understanding of the directional stereophonic listening mechanism. In: Audio engineering society convention 44. Rotterdam
Blauert J (1997) Spatial hearing. The pychophysics of human sound source localization, revised edn. MIT Press, Cambridge, MA
Blauert J (2008) 3-d-Lautsprecher-Wiedergabemethoden. In: Fortschritte der Akustik—DAGA '08. Dresden, pp 25–26
Blauert J, Braasch J (2008) Räumliches Hören. In: Weinzierl S (ed) Handbuch der Audiotechnik, chapter 3. Springer, Berlin, pp 87–122. https://doi.org/10.1007/978-3-540-34301-1_3
Blauert J, Brüggen M, Hartung K, Bronkhorst AW, Drullmann R, Reynaud G, Pellieux L, Krebber W, Sottek R (1998) The AUDIS catalog of human HRTFs. In: Proceedings of the 16th international congress on acoustics, pp 2901–2902. https://doi.org/10.1121/1.422910
Busson S, Nicol R, Warusfel O (2004) Influence of the ears canals location on spherical head model for the individualized interaural time difference. In: CFA/DAGA. Strasbourg
Chowning J (1971) The simulation of moving sound sources. J Audio Eng Soc 19(1):2–6. https://doi.org/10.2307/3679609
Damaske P (2008) Acoustics and hearing. Springer, Berlin. https://doi.org/10.1007/978-3-540-78229-2
Daniel J, Nicol R, Moreau S (2003) Further investigations of high order ambisonics and wavefield synthesis for holophonic sound imaging. In: Audio engineering society convention 114
David jr EE (1988) Aufzeichnung und Wiedergabe von Klängen. In: Winkler K (ed) Die Physik der Musikinstrumente. Spektrum der Wissenschaft, Heidelberg, pp 150–160
Davis MF (2003) History of spatial coding. J Audio Eng Soc 51(6): 554–569. http://www.aes.org/e-lib/browse.cfm?elib=12218
Davis MF (2007) Audio and electroacoustics. In: Rossing TD (ed) Springer handbook of acoustics, chapter 18. Springer, New York, pp 743–781. https://doi.org/10.1007/978-0-387-30425-0_18
Deutsches Institut für Normung (1996) Bild- und Tonbearbeitung in Film-, Video- und Rundfunkbetrieben - Grundsätze und Festlegungen für den Arbeitsplatz
Dickreiter M et al (1978) Handbuch der Tonstudiotechnik, vol 1, 2nd edn. De Gruyter, Munich
Dickreiter M et al (1987) Handbuch der Tonstudiotechnik, vol 1, 5th edn. völlig neu bearbeitete und ergänzte edition. De Gruyter, Munich
Dolby Laboratories Inc (1998) Dolby surround mixing manual, Issue 2. http://www.idea2ic.com/Manuals/dolbySuround.pdf. Last accessed 6 Sept 2016
Dolby Laboratories Inc (2000) Frequently asked questions about dolby digital. http://www.dolby.com/us/en/technologies/dolby-digital.pdf
Dolby Laboratories Inc (2010) Dolby. http://www.dolby.com. Last accessed 30 Sept 2010
Dolby Laboratories Inc (2015) Dolby Atmos®home theater installation guidelines. http://www.dolby.com/us/en/technologies/dolby-atmos/dolby-atmos-home-theater-installation-guidelines.pdf
DTS, Inc (2015a) Welcome to DTS:X - open, immersive and flexible object-based audio coming to cinema and home. http://investor.dts.com/releasedetail.cfm?releaseid=905640
DTS, Inc (2015b) Next-generation object- based codec technology 63(1/2): 130
DTS, Inc (2016) DTS is dedicated to sound. https://dts.com/
Faller C (2009) Spatial audio coding and MPEG surround. In: Luo FL (ed) Mobile multimedia broadcasting standards. Technology and practice, chapter 22. Springer, New York, pp 629–654. https://doi.org/10.1007/978-0-387-78263-8_22

Favrot S, Buchholz JM (2010) LoRA: a loudspeaker-based room auralization system. Acta Acust United Acust 96:364–375. https://doi.org/10.3813/aaa.918285

Friesecke A (2007) Die Audio-Enzyklopädie. Ein Nachschlagewerk für Tontechniker. K G Saur, Munich

Gade A (2007) Acoustics in halls for speech and music. In: Rossing TD (ed) Springer handbook of acoustics, chapter 9. Springer, Berlin, pp 301–350. https://doi.org/10.1007/978-0-387-30425-0_9

Goertz A (2008) Lautsprecher. In: Weinzierl S (ed) Handbuch der Audiotechnik, chapter 8. Springer, Berlin, pp 421–490. https://doi.org/10.1007/978-3-540-34301-1_8

Goertz A (2018) Bowers & wilkins 800 d3. Fidelity. HiFi und Musik 38:070–077

Hubert H (2001) Das Tonstudio Handbuch. Praktische Einführung in die professionelle Aufnahmetechnik, 5. komplett überarbeitete edition. GC Carstensen, Munich

Herre J, Hilpert J, Kuntz A, Plogsties J (2014) MPEG-H audio—the new standard for universal spatial/3D audio coding. In: Audio engineering society convention 137. http://www.aes.org/e-lib/browse.cfm?elib=17418

Herre J, Hilpert J, Kuntz A, Plogsties J (2015) MPEG-H audio—the new standard for universal spatial/3D audio coding. J Audio Eng Soc 62(12): 821–830. http://www.aes.org/e-lib/browse.cfm?elib=17556

Hiebler H (1999) Akustische Medien. In: Hiebel HH, Hiebler H, Kogler K, Wakitsch H (eds) Große Medienchronik. Wilhelm Fink, Munich, pp 541–782

Huber T (2002) Zur Lokalisation akustischer Objekte bei Wellenfeldsynthese. Diloma thesis. http://www.hauptmikrofon.de/diplom/DA_Huber.pdf

International Telecommunication Union (2016) Report ITU-R BS.2159-7: multichannel sound technology in home and broadcasting applications. https://www.itu.int/dms_pub/itu-r/opb/rep/R-REP-BS.2159-7-2015-PDF-E.pdf

ISO/IEC JTC 1/SC 29 (2016) Coding of audio, picture, multimedia and hypermedia information. ISO/IEC 23008-3:2015/amd 1:2016—MPEG-H, 3D audio profile and levels. https://www.iso.org/standard/67953.html

Kan A, Jin C, van Schaik A (2009) A psychophysical evaluation of near-field head-related transfer functions synthesized using a distance variation function. J Acoust Soc Am 125(4):2233–2242. https://doi.org/10.1121/1.3081395

Mäkivirta AV (2008) Loudspeaker design and performance evaluation. In: Havelock D, Kuwano S, Vorländer M (eds) Handbook of signal processing in acoustics, chapter 33. Springer, New York, pp 649–667. https://doi.org/10.1007/978-0-387-30441-0_33

Mores R (2018) Music studio technology. Springer, Berlin, pp 221–258. https://doi.org/10.1007/978-3-662-55004-5_12

Nintendo Co Ltd. (2013) Wii U technical specs. http://www.nintendo.com/wiiu/features/tech-specs/. Last accessed 17 Jan 2014

Owsinski B (2014) The mixing engineer's handbook, 3rd edn. Corse Technology PTR, Boston, MA

Pulkki V (1997) Virtual sound source positioning using vector base amplitude panning. J Acoust Soc Am 45(6):456–466

Pulkki V (2001) Spatial sound generation and perception by amplitude panning techniques. PhD thesis, Helsinki University of Technology, Espoo. http://lib.tkk.fi/Diss/2001/isbn9512255324/

Pulkki V (2008) Multichannel sound reproduction. In: Havelock D, Kuwano S, Vorländer M (eds) Handbook of signal processing in acoustics, chapter 38. Springer, New York, pp 747–760. https://doi.org/10.1007/978-0-387-30441-0_38

Reisinger M (2002) Neue Konzepte der Tondarstellung bei Wiedergabe mittels Wellenfeldsynthese. Diploma thesis, University of Applied Sciences Düsseldorf, Düsseldorf

Rossing TD (1990) The science of sound, 2nd edn. Reading Addison-Wesley, Massachusetts

Schanz GW (1966) Stereo-Taschenbuch. Stereo-Technik für den Praktiker. Philips, Eindhoven

Scheminzky F (1943) Die Welt des Schalls. Das Bergland, zweite ergänzte edition, Salzburg

Schneider Rundfunkwerke-AG (1995) Schneider MP 295 Bedienungsanleitung

Schubert H (2002) Historie der Schallaufzeichnung. http://www.dra.de/rundfunkgeschichte/radiogeschichte/pdf/historie_der_schallaufzeichnung.pdf. Last accessed 23 Aug 2010

Seeber BU, Fastl H (2003) Subjective selection of non-individual HRTF-related transfer functions. In: Proceedings of the 2003 international conference on auditory display. Boston

Slavik KM, Weinzierl S (2008) Wiedergabeverfahren. In: Weinzierl S (ed) Handbuch der Audiotechnik, chapter 11. Springer, Berlin, pp 609–686. https://doi.org/10.1007/978-3-540-34301-1_11

Sodnik J, Susnik R, Tomazic S (2006) Principal components of non-individualized head related transfer functions significant for azimuth perception. Acta Acust United Acust 92: 312–319. https://www.ingentaconnect.com/contentone/dav/aaua/2006/00000092/00000002/art00013

Spors S, Wierstorf H, Raake A, Melchior F, Frank M, Zotter F (2013) Spatial sound with loudspeakers and its perception: a review of the current state. Proc IEEE 101(9):1920–1938. https://doi.org/10.1109/JPROC.2013.2264784

Strube G (1985) Lokalisation von Schallereignissen. In: Bruhn H, Oerter R, Rösing H (eds) Musikpsychologie. Ein Handbuch in Schlüsselbegriffen. Urban & Schwarzenberg, Munich, pp 65–69

Theile G, Plenge G (1976) Localization of lateral phantom-sources. In: Audio engineering society convention 53. Zurich

Theile G (1980) Über die Lokalisation im überlagerten Schallfeld. PhD thesis, University of Technology Berlin. Berlin

Toole FE (2008) Sound reproduction. The acoustics and psychoacoustics of loudspeakers and rooms. Focal Press, Amsterdam

Verheijen E (1997) Sound reproduction by wave field synthesis. PhD thesis, Delft University of Technology. Delft

Vorländer M (2008) Auralization. Fundamentals of acoustics, modelling, simulation, algorithms and acoustic virtual reality. https://doi.org/10.1007/978-3-540-48830-9

Warusfel O, Misdariis N (2004) Sound source radiation syntheses: from performance to domestic rendering. In: Audio engineering society convention 116

Webers J (2003) Handbuch der Tonstudiotechnik. Analoges und Digitales Audio Recording bei Fernsehen, Film und Rundfunk. Franzis, Poing, 8. neu bearbeitete edition

Chapter 8
Wave Field Synthesis

Methods of sound field synthesis aim at physically recreating a natural or any desired sound field in an extended listening area. As discussed in Sect. 5.1.2, sound field quantities to synthesize are mainly sound pressure and particle velocity or sound pressure gradients. If perfect control over these sound field quantities was achieved, virtual sources could be placed at any angle and distance and radiate a chosen source sound with any desired radiation pattern. This way, the shortcomings of conventional audio systems could be overcome: Instead of a sweet spot, an extended listening area would exist. Without the need for panning, lateral and elevated sources could be created and depth in terms of source distance could be implemented. If the sound radiation characteristics of an instrument was captured and synthesized, virtual sources could sound as broad and vivid as their physical counterpart. This means a natural, truly immersive, three-dimensional sound experience including the perception of source width, and motion, listener envelopment, reverberance and alike. Unfortunately, already the theoretic core of most sound field synthesis approaches imposes several restrictions. Current sound field synthesis implementations offer limited acoustic control under certain circumstances with several shortcomings. Still, due to elaborate adaptions of a sophisticated physical core sound field synthesis applications are able to create a realism which is unreachable with conventional audio systems.

In this chapter a short historic overview of sound field synthesis is given. The most prominent sound field synthesis approach includes the spatio-temporal synthesis of a wavefront that propagates through space as desired. In this book this specific approach is referred to as *wave field synthesis* (WFS) or *wave front synthesis* and is treated in detail in this chapter. The term *sound field synthesis* is used as umbrella terms covering several methods which aim at controlling a sound field. These methods include wave front synthesis, ambisonics and alike. The theoretic core of wave field synthesis is derived from several mathematical theorems and physical considerations.

© Springer Nature Switzerland AG 2020

T. Ziemer, *Psychoacoustic Music Sound Field Synthesis*, Current Research in Systematic Musicology 7, https://doi.org/10.1007/978-3-030-23033-3_8

The derivation is explained step by step in the following section.[1] Several constraints make it applicable, as discussed in Sect. 8.3. These constraints lead to synthesis errors which are diminishable by adaptions of the mathematical core. Many sound field synthesis approaches model sound sources as monopole sources or plane waves. So a special treatment is given to the synthesis of the radiation characteristics of musical instruments in Sect. 8.4. Finally, some existing sound field synthesis installations for research and for entertainment are presented.

8.1 Sound Field Synthesis History

As already discussed in Sect. 7.2.2, methods of stereophonic recording and playback arose in the 1930s by Alan Dower Blumlein and others. At the same time Steinberg and Snow (1934a) conceptualized and implemented the *acoustic curtain* to capture and reproduce sound fields.[2] Authors like Ahrens (2012) and Friesecke (2007) consider this as the origin of wave field synthesis.[3] An acoustic curtain is depicted in Fig. 8.1. In principle, one plane wall of a recording room is covered by a mesh of microphones. These capture the auditory scene, e.g., a musical ensemble. In a performance room, one wall is covered by loudspeakers which are arranged in the exact same way as the microphones in the recording room. If now each recording is played back by the co-located loudspeaker, a copy of the auditory scene is created in the performance room. The constellation of the instruments is captured and reproduced this way. Listeners everywhere in the whole performance room have the impression that the instruments are actually there, arranged behind the loudspeaker curtain. Although they implemented the acoustic curtain with a pair or triplet of microphone-loudspeaker pairs only, they report to achieve a perceptually satisfying copy of the auditory scene. Later in this chapter we will see that an infinite acoustic curtain with infinitesimally spaced microphones and loudspeakers is necessary to capture and synthesize a sound field in a half space. This is exactly what the Rayleigh integral describes.

Another sound field recording and synthesis technique was developed in large part by Gerzon (1975) in the 1970s.[4] The sound pressure and the pressure gradients along the spatial dimensions are recorded at one receiver location by means of a microphone array. Two recording setups are illustrated in Fig. 8.2 for the two-dimensional case. These are equivalent if the microphones are proximate to one another compared to the considered wavelength. From these recordings, three channels W, X and Y can be derived as shown in the illustration. These contain the sound pressure W and the sound pressure gradients along two spatial dimensions X and Y. In a three-dimensional

---

[1]Mainly based on Pierce (2007), Williams (1999), Morse and Ingard (1986), Rabenstein et al. (2006), Ziemer (2018).

[2]See Steinberg and Snow (1934a, b).

[3]See Ahrens (2012), pp. 8f and Friesecke (2007), p. 147.

[4]See e.g. Gerzon (1973, 1975, 1981).

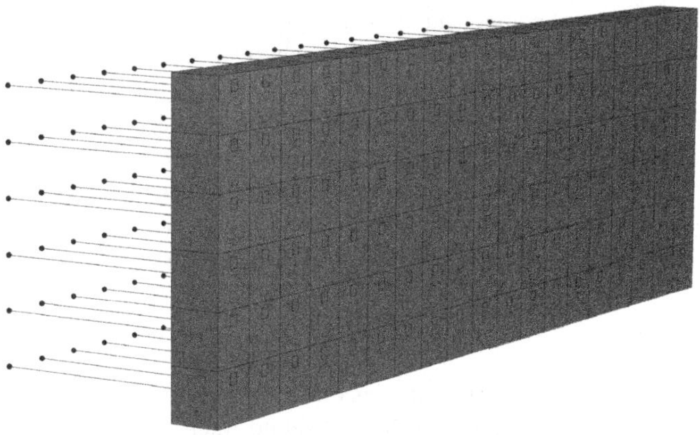

Fig. 8.1 Illustration of the acoustic curtain. After Ziemer (2016), p. 55

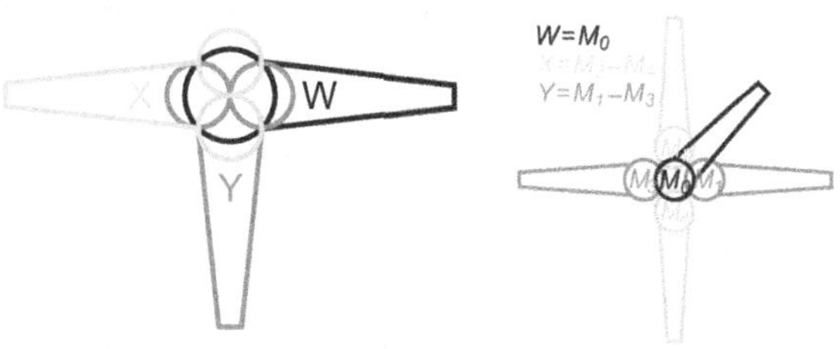

Fig. 8.2 Recording setups for first order ambisonics in two dimensions with different setups. After Ziemer (2017a), p. 315

Fig. 8.3 Ambisonics microphone array in a sound field, after Ziemer (2017a), p. 316

setup an additional channel Z is encoded, containing the pressure gradient along the third dimension. Encoding these channels is referred to as *tetraphony* or *B-Format*. The sound pressure is a scalar and can be recorded by an omnidirectional pressure receiver. The pressure gradients can be recorded by figure-of-eight microphones or approximated by the difference of to opposing omnidirectional microphones which are out of phase. The recreation of the sound field described by these channels by means of a loudspeaker array is referred to as *periphony* or *ambisonics* (Fig. 8.3).

The omnidirectional component W is the spherical harmonic of the zeroth order $\Psi_0^0(\omega, \varphi, \vartheta)$ and X, Y and Z are the three spherical harmonics of the first order $\Psi_1^0(\omega, \varphi, \vartheta)$, $\Psi_0^1(\omega, \varphi, \vartheta)$ and $\Psi_1^1(\omega, \varphi, \vartheta)$ as discussed already in Sect. 5.3.1.1. So the microphone setup performs a spherical harmonic decomposition truncated at the first order. We can write these components in a vector Ψ and try to find loudspeaker signals $\mathbf{A}$ which recreate this encoded sound field. To achieve this, the sound propagation from each loudspeaker to the receiver position needs to be described by means of a matrix $\mathbf{K}$. Then, solving the linear equation system

$$\Psi = \mathbf{KA} \tag{8.1}$$

the loudspeaker signals $\mathbf{A}$ recreate the encoded sound field at the receiver position. The components in Ψ describe the desired sound field. The vector describes the sound pressure and at a central listening position and the pressure gradient along the spatial dimensions whose origin lies at this central position. The B-Format and higher order sound field encoding by means of circular or spherical harmonics are quasi-standardized. In contrast to conventional audio systems, as discussed throughout Chap. 7, the encoded channels are not routed directly to loudspeakers. Only the sound field information is stored. By solving Eq. 8.1 loudspeaker signals approximate the desired physical sound field or the desired sound impression. The solver is the ambisonics decoder. If the desired sound field contains as many values as loudspeakers present, the propagation matrix $\mathbf{K}$ in the linear equation system, Eq. 8.1, is a square matrix. In this case it can be solved directly. This can be achieved by means of an inverse matrix, or by numerical methods like Gaussian elimination. With more loudspeakers than target values the problem is under-determined: we have more known target values than unknown loudspeaker signals. In this case a pseudo inverse matrix can be used to approximate a solution. Unfortunately, this strategy comes along with several issues. First of all this approximate solution does not consider auditory perception. In the Moore Penrose inverse the Euclidean norm, i.e., the squared error, is minimized. This means that small errors in amplitude, phase, and time occur. These may be audible when they lie above the just noticeable difference of level or phase, or above just noticeable interaural level, phase or time difference.[5] The perceptual results of audible errors are a false source localization, especially for listeners that are not located at the central listening position. Other perceptual outcomes are audible coloration effects, spatial diffuseness, or an undesirably high loudness. A psychoacoustical solution to balance these errors would be desirable. Several ambisonics decoders have been proposed.[6] Psychoacoustic evaluation of existing decoders has been carried out.[7] However, psychoacoustic considerations should ideally be carried out already in the development process of the decoder. The radiation method suggested in this book is a physically motivated solution, and

[5]Details about auditory thresholds are discussed in Chaps. 4 and 6.

[6]An overview of ambisonic decoders can be found in Heller (2008). General solutions to inverse problems in acoustics are discussed in Bai et al. (2017).

[7]See e.g. Zotter et al. (2014) and Spors et al. (2013).

not perceptually motivated. But it comes along with a number of psychoacoustic considerations, like the *precedence fade*.[8]

Encoding and reconstructing spherical harmonics of higher order is referred to as *higher order ambisonics* (HOA). With higher order, the sound field is not only synthesized correctly at the very receiver location but in a receiver area which increases with increasing order and increasing wavelength. Originally, the transfer function **K** was modeled as plane waves emanating from the loudspeakers. As discussed in Sects. 5.1.4 and 5.1.6, plane waves are a good approximation of very distant sources. Later, loudspeakers were modeled as monopoles. Monopoles are certainly a better approximation of the actual radiation characteristics of proximate loudspeakers. When their distance to the receiver location has a magnitude of some meters, they do have a relevant amplitude decay and a wave front curvature. These are the main differences between a plane wave and a monopole. When an approximate solution is found, small numeric errors can result in large amplitudes, especially for low frequencies. This is the case because the wave front curvature becomes large compared to the wavelength and the amplitude decay is large compared to the encoded pressure gradient of a plane wave. As a result, **K** has a bad condition number and Eq. 8.1 is ill-conditioned. Compensating these near field effects, e.g., by adoptions of **K** is referred to as nearfield compensated higher order ambisonics (NFC-HOA).[9]

In many applications ambisonics loudspeaker setups are circular or hemispherical. As long as the location of the receiver array and each loudspeaker is known, the necessary loudspeaker signals can be calculated for virtually any constellation. Ideally, the loudspeakers are arranged with regular spacing. Although possible, it is not necessary to encode a sound field by means of a microphone array. One can also freely define or simulate a desired sound field and save it in the B-format. The first approach is referred to as *data based rendering* since it contains measured data. The second approach is called *model based rendering* because source location and sound propagation are calculated. Consequently, the encoded sound field depends largely on the sound propagation model that was used in the calculation. Often, sources are modeled as monopole sources or plane waves. However, models like the *complex point source model* are also conceivable.[10]

Around the late 1980s a wave field synthesis approach was derived, developed and implemented at the Delft University of Technology. The approach was termed *acoustic control* and later *wave front synthesis*.[11] In these works, a mathematical core of sound field synthesis is formulated and interpreted in physical terms. From this core, wave front synthesis, ambisonics and other sound field synthesis methods can be derived. A lot of research and development took place in Delft especially

[8]The phychoacoustic sound field synthesis approach including the radiation method and the precedence fade are introduced in Chap. 9.

[9]More information on HOA and NFC-HOA can be found e.g. in Ahrens and Spors (2008a), Williams (1999), pp. 267ff, Spors and Ahrens (2008), Daniel et al. (2003), Menzies and Al-Akaidi (2007), Daniel (2003) and Elen (2001).

[10]The complex point source model is described in Sect. 5.3.1, applied in Ziemer (2014) and discussed extensively in Ziemer and Bader (2015a), Zimmer (2015a, 2017a).

[11]See e.g. Berkhout (1988) and later Berkhout et al. (1992).

throughout the 1990s.[12] From 2001 to 2003 they were supported by a number of universities, research institutions and industry partners in the CARROUSO research project funded by the European Community.[13] Achievements from this project were market-ready wave front synthesis systems.

Since then, mainly adoptions, extensions, refinements of methods or error compensation[14] and additional features like moving sources and complicated radiation patterns[15] have been implemented. A lot of research is still carried out in the field of wave field synthesis. For example interfaces and techniques for more accessible creation of content and control of wave field synthesis systems are being developed.[16] Another topic is to reduce the number of necessary loudspeakers either by a prioritized sweet area within the extended listening area or by considering psychoacoustic effects.[17] Although sound field synthesis is originally a physically reasoned approach, psychoacoustic considerations are not superfluous. It is the auditory perception that makes sound field synthesis systems sound as desired even though physical synthesis errors are present and easily measurable. An elaborated perceptual evaluation of synthesized sound fields receives more and more attention in the literature.[18]

8.2 Theoretical Fundamentals of Sound Field Synthesis

The general idea of sound field synthesis can be traced back to Huygens' principle. This principle can be described by means of the Kirchhoff–Helmholtz-Integral which is explained in this section. Although often considered as the mathematical core of wave field synthesis, this integral is barely implemented in wave field synthesis. Instead, the Kirchhoff–Helmholtz-Integral is reduced to the Rayleigh-Integral which can be applied rather directly by means of an array with conventional loudspeakers. The adoption process from the mathematical idea to the actual implementation is explained in the subsequent section for wave front synthesis applications.

---

[12] See e.g. papers like Berkhout et al. (1993), de Vries et al. (1994), de Vries (1996), Berkhout et al. (1997) and Boone et al. (1999) and dissertations like Vogel (1993), Start (1997) and Verheijen (1997).

[13] Publications are e.g. Corteel and Nicol (2003), Daniel et al. (2003), Spors et al. (2003), Vaananen (2003) and many more. More information on CARROUSO can be found in Brix et al. (2001).

[14] See e.g. Gauthier and Berry (2007), Menzies (2013), Spors (2007), Kim et al. (2009), Bleda et al. (2005).

[15] See e.g. Ahrens and Spors (2008b), Albrecht et al. (2005) and Corteel (2007).

[16] See Melchior (2010), Fohl (2013) and Grani et al. (2016).

[17] See e.g. Hahn et al. (2016) and Spors et al. (2011), Ziemer (2018) for more information on *local wave field synthesis*, and Chap. 9 and Ziemer and Bader (2015b, 2015c), Ziemer (2016) for details on *psychoacoustic sound field synthesis*.

[18] See e.g. Start (1997), Wierstorf (2014), Ahrens (2016), Wierstorf et al. (2013) and Spors et al. (2013).

8.2.1 Huygens' Principle

Every arbitrary radiation from a sound source can be described as integral of point sources on its surface. In addition, each point on a wave front can be considered as origin of an elementary wave. The superposition of the elementary waves' wavefronts creates the advanced wave front. This finding is called *Huygens' principle* and is the fundament on which wave field synthesis is based on.

Figure 8.4 illustrates the Huygens' principle. Figure 8.5 clarifies this illustration by reducing it to two dimensions and splitting it into states at different points in time. The black disk in Fig. 8.5a represents the source at t_0 which creates a wavefront that spreads out concentrically. This wavefront is illustrated in dark gray in Fig. 8.5b with some points on it. Each point on this wave front can be considered the origin of an elementary source, which again create a wave front, represented by the gray disks in Fig. 8.5c. Together, these wave fronts form the propagated wave front of the original source at a later point in time illustrated in Fig. 8.5d. The distance between those elementary waves has to be infinitesimally small. A monopole-shaped radiation of these elementary waves would create a second wave front at time t_2. This second wave front would be inside the earlier wave front, closer to the original breathing sphere again. This can clearly be seen in both Figs. 8.4 and 8.5c: One half of the elementary waves are located inside the dark gray wave front. This is physically untrue; the elementary waves must have a radiation characteristic which is 0 geared towards the source. This radiation characteristic is described by the Kirchhoff–Helmholtz integral (K-H integral), discussed in the subsequent Sect. 8.2.2.

Fig. 8.4 Illustration of the Huygens' principle. Each point on a wavefront can be considered as the origin of an elementary wave. Together, the elementary waves create the propagated wavefront. From Ziemer (2016), p. 54

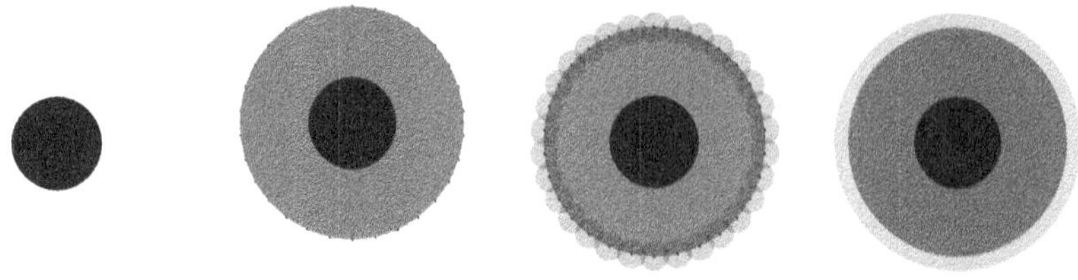

(a) t_0: breathing sphere (black). *(b) t_1: elementary sources (black dots) on emanating wave front (gray).* *(c) t_2: wave fronts from elementary sources.* *(d) t_2: further emanated wave front from breathing sphere.*

Fig. 8.5 Wave fronts of a breathing sphere at three points in time in 2D. The breathing sphere at t_0 (**a**) creates a wave front at t_1 (**b**). Points on this wave front can be considered as elementary sources which also create wave fronts at t_2 (**c**). By superposition these wave fronts equal the further emanated wave front of the breathing sphere (**d**). From Ziemer (2016), p. 55

8.2.2 Kirchhoff–Helmholtz Integral

The Gauss' theorem[19] states that spatial area integrals of a function over a volume V are equal to surface integrals of the normal components of a function over the volume's surface S

$$\int_V \nabla \mathbf{f} \, \mathrm{d}V = \int_S \mathbf{fn} \, \mathrm{d}S \tag{8.2}$$

if it has a piecewise smooth boundary and the function $\mathbf{f}$ is a steady, differentiable vector function.[20] A special case of the Gauss' theorem is described by Green's second theorem[21]:

$$\int_V \mathbf{f}\nabla^2\mathbf{g} - \mathbf{g}\nabla^2\mathbf{f} \, \mathrm{d}V = \int_S \mathbf{f}\nabla\mathbf{gn} - \mathbf{g}\nabla\mathbf{fn} \, \mathrm{d}S \tag{8.3}$$

From Green's second theorem and the wave equations, Eqs. 5.4 and 5.16, the Kirchhoff–Helmholtz integral can be derived, which links the wave field of a source-free volume V with sources $\mathbf{Y}$ on its surface S:

$$-\frac{1}{4\pi}\oiint_S \left(G(\omega, \Delta\mathbf{r}) \frac{\partial P(\omega, \mathbf{Y})}{\partial \mathbf{n}} - P(\omega, \mathbf{Y}) \frac{\partial G(\omega, \Delta\mathbf{r})}{\partial \mathbf{n}} \right) \mathrm{d}S = \begin{cases} P(\omega, \mathbf{X}), & \mathbf{r} \in V \\ \frac{1}{2} P(\omega, \mathbf{X}), & \mathbf{r} \in S \\ 0, & \mathbf{r} \notin V \end{cases} \tag{8.4}$$

Note that Eqs. 8.3 and 8.4 are nonlinear differential equations, which include the sought-after function and its derivative. The K-H integral states that the spectrum

[19] Also called "divergence theorem", see e.g. Pierce (2007), p. 58.

[20] See Merziger and Wirth (2006), p. 551.

[21] See Merziger and Wirth (2006), p. 555.

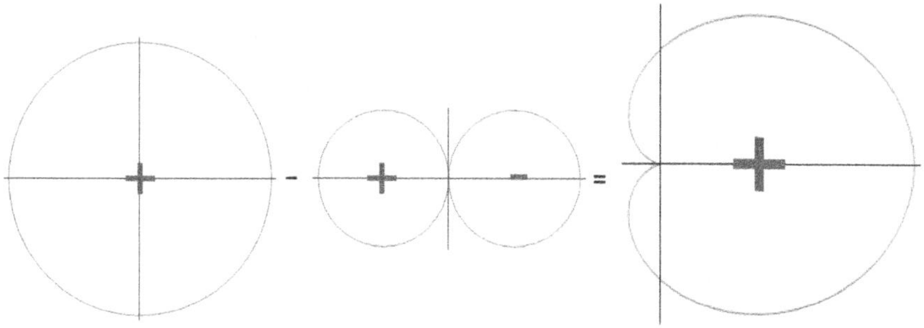

(a) Monopole source. - (b) Dipole source. = (c) Cardioid.

Fig. 8.6 Two dimensional illustration of superposition. Monopole- and dipole-source form a cardioid-shaped radiation. After Ziemer (2018), p. 335. From Ziemer (2016), p. 57

$P(\omega, \mathbf{X})$ at each point $\mathbf{X}$ in a source-free volume V is the integral of the spectra $P(\omega, \mathbf{Y})$ at every point $\mathbf{Y}$ on the bounding surface S and their propagation function $G(\omega, \Delta\mathbf{r})$ in the direction of the normal vector $\mathbf{n}$ pointing inwards. $G(\omega, \Delta\mathbf{r})$ is a Green's function, a solution of the inhomogeneous Helmholtz equation, Eq. 5.22, and $P(\omega, \mathbf{Y})$ is a spectrum, a solution for the homogeneous Helmholtz equation, Eq. 5.9. $\Delta\mathbf{r}$ is the Euclidean distance $||\mathbf{Y} - \mathbf{X}||_2$. The sources $\mathbf{Y}$ on the boundary surface are secondary sources, excited by primary sources $\mathbf{Q}$ which lie in the source volume U. The first term of the closed double contour integral describes a wave which propagates as monopole since the propagation term $G(\omega, \Delta\mathbf{r}) = \frac{e^{-\imath k \Delta\mathbf{r}}}{\Delta\mathbf{r}}$ is a monopole. From the periodic motion equation, Eq. 5.14, it emerges that $\frac{\partial P}{\partial n}$ is proportional to sound particle velocity in normal direction $\mathbf{V}_n$. The second term of the integral is a wave which radiates as dipole, since $\frac{\partial G(\omega, \Delta\mathbf{r})}{\partial \mathbf{n}} = \frac{1 + \imath k \Delta\mathbf{r}}{\Delta\mathbf{r}^2} \cos(\varphi)\, e^{-\imath k \Delta\mathbf{r}}$ is a dipole term. Sound field quantities P and V are convertible into each other after Euler's equation of motion, Eq. 5.1, so the K-H integral is over-determined and several approaches to a solution exist.

As already stated, the secondary sources on the surface of the source-free medium are monopole- and dipole-sources. In phase, they add up and inversely phased they are 0. So the radiation can double inwardly by constructive interference and become 0 outwardly by destructive interference. Combined, they create a *cardioid*, also referred to as *kidney* or *heart*. It is illustrated in Fig. 8.6.

The boundary surface could be the wave front around a source and the source free volume could be the room beyond this wave front. Then, the K-H integral is a quantified formulation of the Huygens' principle. It is illustrated in the two-dimensional Fig. 8.7. In contrast to the earlier illustration of Huygens' principle, Figs. 8.4 and 8.5, this modified version does not create a false wavefront that propagates inwards. This is because the elementary sources on the wavefront are cardioids that face away from the origin of the wave. However, the K-H integral could also describe a wave that propagates inwards. In this case the cardioids would face the focus point of the wavefront. These examples illustrate that both pressure and pressure gradient on a

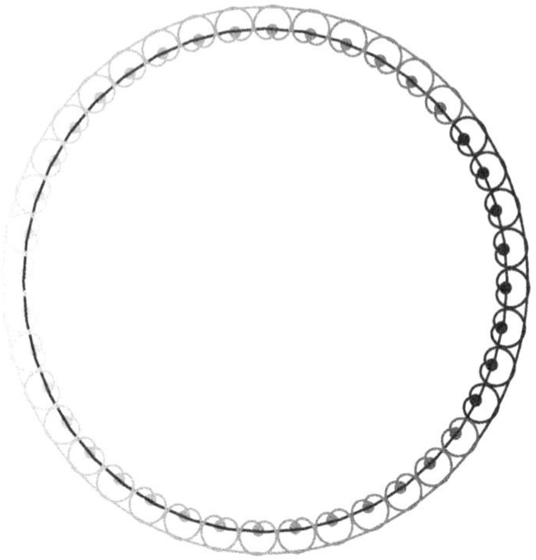

Fig. 8.7 Kirchhoff–Helmholtz integral describing Huygens' principle for an outward propagating wave. From Ziemer (2018), p. 334

surface need to be known do describe the wave propagation direction. The Kichhoff–Helmholtz integral can describe wave fronts of monopole sources or plane waves as well as complex radiation patterns and diffuse sound fields with a random distribution of amplitudes, phases and sound propagation directions. In the illustrated example the elementary waves have different gray levels, indicating different complex amplitudes. So the amplitude and phase are different in any direction, as naturally observed in musical instruments, demonstrated, e.g., for the shakuhachi in Fig. 5.7, in Sect. 5.3.1.

The volume could also be any arbitrary other geometry. It could be the surface of a physically existing or non-existing boundary. This boundary is the separation surface between a source volume, which contains one or more sources, and a source-free volume, which contains the listening area. Any arbitrary closed boundary is conceivable as long as the premises of the Gauss' theorem are observed. Figure 8.8 illustrates three examples for a volume boundary, which will be regarded in later chapters. Two types of setups exist: Surrounding the listener with secondary sources—as in Fig. 8.8a and c—or surrounding the primary source(s), as illustrated in Fig. 8.8b.[22]

The Kirchhoff–Helmholtz integral describes analytically how spectrum and radiation on a volume surface are related to any arbitrary wave field inside a source-free volume. It is therefore the core of wave field synthesis.[23]

[22]Cf. Daniel et al. (2003), p. 3.

[23]See Berkhout et al. (1993), p. 2769.

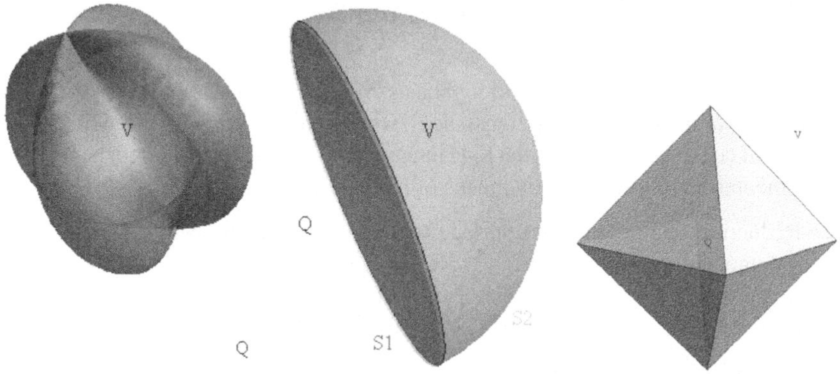

(a) Arbitrary geometry. (b) Hemisphere. (c) Octahedron.

Fig. 8.8 Three volumes V with possible source positions Q. After Ziemer (2016), p. 58

8.3 Wave Field Synthesis

The Kirchhoff–Helmholtz integral is a theoretical construct which cannot simply be put into practice by technical means. It demands control of sound pressure and pressure gradients on a complete surface. That is a continuous distribution of an infinite number of secondary sources with infinitesimal distance, surrounding a volume entirely. Sound pressure and velocity need to be controllable everywhere on the volume surface, which is hardly possible by technical means. However, what we can control is the sound pressure of loudspeakers. But even with an infinite number of infinitesimally distanced loudspeakers completely separating a listening area from a source volume would be insufficient, as long as the pressure gradient cannot be controlled. So for a practical realization the reduction of secondary sources to a finite number of loudspeakers with discrete distances radiating approximately as monopoles or dipoles is feasible.[24] These have to be fed with the correct audio signal, often referred to as "driving function".[25] Surrounding an entire room with speakers is impracticable—as already mentioned in Sect. 4.4.1 and illustrated in Fig. 4.15—and requires enormous technical challenges, computational power, acquisition- and operating-costs. Therefore, concepts with plane arrays[26] and line arrays[27] of the speakers are proposed in the literature and commonly applied.

[24]See e.g. Spors et al. (2008).

[25]See e.g. Spors et al. (2008).

[26]See e.g. Oellers (2010).

[27]One line, see Gauthier and Berry (2007), Baalman (2008), Kolundzija et al. (2009a), Cho et al. (2010), Reisinger (2002, 2003) and Spors (2007), circular array, see Spors (2007), Rabenstein et al. (2006), Reisinger (2002, 2003) and Rabenstein and Spors (2008), and three to four lines surrounding the listening area, see Spors et al. (2003), Reisinger (2002, 2003), Rabenstein et al. (2006).

8.3.1 Constraints for Implementation

For implementing such Wave Field Synthesis (WFS) systems the K-H integral has to be adjusted to the restrictive circumstances, which leads to errors in the synthesis. A number of constraints simplify the K-H integral in a way which allows for a technical implementation of the theory by means of loudspeaker arrays[28]:

1. Reduction of the boundary surface to a separation plane between source-free volume and source volume
2. Restriction to one type of radiator (monopole or dipole)
3. Reduction of three-dimensional synthesis to two dimensions
4. Discretization of the surface
5. Introduction of a spatial border

The particular steps will be successively accomplished in the following subsections.

8.3.2 Rayleigh-Integrals

Imagine a volume V consisting of a circular plane $S1$ closing a hemisphere $S2$, as illustrated in Fig. 8.8b, whose radius converges to ∞. The influence of the radiation from the source on $S2$ becomes 0 for the area in front of $S1$. This coherence satisfies the *Sommerfeld condition*. It remains a separating plane between source free volume and source volume. The K-H integral then consists of an integral over the plane $S1$ and thus fulfills the first simplification criterion from Sect. 8.3.1:

$$-\frac{1}{4\pi}\iint\limits_{S1}\left(G(\omega,\Delta\mathbf{r})\frac{\partial P(\omega,\mathbf{Y})}{\partial\mathbf{n}}-P(\omega,\mathbf{Y})\frac{G(\omega,\Delta\mathbf{r})}{\partial\mathbf{n}}\right)\mathrm{d}S=\begin{cases}P(\omega,\mathbf{X}), & \mathbf{X}\in V\\ 0, & \mathbf{X}\notin V\end{cases}\tag{8.5}$$

This step reduces the area of secondary sources from a three-dimensional surrounding of a source-free volume to a separation plane.

Since the Green's function, Eq. 5.22, is a linear combination of a special solution and a general solution, one term of the integral can be eliminated by adding a deftly chosen general solution to the free-field Green's function. So the radiation can be restricted to one type of radiator. If the Green's function is chosen to be

$$G_D(\omega,\Delta\mathbf{r})=\frac{e^{-\imath k\Delta\mathbf{r}}}{\Delta\mathbf{r}}+\frac{e^{-\imath k\Delta\mathbf{r}'}}{\Delta\mathbf{r}'},\tag{8.6}$$

$G_D(\omega,\Delta\mathbf{r})$ is 0 on the surface S—which satisfies the so-called *homogeneous Dirichlet boundary condition*[29]—and the second term vanishes if $\Delta\mathbf{r}'$ is the mirrored position of $\mathbf{X}$, mirrored at the tangent of point $\mathbf{Y}$ on S. This implicitly models the boundary

[28] These or similar simplifications are also proposed by Rabenstein et al. (2006), p. 529.

[29] See e.g. Burns (1992).

as a rigid surface,[30] leading to the *Rayleigh I integral* for secondary monopole sources as already introduced in Eq. 5.29 in 5.3.3:

$$P(\omega, \mathbf{X}) = -\frac{1}{2\pi} \iint_{S1} \left(G_D(\omega, \Delta\mathbf{r}) \frac{\partial P(\omega, \mathbf{Y})}{\partial \mathbf{n}} \right) \mathrm{d}S. \tag{8.7}$$

Now, considering $\frac{\partial P(\omega,\mathbf{Y})}{\partial \mathbf{n}}$ the desired source signal, an explicit solution can be found e.g. by means of wave field expansion. This approach is called "simple source approach" and is the basis of some sound field reconstruction methods such as HOA.

Since the distance $|\Delta\mathbf{r}|$ between secondary source position $\mathbf{Y}$ and considered position in the source-free volume $\mathbf{X}$ equals the distance between the secondary source position and the mirror position $\left|\Delta\mathbf{r}'\right|$, $G_D(\omega, \Delta\mathbf{r})$ is nothing but a doubling of the free-field Green's function $G(\omega, \Delta\mathbf{r})$:

$$G_D(\omega, \Delta\mathbf{r}) = 2G(\omega, \Delta\mathbf{r}) \tag{8.8}$$

Assuming $\frac{G_N(\omega,\Delta\mathbf{r})}{\partial n}$ to be 0 satisfy the *homogeneous Neumann boundary condition*[31] and the first term of Eq. 8.5 vanishes. This is accomplished by choosing

$$G_N(\omega, \Delta\mathbf{r}) = \frac{e^{-\imath k \Delta\mathbf{r}}}{\Delta\mathbf{r}} - \frac{e^{-\imath k \Delta\mathbf{r}'}}{\Delta\mathbf{r}'}, \tag{8.9}$$

yielding the *Rayleigh II integral* for secondary dipole sources:

$$P(\omega, \mathbf{X}) = -\frac{1}{2\pi} \iint_{S1} \left(P(\omega, \mathbf{Y}) \frac{\partial G(\omega, \Delta\mathbf{r})}{\partial \mathbf{n}} \right) \mathrm{d}S. \tag{8.10}$$

In both cases the second simplification criterion from Sect. 8.3.1 is satisfied. But since the destructive interference outside the source-free volume is missing, $P(\omega, \mathbf{X})$ for $\mathbf{X} \notin V$ is not 0. A mirrored sound field in the source volume is the consequence. In case of monopoles the sound field created by the secondary sources is identical with the one inside the source-free volume. This effect is similar to the earlier illustration of Huygens' principle, Figs. 8.4 and 8.5. In case of dipole sources the phase in the source volume is the inverse of the phase inside the source-free volume. Additionally, the sound pressure or, respectively the particle velocity, duplicate by adding the general solution of the Green's function. Both cases are illustrated in Fig. 8.9 for a one-dimensional loudspeaker array.

Both formulations do not apply for arbitrary volume surfaces but for separation planes only.[32] To ensure that any position around the listening area can be a source position, the listening area has to be surrounded by several separation planes. If Eqs.

---

[30] See Spors et al. (2008), p. 4 and Baalman (2008), p. 27.

[31] See e.g. Burns (1992).

[32] See Spors et al. (2008), p. 5.

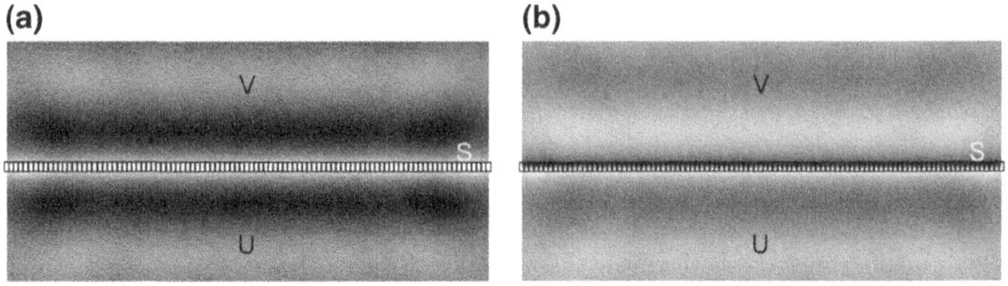

Monopole loudspeakers. *Dipole loudspeakers.*

Fig. 8.9 Desired sound field above and mirrored sound field below a separation plane according to the Rayleigh I integral for secondary monopole sources (**a**) and the Rayleigh II integral for secondary dipole sources (**b**). After Ziemer (2018), pp. 337 and 338

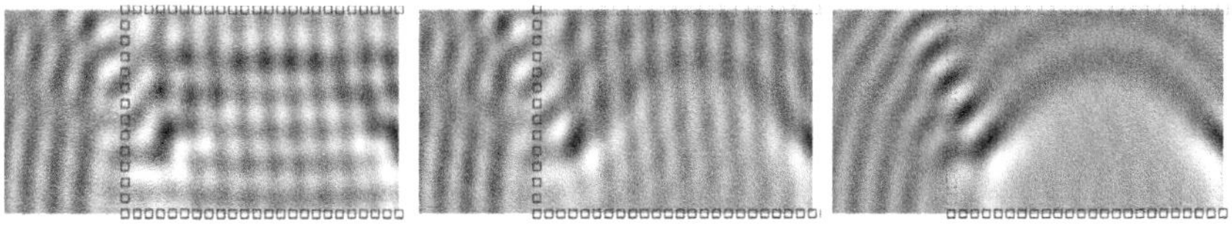

(a) Three active line arrays. *(b) Two active line arrays.* *(c) One active line array.*

Fig. 8.10 Illustration of the spatial windowing effect: A circular wave front superimposes with virtual reflections from two (**a**) or one (**b**) additional loudspeaker array(s). When muting those loudspeakers whose normal direction deviates from the local wave front propagation direction by more than 90° (**c**), the synthesized wave front is much clearer. Here, the remaining synthesis error is a truncation error, resulting from the finite length of the loudspeaker array. After Ziemer (2018), p. 338

8.7 and 8.10 are applied to other geometries, they still deliver approximate results.[33] In any case, the source-free volume has to be convex so that no mirrored sound field lies inside the source-free volume, i.e. volume (a) in Fig. 8.8 is inappropriate.[34] Since $S1$ is implicitly modelled as a rigid surface, several reflections occur when a listening area is surrounded by several separation planes. These unwanted reflections emerge from speakers whose positive contribution to the wave front synthesis lies outside the listening area. The portion of sound that propagates into the listening area does not coincide with the synthesized wave propagation direction. This artifact can be reduced by spatial "windowing"[35] technique applied to the Rayleigh I integral:

$$\begin{aligned} P(\omega, \mathbf{X}) &= d(\mathbf{Y}) \frac{P(\omega, \mathbf{Y})}{\partial \mathbf{n}} 2G(\omega, \mathbf{Y}) \\ d(\mathbf{Y}) &= \begin{cases} 1, & \text{if } \langle \mathbf{Y} - \mathbf{Q}, n(\mathbf{Y}) \rangle > 0 \\ 0, & \text{otherwise} \end{cases} \end{aligned} \tag{8.11}$$

[33] See Spors et al. (2008), p. 5.

[34] See Spors and Ahrens (2008), pp. 4f.

[35] See de Vries et al. (1994), Spors et al. (2008), p. 5 and Gauthier and Berry (2007), p. 3.

Here, $d(\mathbf{Y})$ is the windowing function for spherical waves which is 1 if the local propagation direction of the sound of the virtual source at the position of the secondary source has a positive component in normal direction of the secondary source. If the deviation is $\frac{\pi}{2}$ or more, $d(\mathbf{Y})$ becomes 0 and the speaker is muted. That means only those loudspeakers whose normal component resembles the tangent of the wave front of the virtual source are active. The term $G(\omega, \Delta\mathbf{r})$ describes the directivity function of the secondary source, i.e. of each loudspeaker. The other terms are the sought-after driving functions D of the loudspeakers[36]:

$$D(\omega, \mathbf{Y}) = 2d(\mathbf{Y}) \frac{P(\omega, \mathbf{Y})}{\partial \mathbf{n}} \tag{8.12}$$

An example for the unwanted virtual reflections due to applying the Rayleigh integral although surrounding the listening area from three sides is given in Fig. 8.10. The same wave front is synthesized according to the Rayleigh integral in three ways. In (a), three linear loudspeaker arrays are active. Here, the desired wave front superimposes with virtual reflections from the two additional arrays. In (b), one loudspeaker line array is muted. The contribution of these loudspeakers to the wave front synthesis would lie above the figure, i.e. outside the listening area. Muting them does not decrease synthesis precision in the listening area. In (c), the second line array is muted. Now one can clearly see the desired wave front curvature. No virtual reflections are visible. The remaining sound field synthesis error is the so-called *truncation error*. It will be discussed in detail in Sect. 8.3.3.

Although considered as source- and obstacle-free field, it is to a certain extent possible to recreate the wave field of a virtual source within the source-free volume. This is achieved by assuming an inverse propagation and calculating a concave wave front at the surface which focuses at the position of the virtual source and creates a convex wave front from then on. These sources are called "focused sources".[37] Figure 8.10 already exemplifies a focused source. More examples will be given throughout the chapter. Of course, focused sources will not work for listeners between the active loudspeakers and the focus. For them, the wave front seems to arrive somewhere from loudspeaker array and not from the focus. In contrast, listeners behind the focus do not experience the concave wavefront. They simply hear the convex wave front which seems to originate in the focus point. So focused sources reduce the extent of the listening area.

8.3.2.1 Two Dimensions

For applications in which the audience is organized more or less in plane, it is sufficient to recreate the wave field correctly for that listening plane only, rather than in

[36]See Spors et al. (2008), p. 5.

[37]The derivation of the secondary source signals and further information on these sources can be found e.g. in Kim et al. (2009), Geier et al. (2010), Ahrens and Spors (2009).

the whole listening volume. Furthermore, the best source localization resolution of the human auditory system is in the horizontal plane as discussed in Sect. 4.4. This is the main reason why conventional audio systems mostly focused on horizontal audio setups, as presented in Chap. 7. Luckily, when listening to music, listeners are often organized roughly in plane, like in many concert halls, opera houses, cinemas, theaters, in the car, on the couch in the living room etc. Furthermore, one or several one-dimensional distributions of loudspeakers are easier implementable than covering a complete room surface with loudspeakers. Reducing the three-dimensional wave field synthesis to two dimensions reduces the separation plane $S1$ to a separation line $L1$. In theory, one could simply reduce the surface integral to a simple integral and the Rayleigh integrals would take the forms

$$P(\omega, \mathbf{X}) = \frac{1}{2\pi} \int_{L1} \left(G(\omega, \Delta\mathbf{r}) \frac{\partial P(\omega, \mathbf{Y})}{\partial \mathbf{n}} \right) \mathrm{d}S1 \tag{8.13}$$

and

$$P(\omega, \mathbf{X}) = \frac{1}{2\pi} \int_{L1} \left(P(\omega, \mathbf{Y}) \frac{\partial G(\omega, \Delta\mathbf{r})}{\partial \mathbf{n}} \right) \mathrm{d}S1. \tag{8.14}$$

In these cases $\mathbf{X}$ is two-dimensional

$$\mathbf{X} = \begin{bmatrix} x \\ y \end{bmatrix}. \tag{8.15}$$

This solution was satisfying if no third dimension existed, e.g. if wave fronts of the secondary sources had no spherical but a circular or cylindrical propagation.[38] Then, the propagation function $G(\omega, \Delta\mathbf{r})$ was different, having an amplitude decay of $\frac{1}{\sqrt{r}}$ instead of $\frac{1}{r}$. This is owed to the fact that the surface S of a circle or cylinder doubles with a doubled circle radius r_{circle}

$$S = 2\pi r_{\text{circle}} \tag{8.16}$$

in contrast to the spherical case in which it squares with the doubled radius as already indicated in Eq. 5.24 in Sect. 5.1.6. In this case

$$I \propto \frac{1}{r} \tag{8.17}$$

and thus

$$p \propto \frac{1}{\sqrt{r}}. \tag{8.18}$$

[38] See e.g. Spors et al. (2008) pp. 8f, Rabenstein et al. (2006), pp. 521ff.

So the practical benefit of 8.13 and 8.14 is minor since transducers with a cylindrical radiation in the far field are hardly available.[39] An approximately cylindrical radiation could be achieved with line arrays of loudspeakers.[40] But replacing each individual loudspeaker by a line array of speakers contradicts our goal to reduce the number of loudspeakers. Simply replacing cylindrically radiating speakers by conventional loudspeakers which have a spherical radiation function leads to errors in this wave field synthesis formulation due to the deviant amplitude decay.

The Huygens' principle states that a wave front can be considered as consisting of infinitesimally distanced elementary sources. An infinite planar arrangement of elementary point sources with a spherical radiation could (re-)construct a plane wave, since the amplitude decay which is owed to the 1/r-distance law is compensated by the contribution of the other sources. Imagining secondary line sources with a cylindrical radiation, linear arrangement of sources would be sufficient to create a planar wave front. In a linear arrangement of elementary point sources, the contribution of the sources from the second dimension is missing, resulting in an amplitude decay. Therefore, a "2.5D-operator" including a "far field approximation" which modifies the free-field Green's function to approximate a cylindrical propagation is used.[41] This changes the driving function to

$$D_{2.5\mathrm{D}}(\omega, \mathbf{Y}) = \sqrt{\frac{2\pi\,|\mathbf{Y} - \mathbf{X}_{\mathrm{ref}}|}{\imath k}}\, D(\omega, \mathbf{Y}) \tag{8.19}$$

with $\mathbf{X}_{\mathrm{ref}}$ being a reference point in the source-free volume. This yields the "2.5-Dimensional" Rayleigh integral[42]:

$$P(\omega, \mathbf{X}) = -\int_{-\infty}^{\infty} D_{2.5D}(\omega, \mathbf{Y})\, G(\omega, \Delta\mathbf{r}) \tag{8.20}$$

Taking reference points $\mathbf{X}_{\mathrm{ref}}$ parallel to the loudspeaker array, the wave field can be synthesized correctly along a reference line. Between the speakers and the reference line, the sound pressures are too high, behind it they are too low.

Until now, free-field conditions are assumed. However, if not installed in the free field, reflections may occur and superimpose with the intended wave field created by the loudspeaker system. Under the term "listening room compensation" a variety of methods are proposed to reduce the influence of reflections. The simplest form is passive listening room compensation which means that the room is heavily damped. This is an approved method, applied e.g. in cinemas. However, for some listening rooms, for example living rooms, damping is impractical. Therefore, active solutions are proposed, like adding a filtering function which eliminates the first reflections of

[39]Cf. Spors and Ahrens (2008), p. 6 and Goertz (2008), p. 444.

[40]As often applied in PA systems for concerts, see e.g. Friedrich (2008), pp. 316ff.

[41]See e.g. Spors et al. (2008), pp. 9f or Wittek (2007), p. 58.

[42]See Spors et al. (2008), p. 11, Baalman (2008), pp. 28–46 and Verheijen (1997), pp. 37–49 and pp. 153–156. The derivation of the 2.5D-operator is given in Ahrens (2012), pp. 288f.

the room to the calculated loudspeaker signals.[43] "Adaptive wave field synthesis"[44] uses error sensors which measure errors occurring during WFS of a test stimulus emerging e.g. from reflections. Then any WFS solution is modified by a regularization factor which minimizes the squared error. This is of course a vicious circle since compensation signals corrupt the synthesized wave field and are reflected, too, adding further errors. This problem is related to the error compensation of head-related audio systems. Due to an exponentially increasing reflection density it is hardly possible to account for all higher order reflections. Thus, the approach is limited to first order reflections.

8.3.2.2 Discretization

A discretization of the Rayleigh integrals adopts the continuous formulation to discrete secondary source positions:

$$P(\omega, \mathbf{X}) = \frac{1}{2\pi} \sum_{\mathbf{r}_Y=-\infty}^{\infty} \left(G(\omega, \Delta\mathbf{r}) \frac{\partial P(\omega, \mathbf{Y})}{\partial \mathbf{n}} \right) \Delta\mathbf{r}_Y \tag{8.21}$$

and

$$P(\omega, \mathbf{X}) = \frac{1}{2\pi} \sum_{\mathbf{r}_Y=-\infty}^{\infty} \left(P(\omega, \mathbf{Y}) \frac{\partial G(\omega, \Delta\mathbf{r})}{\partial \mathbf{n}} \right) \Delta\mathbf{r}_Y \tag{8.22}$$

Thereby the Nyquist–Shannon sampling theorem has to be regarded: The sampling frequency has to be at least twice the highest frequency of the signal to be presented for no aliasing to occur. The highest frequency to be represented error-free is the *critical frequency* or *aliasing frequency*. In this case the sampling frequency is spatial; the speaker distance $\Delta\mathbf{Y}$ has to be maximally half the distance of the largest presentable wavelength

$$f_{\text{max}} = \frac{c}{2\Delta\mathbf{Y}} \tag{8.23}$$

between the speakers. The spatial sampling of the secondary source distribution is a process of sampling and interpolation; the interpolator is given by the radiation characteristics of the loudspeakers.[45] For the trace wavelength between the speakers

$$\lambda_{\Delta\mathbf{Y}} = \lambda\,|\sin\alpha| \tag{8.24}$$

---

[43] See Horbach et al. (1999), Corteel and Nicol (2003), Spors et al. (2003, 2004, pp. 333–337, 2007b).

[44] See Gauthier and Berry (2007).

[45] See Spors (2008), p. 1. An adaption of WFS to the radiation characteristic of the loudspeakers is derived in de Vries (1996).

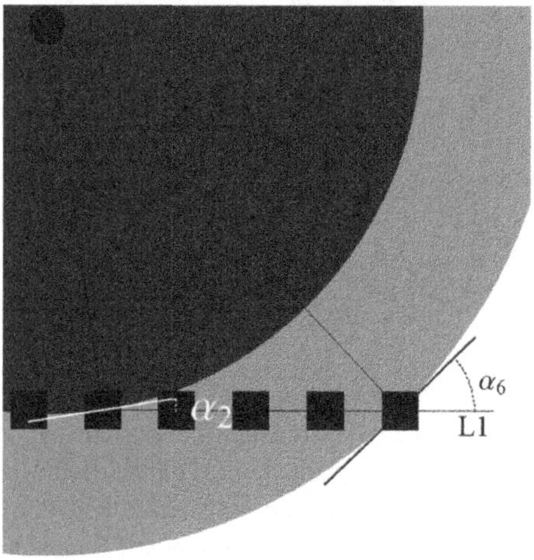

Fig. 8.11 Several incidence angles for one source position. From Ziemer (2016), p. 68

is valid, where α is the angle between the normal direction of a loudspeaker and the wave when striking this loudspeaker. Respectively, it can be considered as angle between separation line $L1$ and the tangent of the wave front when striking the speaker position. This leads to an adjustment of Eq. 8.23 to

$$f_{\max} = \frac{c}{2\Delta\mathbf{Y}\sin\alpha}. \tag{8.25}$$

The angle α may vary depending on position and radiation of the source in a range between $\frac{\pi}{2}$ and $\frac{3\pi}{2}$. Two examples for α are illustrated in Fig. 8.11. to clarify the coherency. The black disk represents the source, the dark and light gray disks the wave front at two different points in time, just as in Figs. 8.4 and 8.5 in Sect. 8.2.1.

Undersampling creates erroneous wavefronts above $f_{\max}$. These erroneous wavefronts contain the frequencies above the critical frequency, cause perceivable changes in sound color and disturb the localization of the virtual source.[46] Two examples of spatial aliasing are illustrated in Fig. 8.12. These illustrations contain an additional error due to the finite number of loudspeakers. It is called truncation error and will be discussed in detail in the subsequent subsection. Aliasing wave fronts create a spatial comb filter effects which colors stationary signals and smear transients. They can be heard as high-frequency echoes following the desired wave front. In the case of focused sources, they create high-frequency pre-echoes preceding the desired wave front. As long as the condition

$$|\sin\alpha(\omega)| < \frac{c}{2\Delta\mathbf{Y} f_{\max}} = \frac{\pi c}{\Delta\mathbf{Y}\,\omega_{\max}} \tag{8.26}$$

is satisfied no aliasing wavefronts will occur.

---

[46]See Spors et al. (2008), p. 14 and Daniel et al. (2003), p. 15.

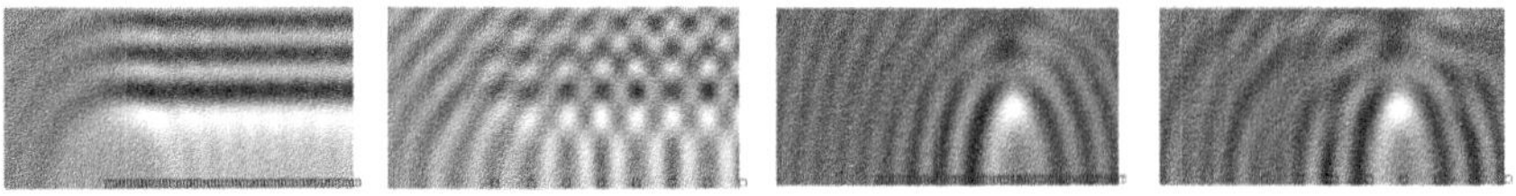

(a) Plane wave without aliasing. (b) Plane wave with aliasing. (c) Focused source without aliasing. (d) Focused source with aliasing.

Fig. 8.12 Virtual sources with (**b** and **d**) and without (**a** and **c**) aliasing. Erroneous wave fronts superimpose with the desired wave fronts. All synthesized wave fronts exhibit a truncation error which has to be compensated. After Ziemer (2016), p. 69

One suggestion to reduce artifacts is to process frequencies above the critical frequency not by means of WFS but by conventional stereophonic sound between two to three loudspeakers. This method is called "Optimized Phantom Source Imaging" (OPSI)[47] and combines WFS with conventional amplitude based panning. Thus, OPSI reintroduces psychoacoustic considerations to WFS. In that manner no aliasing echoes as such occur but the common disadvantages of stereophonic sound become effective: A comb filter effect arises, the display of depth becomes worse and high frequencies are only located correctly on the sweet spot. At other positions two to three wave fronts arrive slightly shifted in time. Also the radiation characteristic of the instrument cannot be displayed. López et al. (2005) suggest a related approach, called "sub band approach", playing frequencies above the aliasing frequency through the one loudspeaker with the most similar direction to the virtual source only.[48] This approach does not bring along the disadvantages of stereo but still a more or less correct localization is only possible in a small part of the listening area. Furthermore, the presentation of the complicated radiation characteristic of high frequencies is not possible. By randomizing the phase of the high frequencies of the loudspeakers artifacts are smeared.[49] An example is illustrated in Fig. 8.13. Due to aliasing, undesired spatially regular amplitude fluctuations occur. By phase randomization the distribution of errors becomes irregular. This reduces the sound coloration but the synthesized wave front still deviates strongly from the desired spherical or plane wave front and the localization accuracy is reduced strongly.[50] A reconstruction of the radiation characteristic of musical instruments is impossible with these methods.

In all three cases the signal is divided by the critical frequency into two frequency regions. For the lower frequency region the theory of WFS is applied. Frequencies in the higher region are not processed to recreate an original wave field but to obtain the natural temporal and spectral properties as well as an approximately correct source position. The methods are based on the same psychoacoustic considerations: Partials of a sound tend to fuse. Higher frequencies tend to be masked by lower frequencies. Altogether the audible portion of sound will be integrated into one auditory stream

[47]See Spors et al. (2008), p. 15, Wittek (2007), pp. 96–105, Reisinger (2002), pp. 42ff, Huber (2002), pp. 20–54.

[48]See López et al. (2005).

[49]See Spors et al. (2008), p. 17.

[50]See Wittek (2007), p. 88.

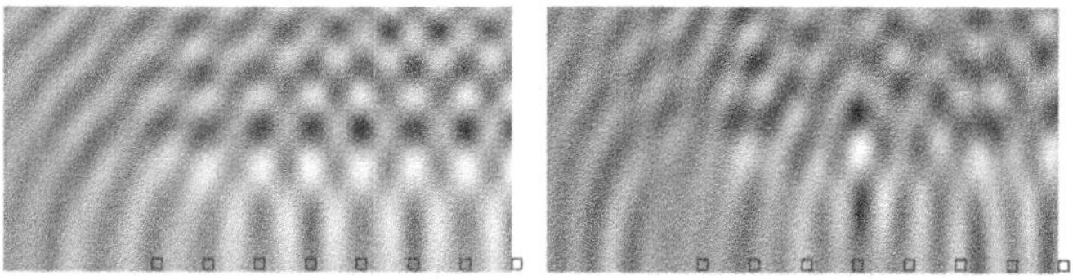

(a) Regular amplitude fluctuations due to spatial aliasing. *(b) Irregular aliasing error due to phase randomization.*

Fig. 8.13 Above the critical frequency, regular amplitude errors occur (**a**). By phase randomization (**b**) the amplitude and phase distribution becomes irregular. After Ziemer (2018), pp. 340 and 341

with one group source position. Then, the lower frequency region—which offers very precise localization cues due to the correct reconstruction of the wavefield—is crucial for a distinct and correct localization and the wrong localization cues of higher frequencies are neglected by the auditory system. All these psychoacoustic phenomenons have been illuminated already in Chap. 4, especially Sects. 4.3 and 4.5.

Of course, these methods work best if the chosen distance between adjacent speakers is so small that the aliasing frequency is as high as possible. Then it can even be speculated that the influence of the frequencies above the critical frequency is weak concerning sound coloration and localization. Spors et al. (2008) confirm this assumption:

> However, the human auditory system seems to be not too sensible to spatial aliasing if the loudspeaker spacing is chosen in the range $\Delta x = 10 \ldots 30$ cm.[51]

Quite a different method is to recreate the wave field not for the discrete loudspeaker positions but for discrete listening positions sampling the listening area. The approach is called "sound field reconstruction" or "sound field reproduction" applying least-squares solution.[52] Sampling positions are chosen under the assumption that if a wave field is reproduced correctly on a grid satisfying the Nyquist–Shannon sampling theorem, the wave field is correct everywhere inside the grid. This approach can be combined with crosstalk cancellation—as discussed in Chap. 7—to create a realistic binaural signal at discrete listening positions.[53]

---

[51] Spors et al. (2008), p. 17. Note that Spors et al. (2008) name the speaker positions "x", in this book they are called **Y**.

[52] Cf. Kolundzija et al. (2009b) and Kirkeby and Nelson (1993).

[53] Proposed and implemented by Menzel et al. (2006).

8.3.3 Spatial Border

A constraint of the discrete Rayleigh integrals, Eqs. 8.21 and 8.22, to a finite number of speaker positions is the 5th simplification of the list in Sect. 8.3.1. This creates two borders from which the created wave front curvatures fade to the wave front of the speaker itself. This effect is called "truncation".[54] It appears like diffraction through a gap and has the effect that the wave field cannot be synthesized in the area beyond the border. Furthermore, a more or less spherical wave front propagates from the border originated in the last speaker,[55] since the compensatory effect of adjacent speakers is missing. The truncation effect can be compensated by reducing the amplitudes of the outermost speakers. This does, however, slightly reduce the listening area extent. Figure 8.14 shows this artifact and its correction by applying a half cosine filter at the left end of the loudspeaker array. This gradual amplitude attenuation is referred to as *tapering*. It can be seen that, due to tapering, the amplitude of the virtual wavefront decays towards the outer positions in the listening area. The truncation error in generally weaker in corners, where two line array meet. An example is illustrated in Fig. 8.15. Compensation sources can compensate truncation by using speakers with antiphased signals at the array ends.

8.3.4 Listening Room

Until now, we assumed a free field and added loudspeaker arrays as secondary sources. So even in a highly damped room or outdoors, the assumption of a free field barely holds. This is especially true if we have actual listeners. Luckily, loudspeakers and listeners cause similar absorption, reflections and diffraction, no matter whether the impinging wave front is natural or synthesized. So the presence of loudspeakers and listeners does not seem to corrupt the wave front synthesis system. But this might change if we consider focused sources. Listeners between the focus point and the loudspeakers not only have trouble localizing the virtual focused source. They also corrupt the concave wave front and, as a consequence, the synthesized wave front is erroneous. This effect seems to be weak, as to the author's knowledge the effect is not addressed in the literature. Probably because the human body barely affects low frequencies. These deflect perfectly round the listener. However, higher frequencies may not deflect perfectly around the listener and create an undesired wave shadow. Even high frequencies are largely absorbed, but the wave fronts of high frequencies are erroneous, anyway, due to spatial aliasing. Furthermore, many wave front synthesis systems are installed slightly above the audience. Consequently, listeners are barely in the direct path between the loudspeakers and other listeners.

Much more critical is the presence of physical borders, i.e., room walls, floor and ceiling. Reflections from these surfaces superimpose with the desired wavefront. If

[54] See Start (1997), pp. 47ff, Verheijen (1997), pp. 50ff and Baalman (2008), pp. 37ff.

[55] See Spors et al. (2008), p. 14.

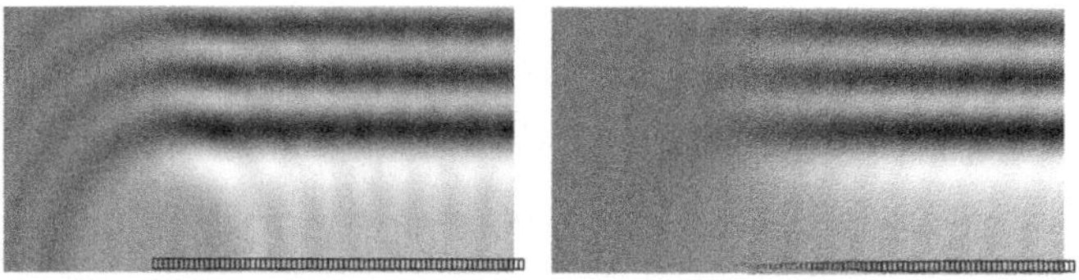

(a) Truncation effect in the case of a planewave.

(b) Reduction of the truncation effect by applying a cosine filter.

Fig. 8.14 Truncation effect of a virtual plane wave (**a**) and its compensation by applying a cosine filter (**b**). The spherical truncation wave emanating from the left end of the loudspeaker array is eliminated. The remaining error occurs from the untapered right end of the array. After Ziemer (2016), p. 71

Fig. 8.15 A virtual point source in the corner. When two linear loudspeaker arrays meet, the truncation error is weak. After Ziemer (2018), p. 343

the wavefronts of direct sound was synthesized correctly, the room acoustics would sound perfectly natural. But as discussed throughout this chapter, this is not the case. First of all, the sound field is typically synthesized in one plane only. And even in this plane, the amplitude decay is too strong, aliasing errors occur and at the ends of the loudspeaker array synthesis errors are produced, may it be due to truncation or due to tapering. Outside this plane, wave fronts are not controlled at all and deviate from natural wave fronts. It follows, that especially reflections from floor and ceiling are unnatural. Synthesizing not only direct sound but additional room acoustics is difficult. These would always superimpose with the reverberation of the listening room. An example of the same synthesized wave front in a free field, in presence of a highly reflective wall, and a highly absorbing wall is illustrated in Fig. 8.16.

Damping the listening room is probably the easiest way to avoid undesired reflections. The downside is that it makes wave field synthesis systems even less flexible. The high number of loudspeakers already affects the interior of the room and so the installation of additional absorbers may be difficult and undesired. Therefore, technical solutions have been developed. Reducing the undesired reverberation of the room by technical means is referred to as *active listening room compensation*.

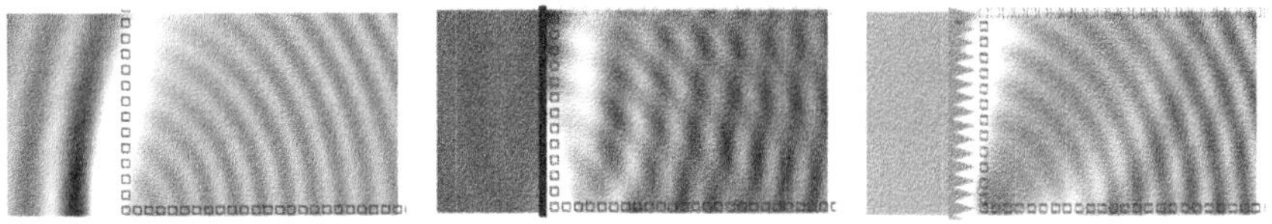

(a) Synthe sized wave front in a free field. *(b) Synthe sized wave front super imposed with wall reflection.* *(c) Synthe sized wave front with highly absorbing wall.*

Fig. 8.16 Wave field in a free field (**a**), in presence of a reflective wall (**b**) and highly absorbing wall (**c**). After Ziemer (2018), p. 343

Unfortunately, this can become a vicious circle, as in cross talk cancellation, discussed in Sect. 7.2.7. Details of active listening room compensation are out of scope of this book but can be found in the literature.[56]

8.4 Sound Field Synthesis and Radiation Characteristics

Radiation synthesis of musical sound is a research topic which receives growing interest.[57] Synthesizing the wave field as propagating from a source with a complex radiation pattern is demanding in many ways. One approach to recreate the natural sound radiation of musical instruments is to use an array of densely spaced loudspeakers. This approach is called "sound radiation synthesis".[58] The idea has been derived and applied by several researchers.[59] They use platonic solids, other regular polyhedrons or a spherical loudspeaker arrangement, as illustrated in Fig. 8.17, to create a high number of spherical harmonics. This approach could be considered as *spherical harmonic composition*, the inverse operation of spherical harmonic decomposition. In the literature it is sometimes referred to as "spatial additive synthesis".[60] The more loudspeakers are used the more complex radiation patterns can be synthesized. Assuming each loudspeaker to radiate as a monopole, the application is straightforward. However, implementing the actual radiation pattern of each loudspeaker and considering diffraction is challenging. An overview of methods is given in Zotter (2009).[61] From auralizations in virtual physical models and from impulse response measurements in actual rooms it has been found that the directivity of musi-

[56]The reader can refer e.g. to Spors et al. (2003, 2007a, b), Corteel and Nicol (2003).

[57]See Ahrens (2012), p. 13.

[58]See e.g. Ahrens (2012), p. 13.

[59]See e.g. Avizienis et al. (2006), Pollow and Behler (2009) and Kassakian and Wessel (2004), Ziemer (2009).

[60]See e.g. Warusfel and Misdariis (2004), p. 3.

[61]See Zotter (2009), pp. 111–152.

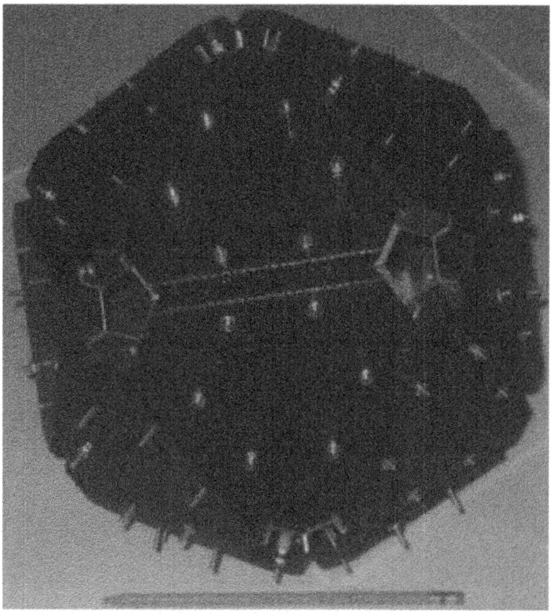

Fig. 8.17 120 loudspeakers mounted on the surface of a dodecahedron for matters of sound radiation synthesis. From Avizienis et al. (2006), with the permission of the Audio Engineering Society

cal instruments significantly affects the room response and leads to changes in the perceived naturalness and loudness.[62]

The high quantity and quality of research in the field of wave field synthesis led to market-ready loudspeaker systems which are able to create impressively realistic sounds with a distinct location of the source. But typically, virtual monopole sources or plane waves are created, which have small perceived dimensions.[63] There have been many attempts already to recreate the sound radiation characteristics of musical instruments via sound field synthesis. Menzel et al. (2006) proposed a WFS method to create binaural signals for a single listening position.[64] Baalman (2008) uses several monopole sources on the body of the virtual sound source to recreate its radiation patterns.[65] This approach is promising but the application is a compromise: A small number of monopole sources does not meet the complexity of many sound sources. A high number of monopole sources on the other hand may lead to an optimal recreation of the radiation characteristic but the computational costs are enormous, as already mentioned in Sect. 5.3.3 about equivalent sources methods in microphone array measurements. However, in more than 70% of the cases subjects of listening tests reported a higher "naturalness" for sources with complex radiation

[62]As already mentioned in Sect. 6.1. See also Martín et al. (2007), p. 395, Otondo and Rindel (2004), p. 1183.

[63]See e.g. Ahrens (2012), p. 198ff.

[64]See Menzel et al. (2006).

[65]See Baalman (2008), p. 97ff.

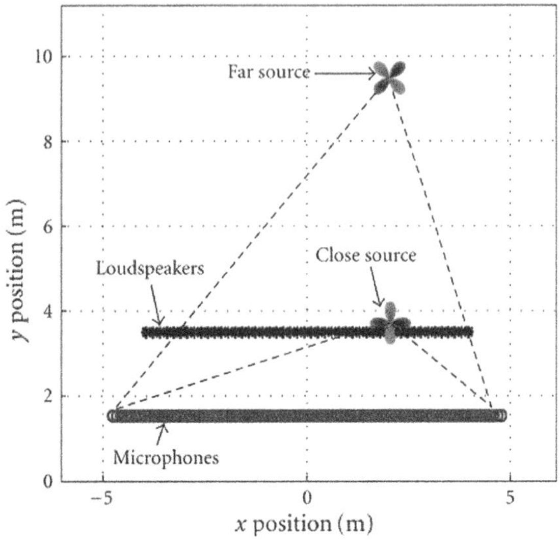

Fig. 8.18 Setup for simulation and actual implementation of synthesizing a complex radiation pattern using wave field synthesis. From Corteel (2007), p. 4, provided under Creative Commons License

patterns compared to virtual monopoles.[66] Corteel (2007) uses a combination of spherical harmonics to calculate driving functions for WFS.[67] He created single and combined spherical harmonics with functions of degree −2 to 2 using closed cabinet loudspeakers and multi-actuator panels (MAPs). A setup for his simulation and actual implementation is illustrated in Fig. 8.18. Although artifacts appear, increasing with decreasing source distance and increasing radiation complexity, he found that such complex source radiations created natural variations while wandering through the listening area as well as an increased perceived "presence" compared to monopole sources.[68] However, no elaborate listening tests have been performed. Böhlke (2016) decomposed the radiation characteristics of a violin to circular harmonics of order 64 and synthesized the radiated sound by means of 128 densely spaced virtual monopole sources in a wave front synthesis system. Listeners observed that the sound field varies when walking through the room, and the source sounds wider and less localizable.[69]

In Ahrens (2012) two formulations to create virtual sound sources with complex radiation patterns via wave front synthesis are given: A finite line sources which is divided into sections vibrating with alternating algebraic sign and spheres vibrating in higher modes. However, he states that these approaches are intermediate steps to a solution to the problem of unnatural spatial radiation characteristics of virtual sources. Firstly, because the computational costs are enormous. And secondly, to him, proper knowledge about the parameters that cause the perception of a certain

[66] See Baalman (2008), p. 19.

[67] See Corteel (2007).

[68] See Corteel (2007), p. 15.

[69] The complete work can be found in Böhlke (2016), a compact version in Böhlke and Ziemer (2017b), and an abstract in Böhlke and Ziemer (2017a).

source extent is needed to enable us to create a sound field that creates the desired spatial impression psychoacoustically, even if the physical wave field is different from a natural wave field emitted by a musical instrument.[70]

As discussed thoroughly in Chap. 5, actual musical instruments may radiate their sound from several vibrating surfaces and through multiple openings. This way wave fronts interfere and create the complicated patterns that make the sound broad and vivid. The radiation characteristics result from the extent of the body. Therefore, it may seem paradox to simplify musical instruments as point sources, especially if the sound radiation characteristics are to be measured, analyzed and synthesized. It is not physically correct but mathematically simple to simplify a musical instrument as a point. Such a point source has a singularity at its origin. From there on the sound wave propagates as a monopole. However, it is possible to define a direction-dependent function that describes a modification of amplitude and phase for each direction. Then, this wave front travels spherically, like a monopole. But this wave front is not necessarily an isobar. Amplitude and phase may vary over the spherical wavefront. Simplifying sound sources and propagation this way is referred to as *complex point source model*.[71] It could be shown that propagating a source sound of musical instruments by means of Eq. 9.1 yields a plausible sound field. The interaural phase- and level differences of a virtual listeners decrease as his distance to the source increases. When applying the complex point source model, the actual source extent of musical instruments could be fairly predicted from propagated sound field quantities. Based on the complex point source model sound radiation characteristics could be measured and synthesized for discrete listening points in space. This can give listeners the impression that the sound radiation characteristics are kept in the loudspeaker playback. The approach has been implemented in an octahedron-shaped loudspeaker array,[72] shown in Fig. 8.19.

8.5 Existing Sound Field Synthesis Installations

Sound field synthesis systems are still at a stage of research and development. Systems are installed in universities and in research and development departments of companies. But in addition to that, several systems are already in practical use. They serve for immersive audio in the entertainment sector, like cinemas, theaters, clubs,

[70]See Ahrens (2012), pp. 198ff.

[71]More details on the complex point source model and investigations on the relationship between musical instruments and the calculated sound field can be found in Ziemer (2014, 2015a, 2017a), Ziemer and Bader (2015a).

[72]The approach is described in Ziemer (2009, 2011a). Demonstrations have been given on the Second International Conference of Students of Systematic Musicology 2009 (SysMus09) conference in Ghent, Belgium.

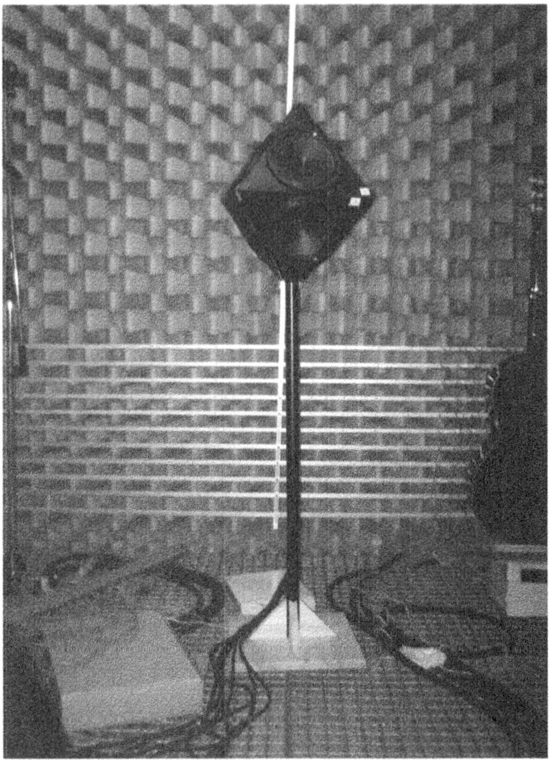

Fig. 8.19 Octahedron-shaped loudspeaker array to synthesize the sound radiation characteristics of musical instruments at 8 discrete locations. From Ziemer (2016), p. 155

and themed environments, or for communication.[73] Some exemplary sound field synthesis installations are presented in the following.

In the field of research, many wave field synthesis systems contain 24 or 32 loudspeakers in front of or around a listening area, which is large enough for about 5 listeners.[74] One example is the installation at the University of Sherbrooke, shown in Fig. 8.20.

A polygonic loudspeaker setup for wave front synthesis and for an alternative approach with a drastically reduced number of loudspeakers had been installed at the Fraunhofer IDMT institute in Ilmenau. Figure 8.21 shows the wave front synthesis layer with densely spaced loudspeakers and the alternative approach with the fourfold spacing of speakers.

[73]Most of all the wave field synthesis system of the Technical University Berlin in cooperation with Deutsche Telekom Laboratories, or IOSONO systems, the wave field synthesis system of Fraunhofer IDMT. Further information on installed wave field synthesis systems can be found e.g. in Baalman (2008), pp. 47ff, Montag (2011), Chaps. 5 and 6, Slavik and Weinzierl (2008), pp. 656f and 664ff and IOSONO GmbH (2008).

[74]See e.g. Gauthier and Berry (2008), p. 1994, Spors et al. (2003), Ahrens et al. (2010), p. 3, Reisinger (2002), pp. 37–39, Reisinger (2003), pp. 40–44, Vogel (1993), pp. 139f, Baalman (2008), p. 48 and Verheijen (1997), p. 103.

Fig. 8.20 Circular wave field synthesis setup for research. Reproduced from Gauthier and Berry (2008, p. 1994) with the permission of the Acoustical Society of America

Fig. 8.21 Wave field synthesis setup for research and development at Fraunhofer IDMT

Fig. 8.22 Psychoacoustic Sound Field Synthesis System at the University of Hamburg. From Ziemer (2016), p. 157

A psychoacoustic sound field synthesis system for music has been developed and tested at the Institute of Systematic Musicology of the University of Hamburg[75] and will be introduced in detail in the subsequent chapter, Chap. 9. It consists of 15 loudspeakers synthesizing a desired sound field in a listening area of around 1 m^2. Just as in many ambisonics systems, the spacing of 0.65 m between the loudspeakers is rather large. In contrast to wave front synthesis and some ambisonics approaches, every loudspeaker is active for every virtual source position. Perceptual mechanisms of the auditory system are considered in the sound field synthesis approach so that a precise localization and a natural and spatial sound impression are created despite physical synthesis errors. Even beyond the listening area, the localization is rather precise. Figure 8.22 is a photo of the installed system.

A full-duplex wave field synthesis system for communication is being developed at the Nippon Telegraph and Telephone (NTT) lab in Tokyo.[76] An individual combination of a loudspeaker- and a microphone array is installed in two separate rooms. In a conference phone call, several subjects can talk on both sides and even move, while all listeners on the other side can localize the speakers well. Of course, the proximity of microphones to loudspeakers on both sides of the line can cause serious problems. So this system focuses on echo-cancellation to suppress feedback loops. Another important topic is real-time implementation on a singe PC, including the signal processing for the microphone array, the wave field synthesis rendering, and

[75]Its developmental progress can be followed by referring to Ziemer (2009, 2011a, b, c, d, 2014, 2015a, b, 2016, 2017a, b, c, 2018), Ziemer and Bader (2015a, b, c, d, 2017).

[76]Details can be found e.g. in Emura and Kurihara (2015).

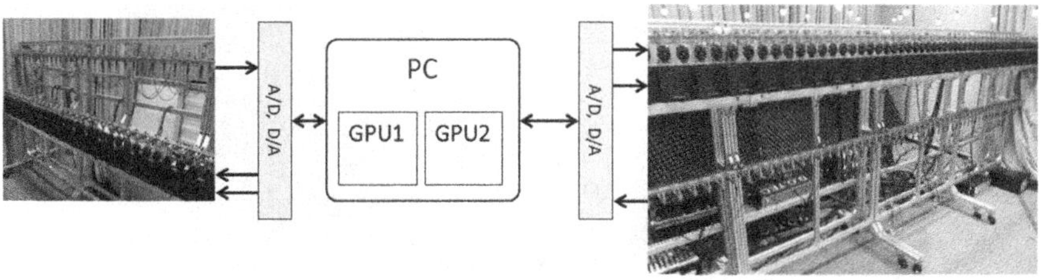

Fig. 8.23 Full duplex wave field synthesis system for communication. From Emura and Kurihara (2015), with the permission of the Audio Engineering Society

the echo-cancellation. This is achieved by fast rendering on two GPUs. The two systems are shown in Fig. 8.23.

The wave front synthesis system at the University of Applied Sciences Hamburg is coupled to a motion capture system.[77] This way, focused sources can be created in such way that they are always between one tracked individual and the loudspeaker array. On the one hand, this brings back the sweet-spot limitation of conventional spatial audio. But on the other hand, a tracked individual can now walk around a virtual source or be surrounded completely by a moving focused source. This offers a new degree of user interaction, which is beneficial, e.g. for virtual reality applications. Furthermore, listeners can control virtual source locations by trackers in their hands. Figure 8.24 is a photo of this system. The wave field synthesis system is linked to a head-mounted display for graphical, three-dimensional virtual reality. With this powerful combination, the WFS system of the University of Applied Sciences Hamburg is used to investigate the potential and limits of *redirected walking*.[78] Here, the translation and/or rotation of a subject can be under- or overemphasized in the virtual auditory and visusal scene. This creates the illusion that the subjects walk paths that exceed the actual physical room.

A wave field synthesis system for research and public events can be found in the auditorium of the Berlin University of Technology,[79] illustrated in Fig. 8.25.

The wave field synthesis system developed in Berlin is also installed at the University of Music and Drama Hamburg. It is in use for both concerts and research, especially in the field of network music performance. The installed system can be seen in Fig. 8.26. The mobile system is transported to event venues like Kampnagel center for performing arts for demonstrations and concerts.

One WFS system containing 832 loudspeakers delivers an immersive sound experience at the Seebühne Bregenz,[80] illustrated in Fig. 8.27. In contrast to conventional PA systems with delay lines, a wave front synthesis system does not create echoes from the rear. These can be annoying to the audience, reduce speech intelligibility and may create conflicting source localization cues. Furthermore, the amplitude

[77] See e.g. Fohl and Nogalski (2013), Fohl (2013), Fohl and Wilk (2015) for details.

[78] See e.g. Nogalski and Fohl (2015, 2016, 2017), Meyer et al. (2016) for details on the approaches.

[79] Some details can be found in Baalman (2008), Chap. 3 and Slavik and Weinzierl (2008), p. 670.

[80] See Slavik and Weinzierl (2008), p. 656.

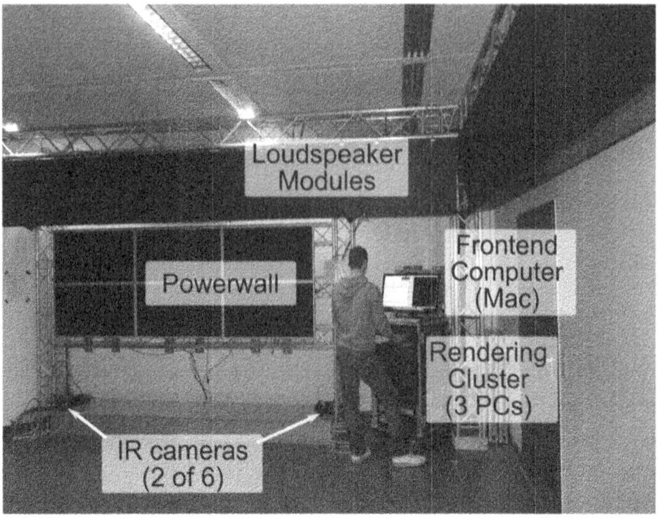

Fig. 8.24 Wave Field Synthesis System at the University of Applied Sciences Hamburg coupled to motion capture technology. Original photo by Wolfgang Fohl, provided under Creative Commons License. The photo is converted to grayscale

Fig. 8.25 Panoramic picture of the WFS loudspeaker system in an auditorium of Berlin University of Technology containing 832 channels and more than 2700 loudspeakers. Pressestelle TU Berlin, with friendly permission by Stefan Weinzierl

decay of line arrays lowers with increasing length. PA systems tend to be too loud in proximity to the loudspeakers. This is necessary to ensure that the sound pressure level is still high enough at the rear seats despite the large amplitude decay over distance. Wave front synthesis systems can create a lower amplitude decay and therefore have a high potential as a PA-alternative for stages with a large audience. Wave field synthesis systems have another advantage for theater. Most large theaters

Fig. 8.26 Wave field synthesis system for music installations and networked music performance at the University of Music and Theater Hamburg

Fig. 8.27 Photo of the WFS loudspeaker system at the Seebühne Bregenz. The speakers are arranged beside and behind the audience. From Slavik and Weinzierl (2008), p. 656

and open air locations for drama create an irritating ear/eye-conflict: The actor is speaking somewhere on the stage but his or her voice will be localized at one of the few PA loudspeaker towers. This can be very confusing in a scene with many actors. A frontal, horizontal WFS line array can give additional localization cues so that the auditory event better fits the visual scene.

In early 2009, a 189-channel wave field synthesis system had been installed at the Casa del Suono in Parma, Italy. This museum is dedicated to the history of audio technology. The loudspeakers are installed behind curtains in a room that is acoustically treated. Visitors can experience the sound without seeing the actual loudspeaker system.[81] Another WFS system had been installed in the Tresor club in Berlin; a famous techno club. Here, the conditions for a wave field synthesis setup are challenging: Standing waves emerge between the solid concrete floor and ceiling. Furthermore, the ceiling is so low above the loudspeaker array that very early reflections superimpose with the desired wave field. Furthermore, the techno music has to be produced specifically for the system so that single tracks can receive their individual virtual source locations or paths. On the other hand, wave field synthesis offers new possibilities for spatial mixing. At the moment, conventional stereophonic dance music tends to make only little use of hard panning.[82] The reason for that is simply that the constellation of loudspeakers and listeners in night clubs is typically far away from the ideal stereo triangle discussed in Sect. 7.2.2. Loudspeakers may be positioned far apart, so a sound hard-panned to either speaker might be inaudible over a large area of the dancefloor. In many night clubs the two stereo channels are mixed together to get rid of hard panning and incoherent loudspeaker signals. Hence, producers of electronic dance music have an eye on mono compatibility. Wave field synthesis systems in night clubs would give music producers and disc jockeys the opportunity to use space as creative and dramaturgical tool instead of trying to stay mono-compatible.

Automotives are a real challenge for both conventional stereophonic audio and wave front synthesis systems. Due to the five seats there is a distribution of listeners, so at least several sweet spots are desired. The exact location of the driver's and passengers' heads may be unknown and may neither be in plane nor static. So sweet spots may even be insufficient as they do not account for head- and torso movement. There is limited space inside a car, so it is not easy to install line arrays of loudspeakers or to surround the interior completely with loudspeakers in one height. Curved loudspeaker arrays on the other hand are challenging in terms of computational effort and synthesis error compensation. Standing waves occur due to the rather small dimensions of cars compared to audible wavelengths. Seats are obstacles that create absorption, deflection and reflections. Their positions are readjusted to the individual needs of the driver and the passengers, so it is challenging to include them in a sound field synthesis calculation. Despite these issues, Audi decided to install Fraunhofer IDMTs wave front synthesis system in the Q7. The system is depicted in Fig. 8.28.

Another commercial system for TV-soundbars is developed and distributed by Sonic Emotion. It promises an enlargement of the sweet spot to a sweet area and an extension of the loudspeaker base by synthesizing plane waves. Figure 8.29 shows a sound bar including multiple speakers that can be used to synthesize wavefronts.[83]

---

[81] A detailed description and a photo can be found in Adriaensen (2010).

[82] See e.g. Owsinski (2014), p. 51 and Stirnat and Ziemer (2017).

[83] More information and advertisements can be found in on the Sonic Emotion website Sonic Emotion (2017) and on their Youtube-channel Sonic Emotion (2012).

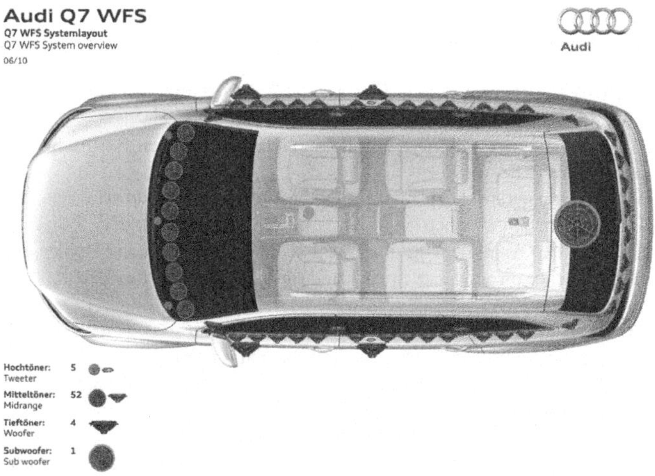

Fig. 8.28 Wave front synthesis installation in a car. Photo from Audi Technology Portal (2011), © Audi

Fig. 8.29 Synthesizing plane waves with multiple loudspeakers in a sound bar enlarges the sweet spot for stereo source signals

Most of these sound field systems aim at reconstructing the spatio-temporal properties of sound waves in terms of wave front synthesis. Psychoacoustic sound field synthesis as installed at the University of Hamburg is an alternative approach which takes auditory perception into account in its derivation. This approach is discussed in Chap. 9.

References

Adriaensen F (2010) The WFS system at La Casa del Suono, Parma. In: Linux audio conference, Utrecht, pp 39–45

Ahrens J (2012) Analytic methods of sound field synthesis. Springer, Berlin. https://doi.org/10.1007/978-3-642-25743-8

Ahrens J (2016) On the generation of virtual early reflections in wave field synthesis. In: Fortschritte der Akustik—DAGA 2016, Aachen

Ahrens J, Spors S (2008a) Analytical driving functions for higher order ambisonics. In: 2008 IEEE international conference on acoustics, speech and signal processing, Las Vegas, NV, pp 373–376. https://doi.org/10.1109/ICASSP.2008.4517624

Ahrens J, Spors S (2008b) Reproduction of moving virtual sound sources with special attention to the doppler effect. In: Audio engineering society convention 124

Ahrens J, Spors S (2009) Spatial encoding and decoding of focused virtual sound sources. In: Ambisonics symposium, Graz

Ahrens J, Geier M, Spors S (2010) Perceptual assessment of delay accuracy and loudspeaker misplacement in wave field synthesis. In: Audio engineering society convention 128

Albrecht B, de Vries D, Jacques R, Melchior F (2005) An approach for multichannel recording and reproduction of sound source directivity. In: Audio engineering society convention 119

Audi Technology Portal (2019) Sound systems. https://www.audi-technology-portal.de/en/electrics-electronics/multimedia_en/sound-systems. Accessed 5 Feb 2019

Avizienis R, Freed A, Kassakian P, Wessel D (2006) A compact 120 independent element spherical loudspeaker array with programable radiation patterns. In: Audio engineering society convention 120. http://www.aes.org/e-lib/browse.cfm?elib=13587

Baalman M (2008) On wave field synthesis and electro-acoustic music, with a particular focus on the reproduction of arbitrarily shaped sound sources. VDM, Saarbrücken

Bai MR, Chung C, Wu P-C, Chiang Y-H, Yang C-M (2017) Solution strategies for linear inverse problems in spatial audio signal processing. Appl Sci 7(6):582. https://doi.org/10.3390/app7060582

Berkhout AJ (1988) A holographic approach to acoustic control. J Audio Eng Soc 36(12):977–995. http://www.aes.org/e-lib/browse.cfm?elib=5117

Berkhout AJ, de Vries D, Vogel P (1992) Wave front synthesis: a new direction in electroacoustics. In: Audio engineering society convention 93, vol 10. https://doi.org/10.1121/1.404755

Berkhout AJ, de Vries D, Vogel P (1993) Acoustic control by wave field synthesis. J Acoust Soc Am 93(5):2764–2778. https://doi.org/10.1121/1.405852

Berkhout AJ, de Vries D, Sonke JJ (1997) Array technology for acoustic wave field analysis in enclosures. J Acoust Soc Am 105(5):2757–2770. https://doi.org/10.1121/1.420330

Böhlke L (2016) Sound radiation of the violin in a virtual acoustic environment

Böhlke L, Ziemer T (2017a) Perception of a virtual violin radiation in a wave field synthesis system. J Acoust Soc Am 141(5):3875. https://doi.org/10.1121/1.4988669

Böhlke L, Ziemer T (2017b) Perceptual evaluation of violin radiation characteristics in a wave field synthesis system. In: Proceedings of meetings on acoustics, vol 30, no 1, p 035001. https://doi.org/10.1121/2.0000524

Bleda S, Escolano J, López JJ, Pueo B (2005) An approach to discrete-time modelling auralization for wave field synthesis applications. In: Audio engineering society convention 118. http://www.aes.org/e-lib/browse.cfm?elib=13141

Boone MM, Horbach U, and de Bruijn WPJ (1999) Virtual surround speakers with wave field synthesis. In: Audio engineering society convention 106, Munich. http://www.aes.org/e-lib/browse.cfm?elib=8252

Brix S, Sporer T, Plogsties J (2001) CARROUSO–a European approach to 3D audio (abstract). In: Audio engineering society convention 110, p 528

Burns TH (1992) Sound radiation analysis of loudspeaker systems using the nearfield acoustic holography (NAH) and the application visualization system (AVS). In: Audio engineering society convention 93

Cho W-H, Ih J-G, Boone MM (2010) Holographic design of a source array achieving a desired sound field. J Audio Eng Soc 58(4):282–298. http://www.aes.org/e-lib/browse.cfm?elib=14607

Corteel E (2007) Synthesis of directional sources using wave field synthesis, possibilities, and limitations. EURASIP J Adv Signal Process Article ID 90509. https://doi.org/10.1155/2007/90509

Corteel E, Nicol R (2003) Listening room compensation for wave field synthesis. What can be done? In: Audio engineering society conference: 23rd international conference: signal processing in audio recording and reproduction, Copenhagen

Daniel J (2003) Spatial sound encoding including near field effect: introducing distance coding filters and a viable, new ambisonic format. In: Audio engineering society conference: 23rd international conference: signal processing in audio recording and reproduction, Copenhagen

Daniel J, Nicol R, Moreau S (2003) Further investigations of high order ambisonics and wavefield synthesis for holophonic sound imaging. In: Audio engineering society convention 114

de Vries D (1996) Sound reinforcement by wavefield synthesis: adaption of the synthesis operator to the loudspeaker directivity characteristics. J Audio Eng Soc 44(12):1120–1131. http://www.aes.org/e-lib/browse.cfm?elib=7872

de Vries D, Start EW, Valster VG (1994) The wave field synthesis concept applied to sound reinforcement restrictions and solutions. In: Audio engineering society convention 96, Amsterdam

Elen R (2001) Ambisonics. the surround alternative. http://www.ambisonic.net/pdf/ambidvd2001.pdf. Accessed 22 Nov 2010

Emura S, Kurihara S (2015) Echo canceler for real-time audio communication with wave field reconstruction. In: Audio engineering society convention 139, New York, NY. http://www.aes.org/e-lib/browse.cfm?elib=17984

Fohl W (2013) The wave field synthesis lab at the HAW Hamburg. In: Bader R (ed) Sound-Perception-Performance. Springer, pp 243–255. https://doi.org/10.1007/978-3-319-00107-4_10

Fohl W, Nogalski M (2013) A gesture control interface for a wave field synthesis system. In: Proceedings of international conference on new interfaces for musical expression, Daejeon + Seoul/Republic of Korea, pp 341–346

Fohl W, Wilk E (2015) Enhancements to a wave field synthesis system to create an interactive immersive audio environment. In: Proceedings of international conference on spatial audio, VDT

Friedrich HJ (2008) Tontechnik für Mediengestalter. Töne hören—Technik verstehen—Medien gestalten. Springer, Berlin

Friesecke A (2007) Die Audio-Enzyklopädie. Ein Nachschlagewerk für Tontechniker. K.G. Saur, Munich

Gauthier P-A, Berry A (2007) Adaptive wave field synthesis for sound field reproduction: theory, experiments, and future perspectives. In: Audio engineering society convention 123

Gauthier P-A, Berry A (2008) Adaptive wave field synthesis for active sound field reproduction: experimental results. J Acoust Soc Am 123(4):1991–2002. https://doi.org/10.1121/1.2875844

Geier M, Wierstorf H, Ahrens J, Wechsung I, Raake A, Spors S (2010) Perceptual evaluation of focused sources in wave field synthesis. In: Audio engineering society convention 128

Gerzon M (1981) Sound reproduction systems. Patent GB 8100018

Gerzon MA (1973) Periphony: with-height sound reproduction. J Audio Eng Soc 21(1):2–10. http://www.aes.org/e-lib/browse.cfm?elib=2012

Gerzon MA (1975) The design of precisely coincident microphone arrays for stereo and surround sound. In: Audio engineering society convention 50, London

Goertz A, Lautsprecher, Weinzierl S (eds) (2008) Handbuch der Audiotechnik. Springer, Berlin, pp 421–490. https://doi.org/10.1007/978-3-540-34301-1_8. (Chap. 8)

Grani F, Di Carlo D, Portillo JM, Girardi M, Paisa R, Banas JS, Vogiatzoglou I, Overholt D, Serafin S (2016) Gestural control of wave field synthesis. In: Proceedings of 13th sound and music computing conference, Hamburg

Hahn N, Winter F, Spors S (2016) Local wave field synthesis by spatial band-limitation in the circular/spherical harmonics domain. In: Audio engineering society convention 140. http://www.aes.org/e-lib/browse.cfm?elib=18294
Heller AJ (2008) Is my decoder ambisonic? In: Audio engineering society convention 125, San Francisco, CA
Horbach U, Karamustafaoglu A, Rabenstein R, Runze G, Steffen P (1999) Numerical simulation of wave fields created by loudspeaker arrays. In: Audio engineering society convention 107. http://www.aes.org/e-lib/browse.cfm?elib=8159
Huber T (2002) Zur Lokalisation akustischer Objekte bei Wellenfeldsynthese. Diloma thesis. http://www.hauptmikrofon.de/diplom/DA_Huber.pdf
IOSONO GmbH (2008) IOSONO—The future of spatial audio. http://www.iosono-sound.com/. Accessed 23 Jan 2011
Kassakian P, Wessel D (2004) Characterization of spherical loudspeaker arrays. In: Audio engineering society convention 117, San Francisco
Kim Y, Ko S, Choi J-W, Kim J (2009) Optimal filtering for focused sound field reproductions using a loudspeaker array. In: Audio engineering society convention 126
Kirkeby O, Nelson PA (1993) Reproduction of plane wave sound fields. J Acoust Soc Am 94:2992–3000. https://doi.org/10.1121/1.407330
Kolundzija M, Faller C, Vetterli M (2009a) Designing practical filters for sound field reconstruction. In: Audio engineering society convention 127
Kolundzija M, Faller C, Vetterli M (2009b) Sound field reconstruction: an improved approach for wave field synthesis. In: Audio engineering society convention 126
López JJ, Bleda S, Pueo B, Escolano J (2005) A sub-band approach to wave-field synthesis rendering. In: Audio engineering society convention 118, Barcelona. https://www.ingentaconnect.com/content/dav/aaua/2006/00000092/00000004/art00013
Martín RS, Witew IB, Arana M, Vorländer M (2007) Influence of the source orientation on the measurement of acoustic parameters. Acta Acust United Acust. 93:387–397. https://www.ingentaconnect.com/contentone/dav/aaua/2007/00000093/00000003/art00007
Melchior F (2010) Wave field synthesis and object-based mixing for motion picture sound. SMPTE Motion Imaging J 3:53–57. https://doi.org/10.5594/j11399
Menzel D, Wittek H, Fastl H, Theile G (2006) Binaurale Raumsyntese mittels Wellenfeldsynthese—Realisierung und Evaluierung. In: Fortschritte der Akustik—DAGA 2006, Braunschweig, pp 255–256
Menzies D (2013) Quasi wave field synthesis: efficient driving functions for improved 2.5D sound field reproduction. In: Audio engineering society conference: 52nd international conference: sound field control-engineering and perception. http://www.aes.org/e-lib/browse.cfm?elib=16930
Menzies D, Al-Akaidi M (2007) Nearfield binaural synthesis and ambisonics. J Acoust Soc Am 121(3):1559–1563. https://doi.org/10.1121/1.2434761
Merziger G, Wirth T (2006) Repetitorium der höheren Mathematik, 5th edn. Binomi, Springe
Meyer F, Nogalski M, Fohl W (2016) Detection thresholds in audio-visual redirected walking. In: Proceedings of 13th sound and music computing conference, SMC
Montag MN (2011) Wave field synthesis in three dimensions by multiple line arrays. Master's thesis. http://www.mattmontag.com/projects/wfs/Montag%20Thesis%202011%20-%20Wave%20Field%20Synthesis%20in%20Three%20Dimensions%20by%20Multiple%20Line%20Arrays.pdf
Morse PM, Ingard KU (1986) Theoretical acoustics. Princeton University Press, Princeton. https://doi.org/10.1063/1.3035602
Nogalski M, Fohl W (2015) Acoustically guided redirected walking in a WFS system: design of an experiment to identify detection thresholds. In: Proceedings of 12th sound and music computing conference, SMC
Nogalski M, Fohl W (2016) Acoustic redirected walking with auditory cues by means of wave field synthesis. In: Proceedings of 23rd IEEE conference on virtual reality. IEEE

Nogalski M, Fohl W (2017) Curvature gains in redirected walking: a closer look. In: Proceedings of 24th IEEE conference on virtual reality. IEEE

Oellers H (2010) Die virtuelle kopie des räumlichen schallfeldes. http://www.syntheticwave.de/. Accessed 27 Sept 2010

Otondo F, Rindel JH (2004) The influence of the directivity of musical instrument in a room. Acta Acust United Acust 90:1178–1184. https://www.ingentaconnect.com/content/dav/aaua/2004/00000090/00000006/art00017

Owsinski B (2014) The mixing engineer's handbook, 3rd edn. Corse Technology PTR, Boston, MA

Pierce AD (2007) Basic linear acoustics. In: Rossing TD (ed) Springer handbook of acoustics. Springer, New York, pp 25–111. https://doi.org/10.1007/978-0-387-30425-0_3. (Chap. 3)

Pollow M, Behler GK (2009) Variable directivity for platonic sound sources based in shperical harmonics optimization. Acta Acust United Acust 95:1082–1092. https://doi.org/10.3813/aaa.918240

Rabenstein R, Spors S (2008) Sound field reproduction. In: Benesty J, Sondhi MM, Huang Y (eds) Springer handbook of speech processing. Springer, Berlin, pp 1095–1114. https://doi.org/10.1007/978-3-540-49127-9_53. (Chap. 53)

Rabenstein R, Spors S, Steffen P (2006) Wave field synthesis techniques for spatial sound reproduction. In: Hänsler E, Schmidt G (eds) Topics in acoustic echo and noise control. Selected methods for the cancellation of acoustical echoes, the reduction of background noise, and speech processing. Signals and communication technology. Springer, Berlin, pp 517–545. (Chap. 13)

Reisinger G (2003) Einsatz von stereophonen Aufnahmetechniken für die räumliche Übertragung ausgedehnter Schallquellen mit Hilfe der Wellenfeldsynthese. Diploma thesis, University of Applied Sciences Düsseldorf, Düsseldorf

Reisinger M (2002) Neue Konzepte der Tondarstellung bei Wiedergabe mittels Wellenfeldsynthese. Diploma thesis, University of Applied Sciences Düsseldorf, Düsseldorf

Slavik KM, Weinzierl S (2008) Wiedergabeverfahren. In: Weinzierl S (ed) Handbuch der Audiotechnik. Springer, Berlin, pp 609–686. https://doi.org/10.1007/978-3-540-34301-1_11. (Chap. 11)

Sonic Emotion (2012) Sonic emotion absolute 3D sound in a nutshell/stereo VS WFS. https://www.youtube.com/user/sonicemotion3D/videos

Sonic Emotion (2017) Sonic emotion absolute 3D. https://www.youtube.com/user/sonicemotion3D/videos

Spors S, Kuntz A, Rabenstein R (2003) An approach to listening room compensation with wave field synthesis. In: Audio engineering society conference: 24th international conference: multichannel audio, the new reality

Spors S, Helwani K, Ahrens J (2011) Local sound field synthesis by virtual acoustic scattering and time reversal. In: Audio engineering society convention 131

Spors S (2007) Extension of an analytic secondary source selection criterion for wave field synthesis. In: Audio engineering society convention 123

Spors S (2008) Investigation of spatial aliasing artifacts of wave field synthesis in the temporal domain. In: Fortschritte der Akustik—DAGA 2008, Dresden

Spors S, Ahrens J (2008) A comparison of wave field synthesis and higher-order ambisonics with respect to physical properties and spatial sampling. In: Audio engineering society convention 125

Spors S, Teutsch H, Kuntz A, Rabenstein R (2004) Sound field synthesis. In: Huang Y, Benesty J (eds) Audio signal processing. For next-generation multimedia communication systems. Springer, New York, pp 323–344. https://doi.org/10.1007/1-4020-7769-6_12. (Chap. 12)

Spors S, Buchner H, Rabenstein R, Herbordt W (2007a) Active listening room compensation for massive multichannel sound reproduction systems using wave-domain adaptive filtering. J Acoust Soc Am 122(1):354–369. https://doi.org/10.1121/1.2737669

Spors S, Buchner H, Rabenstein R, Herbordt W (2007b) Active listening room compensation for massive multichannel sound reproduction systems using wave-domain adaptive filtering. J Acoust Soc Am 122(1):354–369. https://doi.org/10.1121/1.2737669

Spors S, Rabenstein R, Ahrens J (2008) The theory of wave field synthesis revisited. In: Audio engineering society convention 124

Spors S, Wierstorf H, Raake A, Melchior F, Frank M, Zotter F (2013) Spatial sound with loudspeakers and its perception: a review of the current state. Proc IEEE 101(9):1920–1938. https://doi.org/10.1109/JPROC.2013.2264784

Start EW (1997) Direct sound enhancement by wave field synthesis. PhD thesis, Delft University of Technology, Delft

Steinberg JC, Snow WB (1934a) Symposium on wire transmission of symphonic music and its reproduction in auditory perspective. physical factors. Bell Syst Tech J XIII

Steinberg JC, Snow WB (1934b) Auditory perspective–physical factors. Electr Eng 12–17

Stirnat C, Ziemer T (2017) Spaciousness in music: the toneister's intention and the listener's perception. In: Proceedings of the klingt gut! symposium, Hamburg

Vaananen R (2003) User interaction and authoring of 3D sound scenes in the Carrouso EU project. In: Audio engineering society convention 114. http://www.aes.org/e-lib/browse.cfm?elib=12483

Verheijen E (1997) Sound reproduction by wave field synthesis. PhD thesis, Delft University of Technology, Delft

Vogel P (1993) Applications of wave field synthesis in room acoustics. PhD thesis, Delft University of Technology, Delft

Warusfel O, Misdariis N (2004) Sound source radiation syntheses: from performance to domestic rendering. In: Audio engineering society convention 116

Wierstorf H (2014) Perceptual assessment of sound field synthesis. PhD thesis, University of Technology Berlin, Berlin

Wierstorf H, Raake A, Geier M, Spors S (2013) Perception of focused sources in wave field synthesis. J Audio Eng Soc 61(1/2):5–16. http://www.aes.org/e-lib/browse.cfm?elib=16663

Williams EG (1999) Fourier acoustics. Sound radiation and nearfield acoustical holography. Academic Press, Cambridge

Wittek H (2007) Perceptual differences between wavefield synthesis and stereophony. PhD thesis, University of Surrey, Guilford

Ziemer T (2009) Wave field synthesis by an octupole speaker system. In: Naveda L (ed) Proceedings of the second international conference of students of systematic musicology (SysMus09), pp 89–93. http://biblio.ugent.be/publication/823807/file/6824513.pdf#page=90

Ziemer T (2011a) Wave field synthesis. Theory and application. Magister thesis, University of Hamburg

Ziemer T (2011b) A psychoacoustic approach to wave field synthesis. In: Audio engineering society conference: 42nd international conference: semantic audio, Ilmenau, pp 191–197. http://www.aes.org/e-lib/browse.cfm?elib=15942

Ziemer T (2011c) Psychoacoustic effects in wave field synthesis applications. In: Schneider A, von Ruschkowski A (eds) Systematic musicology. Empirical and theoretical studies. Peter Lang, Frankfurt am Main, pp 153–162. https://doi.org/10.3726/978-3-653-01290-3

Ziemer T (2011d) A psychoacoustic approach to wave field synthesis. J Audio Eng Soc 59(5):356. https://www.aes.org./conferences/42/abstracts.cfm#TimZiemer

Ziemer T (2014) Sound radiation characteristics of a shakuhachi with different playing techniques. In: Proceedings of the international symposium on musical acoustics (ISMA-14), Le Mans, pp 549–555. http://www.conforg.fr/isma2014/cdrom/data/articles/000121.pdf

Ziemer T (2015a) Exploring physical parameters explaining the apparent source width of direct sound of musical instruments. In: Jahrestagung der Deutschen Gesellschaft für Musikpsychologie, Oldenburg, pp 40–41. http://www.researchgate.net/publication/304496623_Exploring_Physical_Parameters_Explaining_the_Apparent_Source_Width_of_Direct_Sound_of_Musical_Instruments

Ziemer T (2015b) Spatial sound impression and precise localization by psychoacoustic sound field synthesis. In: Deutsche Gesellschaft für Akustik e.V., Mores R (eds) Seminar des Fachausschusses Musikalische Akustik (FAMA): "Musikalische Akustik zwischen Empirie und Theorie", Hamburg. Deutsche Gesellsch. f. Akustik, pp 17–22. https://www.dega-akustik.de/fachausschuesse/ma/dokumente/tagungsband-seminar-fama-2015/

Ziemer T (2016) Implementation of the radiation characteristics of musical instruments in wave field synthesis application. PhD thesis, University of Hamburg, Hamburg

Ziemer T (2017a) Source width in music production. Methods in stereo, ambisonics, and wave field synthesis. In: Schneider A (ed) Studies in musical acoustics and psychoacoustics. Current research in systematic musicoogy, vol 4. Springer, Cham, pp 299–340. https://doi.org/10.1007/978-3-319-47292-8_10. (Chap. 10)

Ziemer T (2017b) Perceptually motivated sound field synthesis for music presentation. J Acoust Soc Am 141(5):3997. https://doi.org/10.1121/1.4989162

Ziemer T (2017c) Perceptual sound field synthesis concept for music presentation. In: Proceedings of meetings on acoustics, Boston, MA, 015016. https://doi.org/10.1121/2.0000661

Ziemer T (2018) Wave field synthesis. In: Bader R (ed) Springer handbook of systematic musicology. Springer, Berlin, pp 175–193. https://doi.org/10.1007/978-3-662-55004-5_18. (Chap.18)

Ziemer T, Bader R (2015a) Complex point source model to calculate the sound field radiated from musical instruments. In: Proceedings of meetings on acoustics, vol 25. https://doi.org/10.1121/2.0000122

Ziemer T, Bader R (2015b) Implementing the radiation characteristics of musical instruments in a psychoacoustic sound field synthesis system. J Audio Eng Soc 63(12):1094. http://www.aes.org/journal/online/JAES_V63/12/

Ziemer T, Bader R (2015c) Implementing the radiation characteristics of musical instruments in a psychoacoustic sound field synthesis system. In: Audio engineering society convention 139, New York, p 9466. http://www.aes.org/e-lib/browse.cfm?elib=18022

Ziemer T, Bader R (2015d) Complex point source model to calculate the sound field radiated from musical instruments. J Acoust Soc Am 138(3):1936. https://doi.org/10.1121/1.4934107

Ziemer T, Bader R (2017) Psychoacoustic sound field synthesis for musical instrument radiation characteristics 65(6):482–496. https://doi.org/10.17743/jaes.2017.0014

Zotter F (2009) Analysis and synthesis of sound-radiation with spherical arrays. PhD thesis, University of Music and Performing Arts, Graz

Zotter F, Frank M, Kronlachner M, Choi J-W (2014) Efficient phantom source widening and diffuseness in ambisonics. In Proceedings of EAA joint symposium on auralization and ambisonics, Berlin

Chapter 9
Psychoacoustic Sound Field Synthesis

For listeners, perceptual precision of sound is more important than the physical precision of waves. The success of conventional audio systems as well as of psychoacoustic audio compression methods can be explained by this fact. Here, major physical differences are perceived as minor deviations. As stated already in Sect. 8.1, this is even true for sound field synthesis. Physical errors occur and are measurable. Still, the auditory system is not too sensitive for these errors. This raises the question: Why not combine physical and perceptual reasoning to derive a new approach to a spatial audio system? Although there is an increasing interest in perceptual assessment of spatial audio systems,[1] rarely any audio system considered perception thoroughly during the derivation and development process. Taking perception into account could be a powerful approach towards a new type of spatial audio systems and other audio technology, e.g., in the field of audio compression and music information retrieval. Conventional wave field synthesis approaches are derived from a physical core as discussed in Chap. 8. The original objective was "acoustic control", i.e., control over the physical wave.[2] To put it into practice a number of issues are solved because the real audio equipment and setup deviate from the idealized physical concept. Solutions tend to come from the field of engineering. They are solved by technical means. Tools, like discretization, windowing and tapering, have proven their value in many applications in electrical engineering, digital signal processing and communication technology. The impressive sound experience that many wave field synthesis systems provide demonstrates the benefit of these methods once again. An unconventional, yet promising, approach is *psychoacoustic control*. This implies controlling the audible portion of the physical sound field with a precision that is

[1] See e.g. Wittek (2007), Wittek et al. (2004), Wierstorf et al. (2017), Ahrens et al. (2010), Geier et al. (2010), Wierstorf et al. (2013), Spors et al. (2013), Lindau et al. (2014), Ahrens (2015), Wierstorf (2014), Böhlke and Ziemer (2017a, b).

[2] See Berkhout (1988).

© Springer Nature Switzerland AG 2020

T. Ziemer, *Psychoacoustic Music Sound Field Synthesis*, Current Research in Systematic Musicology 7, https://doi.org/10.1007/978-3-030-23033-3_9

sufficient for a human listener. To achieve this we have to consider the resolution of the auditory system as described in detail in Chap. 4.

This idea of a psychoacoustic sound field synthesis system is the main motivation of this book. An approach to psychoacoustic sound field synthesis is given in this chapter. The chapter starts with psychoacoustic reasoning. After that, the physical framework of the presented psychoacoustic sound field synthesis approach is derived in successive subsections. Finally, the psychoacoustic considerations are implemented in the physical framework and evaluated in calculations, simulations and listening experiments. After a short summary, the chapter closes with prospects on further potential developmental steps for psychoacoustic sound field synthesis.

9.1 Psychoacoustic Reasoning

When taking auditory perception into consideration, it is important to identify the sound qualities that shall be created by the audio system. For musical applications, an audio system should be able to create a natural instrumental sound. As mentioned in Sect. 5.2, main acoustic features of musical instruments include musical scale, dynamics, time envelope and radiation characteristics. Perceptually, we hear the musical instrument's source location and width, loudness, pitch and timbre, as well as their changes in time. These are considered the most important aspects of direct sound in the present psychoacoustic sound field synthesis approach. In this book, these impressions are summarized as a *natural, spatial sound impression*. The auditory system processes these impressions with limited time and frequency resolution. Further sound impressions are related to indirect sound, as discussed extensively in Chap. 6. These are out of scope of this psychoacoustic sound field synthesis chapter but they are certainly highly interesting for further development of psychoacoustic sound field synthesis for music.

9.1.1 Integration Times

The auditory system has several integration timeintegration times. These are partly controversial. However, we can clearly distinguish two orders of magnitude. Source localization in the horizontal plane is a matter of interaural time differences of less than one millisecond. Other aspects of instrumental sound are a matter of range between 5 and 300 ms.

As discussed in Sect. 4.1, the perception of pitch and timbre need at least 5 ms to build up. But pitch strength increases until a duration of 300 ms is reached.[3] Likewise, loudness and features of timbre—such as brightness and sharpness—are

[3]See e.g. Zwicker and Fastl (1999), p. 138 and Bader (2013), p. 325.

not a matter of temporal fine-structure but rather an integration over 50–200 ms.[4] As discussed in Sect. 4.1, it takes at least 7–30 ms—depending on frequency—to distinguish alternating complex tones from simultaneous complex tones. And as mentioned in Sect. 4.5.4, sequential grouping creates the perception of melodies and established auditory streams can sustain for seconds and more. Perceived source extent is generally considered as a matter of ear signal (in-)coherence during direct sound and early reflections.[5] As discussed in Sect. 6.2, the integration timeintegration time of 50–80 ms separates the early from the late reflections in the field of subjective room acoustics. The late reflections have a larger influence on listener envelopment than on apparent source width. As discussed in Sect. 4.3.1, pre-masking before note note onsets can be effective between 20 and 100 ms, followed by an overshoot of up to 50 ms. After the note offset the masking effect sustains for about 5 ms before a gradual decrease that takes up to 200 ms. As discussed in Sect. 4.3.2, the overshoot phenomenon is even evident in binaural masking. Here, the masking threshold is up to 10 dB higher than during steady state.

This integration timeintegration time of several milliseconds implies that signal properties do not need to be synthesized correctly the whole time. A good example for this fact is the singing voice. Single notes of sung melodies tend to be no separate stable notes. Instead, pitch glides from one note to the other. This subtle portamento does not sound detuned. It supports the natural auditory scene analysis principle of continuity and thereby strengthens the temporal grouping of notes to form a melody. Due to the inertia of the auditory system, i.e., the integration timeintegration time, the quick correction of the sung fundamental frequency creates the impression of a stable pitch. Another example is audio compression based on temporal masking. Codecs such as MP3 tend to encode soft passages right before loud note onsets with a low dynamic resolution. Even though this may create a pre-echo, i.e., additional sound energy right before the note onset, the sound does not appear louder. Due to backwards making, the pre-echo is inaudible and does not contribute to loudness perception. Technically speaking, the integration timeintegration time is a time margin in which we can inaudibly add, subtract or degrade signal information.

Source localization in the horizontal plane depends on signal properties, like frequency, but is mainly a matter of ILD and ITD of the first wave front and very early reflections that arrive at a listener. As discussed in Sect. 4.4.2, an ITD of 27 μs or an ILD of slightly over 1 dB may already be interpreted as lateral source deflection. When several wave fronts arrive, the first wave front tends to dominate localization mechanisms. But, e.g., with a stereo setup, the first arriving wave front does not always draw the auditory event towards the loudspeaker location. With an inter channel time delay of a few milliseconds, the second wave front affects the phantom source angle, allowing for time based panning. For the detection of source motion,

---

[4]For a discussion of integration timeintegration times, see e.g. Ziemer (2017c) and psychoacoustic studies by Zwicker and Fastl (1999), von Bismarck (1974), Aures (1985) and Daniel and Weber (1997).

[5]An extensive review of physical measures to explain the perceived width of sound sources width can be found in Ziemer (2017a).

the integration time is much higher,[6] lying around 150–300 ms. Since the integration timeintegration time for source localization lies in the order of microseconds, signal manipulations can easily affect perceived source location. The challenge is to ensure a robust source localization even though the signal synthesized by the audio system is not physically accurate.

9.1.2 *Frequency Resolution*

Not all spectral components are equally important for sound perception. As discussed in Sect. 4.2, the auditory system does not process sound as a whole but in critical frequency bands. This is true for the perception of loudness and masking effects, aspects of timbre and of spatial hearing and of auditory stream segregation. The loudest frequency within each critical band rate is dominant over other partials. Psychoacoustic audio compression algorithms like MP3 make use of this fact by lowering the dynamic precision of partially or completely masked partials.[7]

From the different integration times it is obvious that source localization is very sensitive to small temporal changes, whereas musical scale, loudness, timbre and even source width are a matter of larger temporal magnitudes. The auditory system has a focus on the loudest partial within a critical band and is less sensitive to inaccuracies of other partials. These fundamental observations are the psychoacoustic core of the psychoacoustic sound field synthesis approach described below. In contrast to other sound field synthesis approaches, such as wave field synthesis and ambisonics, it accounts for auditory perception in the complete chain from measurements over calculations to the actual sound field synthesis. The complete derivation of the psychoacoustic sound field synthesis approach as well as further details and listening test results can be found in the literature.[8]

9.2 Physical Fundamentals

The core of the present psychoacoustic sound field synthesis approach can be considered as physical. In fact, this core already includes some psychoacoustic considerations, as will become clear in the section about the implementation of psychoacoustics, Sect. 9.3. The physical fundamentals will be introduced in this section.

---

[6] Also discussed in Sect. 4.1, after Grantham (1986).

[7] See e.g. International Organization for Standardization/International Electrotechnical Commission (1993), Brandenburg and Stoll (1994).

[8] See Ziemer (2009, 2011a, b, c, d, 2014, 2015b, 2016, 2017b, 2018) Ziemer and Bader (2015a, b, c, d, 2017).

9.2.1 Radiation of the Loudspeakers

A loudspeaker spectrum $P_{\mathbf{Y}}(\omega)$ is the solution to the homogeneous Helmholtz equation, Eq. 5.9, discussed in Sects. 5.1.3 to 5.1.4. Its radiation from the source point $\mathbf{Y}$ is described by a complex transfer function, the free-field Green's function $G(\omega, r)$. It was introduced in Eq. 5.22 and described from Sects. 5.1.5 to 5.1.6. In the case of a monopole source the radiation is independent of the propagation direction. Higher-order radiators have complex angular factors $\Psi(\omega, \varphi, \vartheta)$—which comprise of an azimuth and a polar factor $\Gamma(\omega, \varphi)$ and $\Theta(\omega, \vartheta)$—that describes an amplitude- and phase shift for each direction as explained in Sect. 5.3.1.1. Reducing the considered wave field to two dimensions for the measurement and reconstruction, as commonly done in wave field synthesis applications, eliminates the term $\Theta(\omega, \vartheta)$. Then, $\Gamma(\omega, \varphi)$ is the only angle-dependent term which modifies pressure and phase distribution on the circular wave front. The term describes the sound radiation characteristics. It can be any arbitrary complex function of azimuth angle as long as it is defined from 0–2π. Theoretically, the radiation pattern could even be a Dirac delta impulse. This corresponds to a ray as used in Sect. 6.1 for geometrical room acoustics. It could be composed of an infinite series of circular harmonics, described in Sect. 5.3.1.1. In principle, circular harmonics decomposition is a Fourier transform over the horizontal angle as explained in Sect. 5.3.1.1. Since $\Gamma(\omega, \varphi)$ is the only term which is dependent on azimuth angle, it is sufficient to determine the radiation characteristics of a source if we consider it as a complex point source. The relationship between frequency spectrum of a loudspeaker $P_{\mathbf{Y}}(\omega)$, its complex angular factor $\Gamma_{\mathbf{Y}}(\omega, \varphi)$, the spatial wave propagation $G(\omega, \Delta\mathbf{r})$ and the resulting signal at a receiver location $P_{\mathbf{M}}(\omega)$ for a free field with one complex point source can be written as

$$P_{\mathbf{M}}(\omega) = P_{\mathbf{Y}}(\omega)\, \Gamma_{\mathbf{Y}}(\omega, \varphi)\, G(\omega, \Delta\mathbf{r}) \ . \tag{9.1}$$

Recording a signal at any point in the far field $\mathbf{M}$ and assuming the source to originate in exactly one point $\mathbf{Y}$, the measured signal $P_{\mathbf{X}}(\omega)$ equals the source signal $P_{\mathbf{Y}}(\omega)$, modified by the amplification towards the direction $\Gamma(\omega, \varphi)$ and distance $G(\omega, \Delta\mathbf{r})$ between source point and receiver points for any frequency ω. If the distance between the source point and an array of receiver positions is kept constant, the receivers lie on a circle with the source in the center. In polar coordinates, like Eq. 5.18 in Sect. 5.1.6, the position of the loudspeaker and the receiver points can be written as

$$\begin{aligned} \mathbf{Y} &= \begin{bmatrix} 0 \\ 0 \end{bmatrix} \\ \mathbf{M}_m &= \begin{bmatrix} 1 \\ \frac{2\pi(m-1)}{M} \end{bmatrix}, \qquad m = 1, 2, 3, \ldots, M \end{aligned} \tag{9.2}$$

with M microphones located at positions $\mathbf{M}_m$. An exemplary setup for far field recordings with 128 microphones is illustrated in Fig. 9.1. The loudspeaker driving

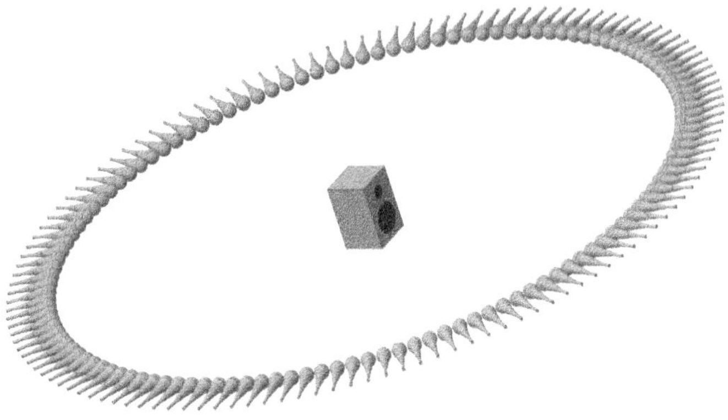

Fig. 9.1 Circular microphone array recording the radiation characteristics of a loudspeaker

signal $P_Y(\omega)$ can be chosen to be 1. At a distance of 1 m from the loudspeaker the distance function $G(\omega, \Delta \mathbf{r}) = e^{\imath kr}/r = e^{\imath k}$ implies nothing but a phase shift which is equal for each microphone location. Consequently, each recorded spectrum $P_X(\omega)$ is directly proportional to the directivity factor at the corresponding angle $\Gamma_Y(\omega, \varphi)$. So the measurement implicitly samples the complex radiation characteristics of a loudspeaker which is represented as a complex point source. So the RC of a loudspeaker can be stored in a database by saving the discrete complex, frequency-dependent amplitudes per angle. Note that a loudspeaker can play the same driving signal over and over. So if no microphone array is available, the loudspeaker could be placed on a turn table. Then, one microphone can record the radiation pattern of one angle. After that, the loudspeaker is rotated by x degrees and the measurement is repeated, until all 360° are covered.

9.2.2 Radiation of Musical Instruments

In Eq. 9.1 the radiation characteristic $\Gamma(\omega, \varphi)$ of a loudspeaker can be calculated because the original spectrum $P_Y(\omega)$ can be controlled. So if a source is excited by a Dirac delta impulse, $\Gamma(\omega, \varphi)$ is the only unknown term if $P_M(\omega)$ is measured with a circular microphone array in the far field that surrounds the source concentrically. If the source is excited by another driving signal, a relative $\Gamma(\omega, \varphi)$, $\Gamma' = \frac{\Gamma_Q(\omega\varphi, \vartheta)}{P_Q(\omega)}$, can be calculated. As Γ' is proportional to Γ, it is an adequate measure to describe and compare the radiation characteristic of any complex point source even though the source amplitude is not known. Γ' only lacks normalization. Standard works, like (Meyer 2009) and (Fletcher 2008) use this same method to describe and compare the radiation characteristic of musical instruments, even for instruments that are much

Fig. 9.2 Photo of the measurement setup recording the radiation characteristic of a shakuhachi. The microphones stick out of the circular rim that can be seen behind the instrumentalists

larger than the considered wavelengths.[9] This simplification is physically untrue and neglects diffraction around the instrumental body and the instrumentalist, interference between radiations from various parts of the instrument, different arrival times and amplitude decays due to different paths from different body parts to each receiver. But besides the immense advantage of a uniformly defined measure, which makes the RC describable and comparable, the complex point source simplification has another very important benefit: its simplicity in application. Due to the reduction of the radiation characteristic to one function of angle only, complicated measurement techniques like acoustical holography are dispensable. Furthermore, calculating the wave field emerging from such a complex point source is straight forward for any source- and receiver location and practical implementations showed that "a circular array of microphones leads to the best extrapolation results".[10] Spherical and dual-spherical microphone arrays have been suggested to record and reconstruct the radiation characteristics of sound sources.[11] It could be demonstrated that the complex point source method yields a plausible sound field in terms of interaural level and phase differences and other measures of interaural coherence.[12] An arbitrary source signal $P(\omega)$ can be multiplied by the relative angular factor for the different frequency regions for the desired direction and forward-propagated according to the free-field Green's function for the desired distance. No superposition of spherical harmonics or integral of point sources is necessary.

Figure 9.2 shows the same setup as Fig. 9.1. But this time an instrumentalist is placed in the center of the microphone array. In the photo the white rim contains 128 synchronized microphones. By playing different notes successively, a large portion of the audible frequency range can be created. These can be measured, normalized

[9] See Fletcher and Rossing (2008), p. 395, showing the plots from Meyer (2009), pp. 163ff.

[10] Ajdler et al. (2008), p. 157.

[11] See Bader (2014), pp. 1181f.

[12] This is demonstrated e.g. in Ziemer (2014), Ziemer and Bader (2015a), Ziemer (2015a) and Ziemer (2017a).

and stored in a Γ'-database. One second of quasi-stationary sound is recorded and transferred into frequency domain to achieve a frequency precision of 1 Hz.

9.2.3 Sound Field Extrapolation

Assuming a point source Q in a free field having an arbitrary source spectrum $P_{\mathrm{Q}}(\omega)$, one can easily calculate the pressure distribution in an extended listening area. To achieve this, the listening area is sampled into discrete listening points. For each listening point

$$P_{\mathrm{X}}(\omega) = P_{\mathrm{Q}}(\omega)\,\Gamma_{\mathrm{Q}}(\omega, \varphi)\,G(\omega, \Delta\mathbf{r}) \tag{9.3}$$

is valid. Here, either $\Gamma_{\mathrm{Q}}(\omega, \varphi)$ can be chosen freely or Γ' is measured by means of a circular microphone array. Choosing $\Gamma_{\mathrm{Q}}(\omega, \varphi) = 1$ equals a monopole source. This is certainly the simplest case. The forward-propagation in Eq. 9.3 is an extrapolation of the source spectrum $P_{\mathrm{Q}}(\omega)$ from the source $\mathbf{Q}$ to the listening points $\mathbf{X}$ which sample an extended listening area. $\Gamma_{\mathrm{Q}}(\omega, \varphi)$ and $G(\Delta\mathbf{r})$ constitute the transfer function from the source to the listening position, which can be combined to one function

$$K(\mathbf{Q}, \mathbf{r}) = \Gamma_{\mathrm{Q}}(\varphi)\,G(\Delta\mathbf{r}). \tag{9.4}$$

Summarizing the transfer function to one term K, function 9.1 can be written as matrix equation

$$\mathbf{P_X} = P_{\mathrm{Q}}\mathbf{K}. \tag{9.5}$$

The amplitudes $\mathbf{P_X}$ describe the wave field at the listening points $\mathbf{X}$ under free field conditions. They are created by the propagated source term P_{Q} and are to be reproduced by the loudspeaker system. Calculating the desired wave field this way is a simple forward propagation of a known source signal. An example is illustrated in Fig. 9.3. It shows a freely chosen source location, freely chosen listening points and a measured radiation factor Γ'. The measured radiation factor modifies the amplitude for each angle between the facing direction of the source and each listening point. Towards $\mathbf{X}_4$ the amplitude Γ' in $\mathbf{K}$ is rather low. It is much higher towards $\mathbf{X}_1$ and $\mathbf{X}_3$. The illustration assumes a purely real Γ' and only depicts the real part of $\mathbf{K}$. However, finding the necessary loudspeaker signals which create this calculated sound field is an inverse problem. This will be discussed in the following subsection.

9.2.4 Sound Field Reconstruction

The calculated wave field at the listening points are modified versions of the original source signal. The modification is caused by the propagation through the free field including amplitude decay and phase shift per distance and the different complex

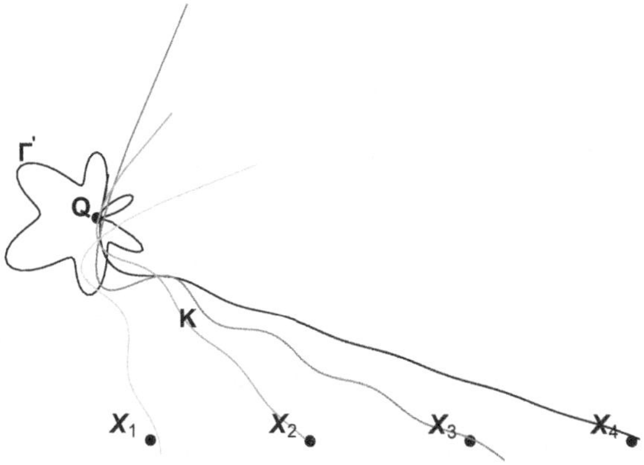

Fig. 9.3 Forward propagation from a source **Q** to receivers $\mathbf{X}_m$ by means of the propagation matrix **K** which includes the angular amplitude factor Γ'

radiation factors per angle. This calculated wave field is the desired wave field to be reproduced by a loudspeaker system. In other words, the speakers have to be fed by versions of the source signal which are manipulated in a way that results in exactly the calculated wave field at the listening points. The loudspeaker driving signals have to be calculated for the individual constellation of virtual source location, loudspeaker distribution and listening points. The wave front synthesis approach described in the previous chapter assumes loudspeakers to radiate like monopoles. This is a reasonable simplification which allows us to apply the Rayleigh integral and find an analytic solution. Loudspeaker signals are delayed and attenuated according to the distance between the virtual monopole source and the loudspeaker. However, due to the finite number of secondary sources, and their discrete placement in one or several one-dimensional arrays, all the above-mentioned adoptions are necessar, like tapering and the 2.5D-operator. These are not analytically correct but they work in practice. There are also approaches to account for the radiation characteristics of the secondary sources.[13] It has been found that the deviant radiation pattern of the loudspeakers contained in the array are transferred to the virtual point source.[14] But as long as all loudspeakers exhibit the same radiation characteristics, it can be sampled and implemented in the wave front synthesis calculation with little synthesis errors. Likewise, (Ahrens and Spors 2009) deliver a formulation to implement the actual radiation characteristics of loudspeakers as measured with a microphone array under anechoic conditions. The authors of both papers agree that all loudspeakers used for the sound field synthesis should exhibit the same radiation characteristics.

In ambisonics, loudspeakers are also typically assumed to either like radiate sound like monopoles or to radiate plane waves. This distinction is the main difference between ambisonics and nearfield compensated ambisonics. With this assumption, one could reconstruct the desired sound field by solving the linear equation system

---

[13] See e.g. de Vries (1996) and Ahrens and Spors (2009).

[14] See de Vries (1996), p. 1123.

$$\mathbf{P}_{\mathbf{X}_n} = \mathbf{P}_{\mathbf{Y}_l}\mathbf{G},\ n = 1, \ldots, N,\ l = 1, \ldots, L. \tag{9.6}$$

Here, L loudspeakers are simplified as monopole sources and the transfer function from each loudspeaker $\mathbf{Y}_l$ to each of the N listening point $\mathbf{X}_n$ is fully described by the free field Green's function $\mathbf{G}$. As mentioned earlier, Eq. 9.6 is ill-posed because the sound field is known but not the loudspeaker signals which create it. However, when choosing as many listening points as loudspeakers available, i.e. $L = N$, an exact and unique solution can be found. According to the Nyquist-Shannon sampling theorem, the solution should not only create the correct sound field at the discrete listening points but in the whole listening area, as long as the considered frequency lies below the critical spatial frequency. This critical frequency is determined by the distance between neighboring listening points.

Problematic here is the fact that Eq. 9.6 may be ill-conditioned. If two loudspeakers are proximate to one another compared to the considered wavelength, their influence on one listening point is almost the same. Likewise, the influence of one loudspeaker on two listening points that are close to each other is almost the same. In these cases, the rows of the transfer matrix are not mutually independent and the matrix is ill-conditioned. The condition $\kappa\,(\mathbf{K})$ of a matrix $\mathbf{K}$ can be described by the ratio of its highest (max) to its lowest (min) eigenvalue $\text{eig}(\mathbf{K})$

$$\kappa\,(\mathbf{K}) = \frac{\text{eig}_{\text{max}}\,(\mathbf{K})}{\text{eig}_{\text{min}}\,(\mathbf{K})}. \tag{9.7}$$

A high κ indicates a badly conditioned matrix. So when the condition of $\mathbf{G}$ in 9.6 is bad, the solution may have huge loudspeaker signal amplitudes, caused e.g., by subtle quantification noise. That means the reconstruction energy is very high. A relaxation method which improves the condition number is the minimum energy method which comes from the field of nearfield microphone array technology and is discussed in Sect. 5.3.3. The method slightly changes the assumed loudspeaker directivity from a monopole to a more directive sound source. Even subtle changes decrease the condition number drastically. For this method the optimal solution creates the lowest reconstruction energy, i.e. the lowest loudspeaker amplitudes. Another relaxation approach is to stick to the complex point source assumption as in Eqs. 9.1 and 9.5. In this case Eq. 9.6 changes to

$$\mathbf{P}_{\mathbf{X}} = \mathbf{P}_{\mathbf{Y}}\mathbf{K}. \tag{9.8}$$

Here, the propagation matrix

$$\mathbf{K} = \boldsymbol{\Gamma}\mathbf{G} \tag{9.9}$$

includes the loudspeaker radiation as measured in Sect. 9.2.1. This relaxation method is referred to as *radiation method* or *radiation methodr-method*. Loudspeakers have at least subtle variations of amplitude and phase over angle. For low frequencies, they are often close to a monopole. Higher frequencies are radiated even more directive. An example of one low and one high frequency is illustrated in Fig. 9.4. The loudspeaker membrane is facing towards the right hand side. The radiation pattern of the low

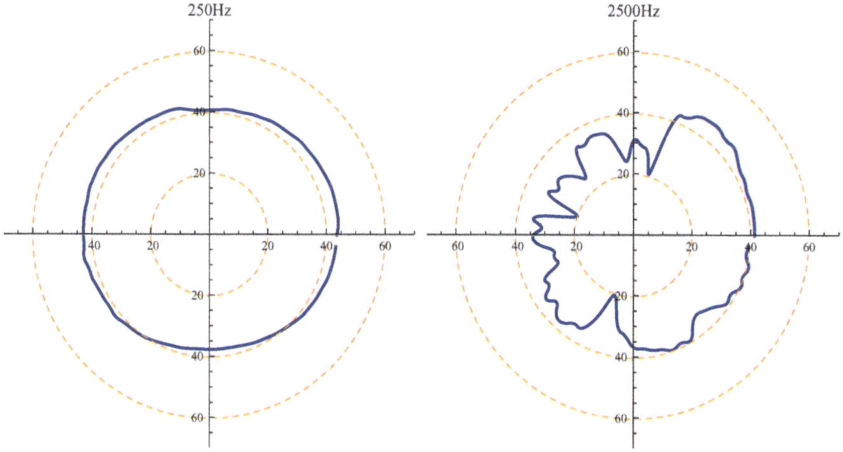

Fig. 9.4 Measured radiation characteristics of a loudspeaker at frequencies of 250 Hz (left) and 2.5 kHz (right). From Ziemer (2016), pp. 164–165

frequency resembles a monopole. But in the front and the back of the loudspeaker, the amplitude is slightly higher than towards the sides. The radiation pattern of the high frequency looks more complicated and asymmetric. The front-left is emphasized and deep amplitude dips can be seen at the sides. When including the measured radiation characteristics, the rows in **K** of Eq. 9.8 become independent from one another. Loudspeakers at different angles from one receiver point exhibit individual radiation factors. Chances are low that they exhibit exactly the same numerical value. But even if they do, it is very unlikely that they share the same complex amplitude for every single listening point. So by including the radiation characteristics, the matrix **K** in Eq. 9.8 has a better condition κ than the matrix **G** in Eq. 9.6. So the radiation method relaxes the linear equation system just as the minimum energy method. The propagation matrix exhibits a lower $\kappa(\mathbf{K})$ and the loudspeaker amplitudes in the solution are lower. But in contrast to the MEM, the radiation method is based on measurements and may represent the actual physics of the loudspeaker radiation better.

An exemplary scenario is simulated to demonstrate the benefit of the regularization methods and to compare their performance. The scenario is illustrated in Fig. 9.5. A source with a shakuhachi radiation is placed in a free field at position **Q**. It radiates a stationary signal containing 25 frequencies. In this example, the frequencies are the center frequencies of the 25 critical bands according to the Bark scale as listed in Table 4.1 in Sect. 4.2. 15 listening points $\mathbf{X}_{1-15}$ are chosen to lie at a distance of a few meters away from the source. These points sample a listening area. 15 loudspeakers $\mathbf{Y}_{1-15}$ surround the listening area from three sides. The sound field can be calculated for these 15 points as well as for additional points between and besides these listening point according to Eq. 9.5. The sound field can now be synthesized by considering the loudspeakers as monopole sources, Eq. 9.6, by applying the minimum energy

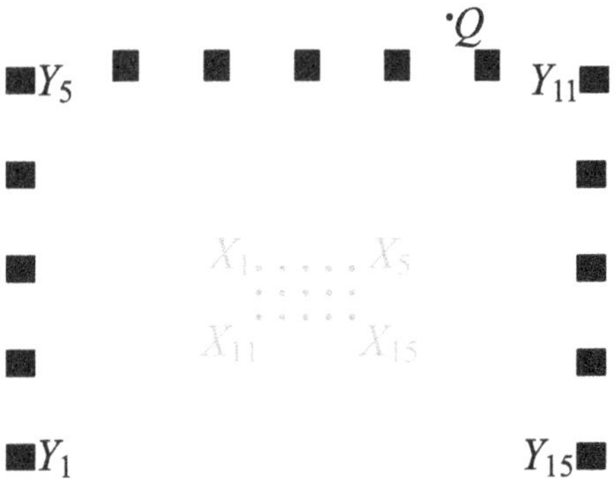

Fig. 9.5 Setup of the simulated scenario which demonstrates the performance of the regularization techniques

method, Eq. 5.31 in Sect. 5.3.3, and by applying the radiation method, Eq. 9.8. For the radiation method, the measured directivity function of actual loudspeakers is used.

Without regularization small numerical errors cause massive increases of sound pressure levels in the solution. The worse the condition the higher the amplitudes calculated by a solver. Such a solution may be mathematically correct but unphysical. Many regularization methods improve the mathematical condition[15] However, a plausible regularization method takes the natural physics of the considered problem into account. Two regularization methods serve for this purpose.

The philosophy behind the minimum energy method is convincing for the given inverse problem. The idea is that nature regulates itself. There are countless examples of self-organization in nature. In musical acoustics the self-sustained oscillation in organ pipes and the aeroacoustical synchronization between organ pipes result from self-organization of a nonlinear system. It is assumed that nonlinearities like mode locking are essential to produce nearly harmonic overtone series in musical instruments. Synchronization of musicians, dancers, and music listeners are a matter of self-organization, referred to as entrainment. It helps to keep track of the tempo and bond a group, which becomes obvious, e.g., in work songs.[16] This self-organization is energy-efficient. Without it, many musical instruments would not be able to sound at all, and playing in an ensemble would be much more demanding. The idea in the MEM is that the most energy-efficient solution is the natural one.

The radiation method makes a similar assumption. However, it is more directly linked to the problem statement. As nature regulates itself, it may be beneficial to consider the loudspeaker physics more closely. Instead of simplifying loudspeakers as ideal monopole sources their sound radiation characteristics are measured and included in the calculation. For low frequencies the only small deviations from a monopole can be observed. For high frequencies the difference is huge. However,

[15]A review of regularization methods in inverse problems in spatial audio can be found in Bai et al. (2017).

[16]Refer to Fischer et al. (2016), Bader (2013), pp. 253f and pp. 159–163, Fletcher (1978), Dubnov and Rodet (2003), Jensenius et al. (2010), Patel (2002), Chap. 3 for details on the given examples.

over the whole audible frequency range the natural deviations of the sound radiation properties turn out to contribute a lot to the regularization of the linear equation system.

The benefit of the regularization methods is demonstrated in an example. Fig. 9.6 shows the condition number κ of the three solutions for each frequency according to Eq. 9.7. It can be observed that without regularization the condition number decreases as the frequency increases. This is expected: When wavelengths are large compared to the distance between listening points and between the loudspeakers, the rows in the propagation matrix **G** in Eq. 9.6 become more similar. The MEM exhibits much lower condition numbers by assuming the loudspeakers to be slightly more directive than monopole sources. This way the propagation between all loudspeakers and two neighboring listening points are more diverse. However, the method does not choose the solution with the lowest condition number but with the lowest reconstruction energy. These are not necessarily the same. Their relation can be observed in Fig. 9.7 for one frequency in the scenario. With different values for Ω, both reconstruction energy and condition number change. Both curves look similar. Already a slight increase of Ω reduces the reconstruction energy and the condition number drastically. But several other values of Ω lead to a slightly lower condition number. One could choose the value for Ω with the lowest condition, to have the mathematically ideal solution. However, choosing the solution with the lowest reconstruction energy has a certain advantage. For all solutions the sound field at the very listening points $\mathbf{P_X}$ is assumed to be the same. Lower reconstruction energy means lower loudspeaker amplitudes. The lower the loudspeaker amplitudes, the more constructive the waves interfere in the listening area. If the loudspeaker amplitudes were higher, the sound field was generated by destructive interference. In an actual implementation we usually do not have a free field. As already illustrated in Fig. 8.16, waves reflect from room surfaces and interfere with the desired sound field. Beyond the listening area the propagated loudspeaker signals do not superimpose in a controlled way but rather in random phase relations. So the lower the loudspeaker gains, the higher the degree of constructive interference inside the listening area and, consequently, the lower the amplitudes outside the listening area. This is true for the direct sound and the reflections. This makes the minimum energy approach very valuable for a the synthesis of a wave field. When applying the radiation method, an even lower condition number can be observed for all frequencies, which results in lower loudspeaker amplitudes. Note that the ordinate in the graph in Fig. 9.6 is in logarithmic scale. Compared to the unregulated solution, the MEM reduces the condition of the propagation matrix by an order of 1 to almost 8 decimals, the radiation method even by 5–15 orders of magnitude. A similar trend can be observed for the reconstruction energy, illustrated in Fig. 9.8. So, although derived from physical measurements, the radiation method seems to be superior over the iterative minimum energy method for this sound field synthesis task. This is certainly owed to the fact that the actual radiation patterns of loudspeakers tend to be much more diverse than the slight adjustments that the MEM introduces. The pattern in Fig. 9.4 is much more complicated and deviant from a monopole than the pattern in Fig. 5.13 in Sect. 5.3.3.

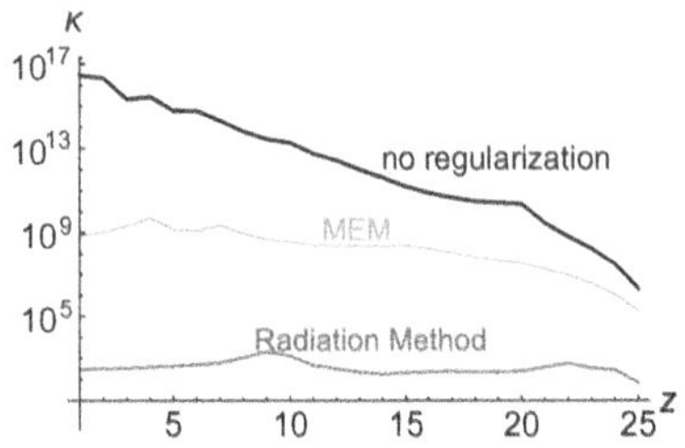

Fig. 9.6 Condition numbers κ for each frequency band without regularization (black) and when applying the MEM (light gray) and r-method (gray). From Ziemer and Bader (2017), p. 486, with the permission of the Audio Engineering Society

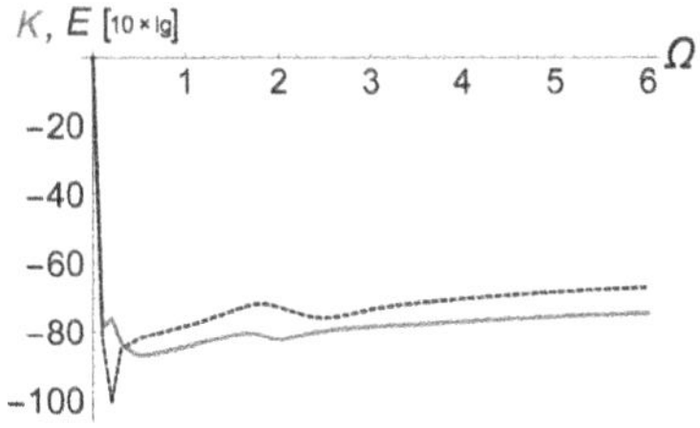

Fig. 9.7 Exemplary reconstruction energy E (black) and condition number κ (gray) for different frequency bands in the given scenario. Both are calculated as $10\,\text{lg}\,\frac{\text{value}}{\text{max}}$

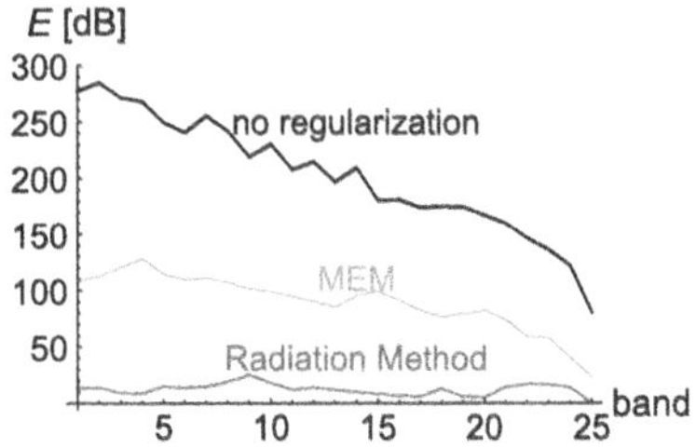

Fig. 9.8 Reconstruction energy E for each frequency band without regularization (black) and when applying the MEM (light gray) and r-method (gray). From Ziemer and Bader (2017), p. 486, with the permission of the Audio Engineering Society

The solution of a linear equation system with a badly conditioned matrix can change dramatically when changing the initial conditions— i.e. the constellation of source, receivers and loudspeakers—only slightly. That means, the solution is not robust. Subtle changes of the source spectrum or a little shift of the virtual source position would yield a very different solution, i.e. a strongly varied constellation of loudspeaker signals. In theory, this would be no problem, as the sound field at the listening points is mathematically correct. But in fact equations are already a simplification of the physical scene, and numeric quantification, as well as the DFT over a finite time window, introduce small errors. So in the actual implementation,

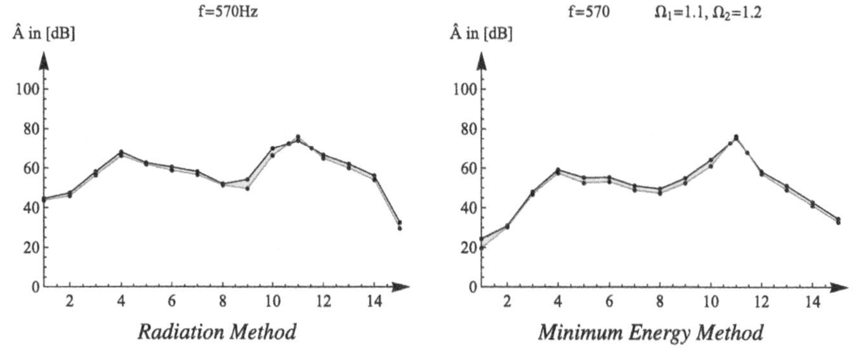

Fig. 9.9 Loudspeaker amplitudes for two proximate virtual sources with the same source signal, solved by the radiation method (left) and the minimum energy method (right). From Ziemer (2016), p. 296

such strong changes may create audible discontinuities. These discontinuities may be stronger outside the listening area but become audible inside the listening area due to reflections. A good relaxation method yields robust solutions. This is the case for both presented methods,[17] as can be seen in Fig. 9.9. It shows the 15 loudspeaker amplitudes of one frequency that is radiated from two virtual sources which have a distance of 5 cm from one another. The distance between the two plots is indicated as a gray area. It indicates the difference between the solution for the original and the slightly misaligned source. It can be observed that the difference tends to lie below 3 dB. This is plausible, as the desired sound field will also change when slightly moving a source with a complex radiation pattern.

9.3 Implementation of Psychoacoustics

Until now, the psychoacoustic sound field synthesis approach is described in a mathematical way which approximates a musical sound source and loudspeakers by means of complex point sources. It is formulated in frequency domain. In the strict sense, a Fourier transform is only valid for stationary signals. It yields the correct spectrum if we assume that the considered time window contains exactly one period of a stationary signal that started at $t_{\text{start}} = -\infty$ and will repeat until $t_{\text{end}} = \infty$. In this case, the wave front has passed the listening area already, since we are looking at one time window at $t_0 = 0$. So the wave front has already traveled a distance of $|(t_0 - t_{\text{start}})\, c| = \infty$. So in the case of stationary sound, the equations in frequency domain are valid for any point in time. However, music only consist of quasi-stationary phases which are interrupted by transients, like note-onsets, offsets and modulations.

[17]This has been demonstrated in simulations in Ziemer (2011a).

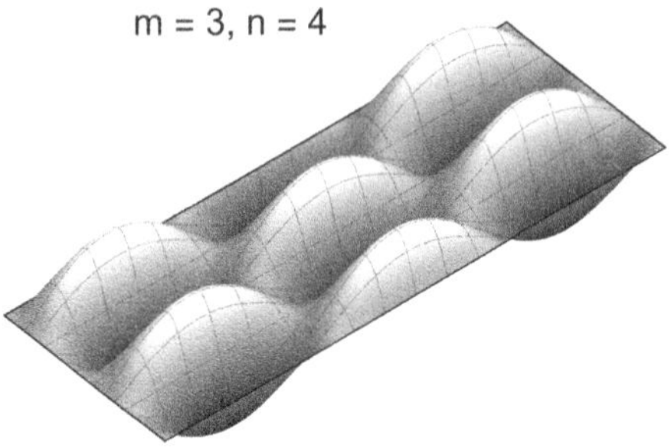

Fig. 9.10 Eigenmode of a rectangular membrane, as a demonstration of a two-dimensional standing wave. No wave front can be identified. Still, two points can have sound pressure level and phase differences

The travel time and direction of a wave front deliver the cues that the auditory system uses to localize a sound source. One cue is the ITD caused by the individual paths from the source to each ear. Another cue is the ILD, which is also partly caused by the individual path length between the source and the two ears as well as the acoustic wave shadow. Also caused by the head related transfer function, is an individual spectral shaping of both ear signals which depends on the source direction. When entering a room in which a quasi-stationary sound is already playing, the direct sound interferes with reflections of first and higher order. In such a diffuse field, source localization is also diffuse. Likewise, standing waves do not seem to have a source location. Their regular interference patterns creates ear signals that do not point at a source location. An example for a standing wave is given in Fig. 9.10. Even by vision, neither wavefronts nor source locations can be recognized. So there is no doubt that the travel time and direction of a wave front is important for a proper localization. However, it barely affects the perception of level, pitch and timbre. Also, the impression of auditory source width seems to arise after an integration of the first wave front and the following 50–80 ms.[18] Furthermore, a low temporal, dynamic and spatial precision of masked partials does not deteriorate the sound perception. All these facts about auditory perception are considered to make the psychoacoustic sound field synthesis approach suitable for music presentation.

9.3.1 Implementation of Critical Bands

The auditory system does not process all impinging sound as a whole. As one of the first steps, it carries out a frequency analysis, right in the cochlea. From here on frequencies are processed in dynamic frequency bands. In each frequency band the frequency with the amplitude tends to mask the other frequencies within that band. Or the frequencies in the band are processed together, creating a common impression, like beating or roughness. In psychoacoustic audio processing this is typically modeled by static bands like the critical bands according to the Bark scale. We can use this scale throughout the complete psychoacoustic sound field synthesis procedure,

[18]As briefly discussed in Sect. 6.2 and more deliberately demonstrated in Ziemer (2017a).

from the measurement of the loudspeakers' and sources' radiation characteristics over the sound field extrapolation to the calculation of the loudspeaker signals, i.e. the reconstruction of the calculated sound field.

Note that all formulations in this chapter are in frequency domain. To be processed in frequency domain, a discrete Fourier transform over a time window of the source signal is carried out. A common sample rate of musical material is 44.1 kHz. Then, a time window of 512 samples length corresponds to a duration of around 11 ms. This yields the amplitude and phase of 256 positive frequencies. Processing all frequencies correctly would imply to measure the RC of the loudspeakers and the musical instruments for these 256 frequencies. These need to be stored in a database and included in the wave field extrapolation. Then, the linear equation system needs to be solved 256 times when applying the radiation method. When applying the minimum energy method the 256 equation systems are solved 18 times if Ω is chosen between 0 and 10 and approximated to a precision of 0.1, i.e. iterated two times. So the linear equation system was solved 4608 times for each time window. This can imply high computational demands if the number of loudspeakers is large. Furthermore, amplitude and phase of each frequency bin are manipulated. This massive filtering can create a massive deviation from the original source signal and a strong decorrelation of the loudspeaker signals. For an actually stationary signal this is not a problem. The propagated signals superimpose in the listening area and create the desired sound field. But music is typically characterized by the interplay of transients and quasi-stationary periods. Here, the Fourier transform already implies a time integration and introduces spectral leakage effects. Then, manipulating phase information of all frequency bins modifies the temporal signal envelope. For stationary signals, this would not make a difference. Each calculated loudspeaker signal was just differently looking periods of a static sound. But in transient signals a phase modification results in an audible temporal modification. Attacks might get smeared and clicks might occur at the start and the end of each window. To reduce these artifacts, several windowing methods exists. These are out of scope of this book.[19] However, the more frequencies are manipulated, the stronger also the effect on the time signature of the signal. This can have the consequence that some loudspeaker signals cannot be recognized as modifications of a common source signal. This becomes obvious if we consider a typical example. When a source signal does not contain a specific frequency, the amplitude of this frequency component should be 0. A continuous, Fourier transform over an infinite time window would yield 0 for this frequency. However, when applying a discrete Fourier transform, its amplitude could be higher due to the leakage effect that can occur in a finite time window or due to numeric errors. However, after an inverse discrete Fourier transform, the frequency will not be contained is the signal because the Fourier transform has a one-to-one correspondence. But the solution of the linear equation system will contain other non-zero amplitudes for the loudspeakers. These amplitudes can be much higher than the original amplitude in the source spectrum. In this case, this frequency may

[19] An overview about windowing can be found e.g. in Nuttal (1981). Exemplary windowing functions can be found in a YouTube video on https://tinyurl.com/y7nfgfd8.

be audible in the time signal after an inverse Fourier transform. Depending on the size of the time window and the spectral density of the source signal, this can happen to hundreds of frequencies. This can cause two effects. First, the loudspeaker signals exhibit strong numeric noise or they are so deviant from the original source signal that they are not recognized in the synthesized sound field except for the spectral contour of the quasi-stationary periods. Second, the loudspeaker signals are so incoherent that they are not recognized as different versions of the same source signal. In this case they are not integrated into a common auditory stream but are segregated. Then, they are processed individually by the auditory system, resulting in the perception of different streams with individual locations and timbres. Or the high number of incoherent signals are perceived as overwhelming, chaotic ambient sound. Such effects are unacceptable for an audio system. However, considering critical frequency bands can prevent from these effects.

As mentioned earlier, the RC of a musical instrument does not only vary with frequency but also with the played fundamental, fingering, or playing technique. Storing all possible radiation patterns would create a huge database. Yet, it is unknown to what degree a listener can differentiate between radiation patterns or even recognize a specific pattern. It is certainly meaningful to represent the radiation characteristics of musical instruments by a subset of radiation patterns. Here, the critical bandwidth can serve as a compromise between precision and database size. 25 radiation patterns, one for each critical frequency band, can easily be measured, stored in a database and accessed from a lookup table. This is a meaningful subset of radiation patterns from a perceptual point of view, for musical instruments as well as for loudspeakers.

In the next step, when the desired sound field is calculated, one protruding frequency from each frequency band is extrapolated to the listening region. So the desired sound field in the listening area only consists of 25 frequencies; the loudest frequency of each band. Then, the linear equation system is solved only 25 times to synthesize a source sound with the full audible bandwidth, applying the *radiation method*r-method. When applying the MEM, the linear equation system is solved 450 times per time window. All other frequencies contained in the source signal stay untouched. Consequently, all but 25 frequencies in the loudspeaker spectra equal the original source spectrum. The computational advantage of taking critical bands into account is obvious. The number of calculations is reduced by a factor of 10. But in addition to the computational demands the perceptual advantage is even more important. As stated earlier, manipulating amplitude and phase of all spectral components can cause audible noise and decorrelated signals, which may be segregated into different auditory streams. However, the less frequencies are manipulated, the lower the degree of decorrelation. Furthermore, to maintain transient properties of the source signal, it is important to introduce only minimum phase manipulations. Fig. 9.11 makes this relationship clear. It shows the time signal and spectrum of an original electronic bass drum sound, one version with 25, and another version with 2048 manipulated frequencies. Manipulating 25 frequencies has moderate effects on the time series and the frequency spectrum. Manipulating all frequencies contained in the has a stronger effect. The contour of the spectrum looks rather similar to the original version. However, it does look a little noisy. The spectral peaks are less

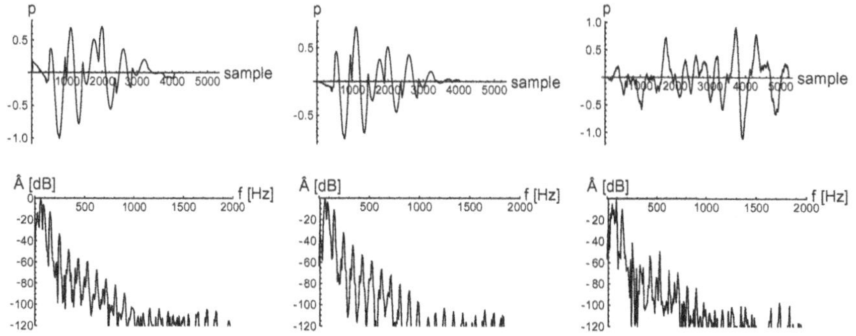

Fig. 9.11 Time series (top) and spectra (bottom) of an original electronic bass drum sample (left) and a version with 25 (center) and 2048 (right) frequencies whose amplitude and phase were manipulated. Especially phase manipulations degrade affect the overall contour of the time series

distinct. This is because many frequencies that used to have a very low amplitude in the original version are now much louder. More obvious is the effect of the spectral manipulation on the time series. Especially due to the high number of phase manipulations the time series has changed dramatically. Neither the fine structure nor the envelope of the original bass drum are kept. Despite a similar frequency spectrum, the steep attack, the periodicity and the decay are missing. This manipulated version barely sounds like a bass drum anymore.[20]

Both the overall spectral contour and the impulsive attack get lost when manipulating too many frequencies. Furthermore, as we only consider the frequencies with the highest amplitudes, the solutions will certainly not amplify quantification noise. This reduces the risk that the solutions contain noise or that they are not recognized as modified versions of the source signal. Still, these 25 frequencies have the strongest effect on the sound impression, especially timbre.

9.3.2 Implementation of Interaural Coherence

When it comes to the spatial precision of the radiation RC measurements, it is also meaningful to consider perceptual aspects. For source localization interaural level and phase differences are of major importance. This is also true for the ASW. The ears of a human listener have a distance of about 0.15 m from each other. So at least one microphone every 0.15 should be used to measure the radiation characteristics. This way, the recordings contain information comparable to binaural data. For the same reason, it may be useful to sample a listening area by listening points which

[20]Note that the very steep attack transient at the very beginning of the original time series is lost in both manipulated versions. This is mainly due to the *precedence fade*. The precedence fade is a method to improve source localization in the psychoacoustic sound field synthesis approach. It is explained in detail in Sect. 9.3.3.

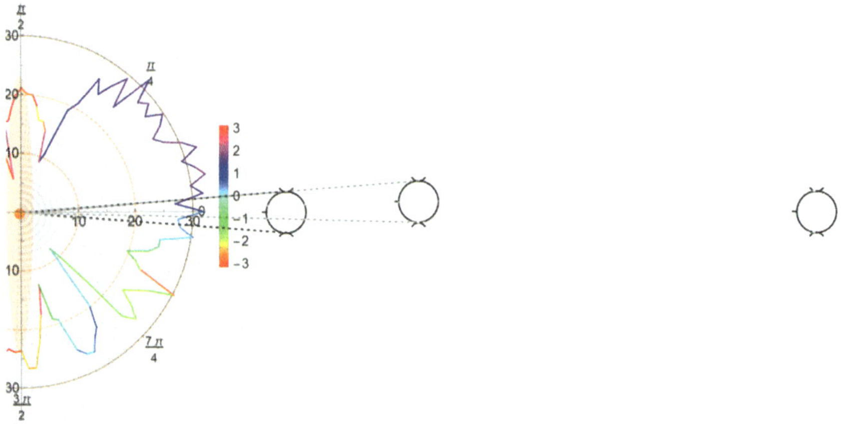

Fig. 9.12 Radiation pattern and extrapolation paths from a virtual complex point source to 3 listeners at a distance of 1, 1.5 and 3 m. From Ziemer (2017a), p. 323

are 0.15 m apart. If the angles of two listening point lie between the angles of two microphones, an interpolation is carried out to receive a meaningful approximation. However, if both ears lie between the same pair of microphones, they will receive a similar radiation factor due to the interpolation. Their difference depends only on the type of interpolation. Hence, 0.15 m should be considered the least distance of neighboring microphones. In this case, one individual radiation factor is provided for each ear. This is true as long as the listener has a distance to the virtual source which is smaller or equal to the distance of the microphones from the original source. When choosing a distance of 0.05 m for neighboring microphones in a circular array with 128 microphones and a radius of 1 m, the radiation characteristics can be extrapolated up to 3 m before the two ears of a listener facing the source may lie between the same pair of extrapolated microphone signals. This is demonstrated in Fig. 9.12. It shows a radiation pattern of one violin frequency. Three listeners are at a distance of 1, 1.5 and 3 m are facing the source. Each of their ears lies on the propagation path of one microphone angle. Only at a distance of > 3 m the two ears may lie between two neighboring microphone angles.

It is commonly accepted that musical instruments can be considered as point source sources if receiver positions are far away compared to the size of the source and compared to the distance between the receivers themselves. The same is true for vision. Details of shape, color and contour of proximate objects are visible and both eyes may have very different perspectives. From a large distance, even broad objects appear narrow and complicated contours become blurred. A proximate and a remote harpsichord are illustrated as an example in Fig. 9.13. An example from the field of audio applications are head related transfer functions. These are typically measured

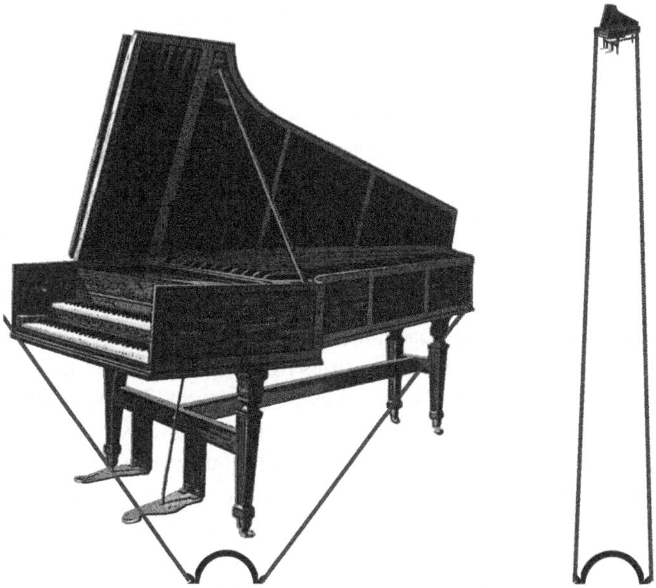

Fig. 9.13 Example for width and detail of a near object compared to a remote object. The near harpsichord looks and sounds broad and has rich detail. The distant harpsichord in a free field looks and sounds narrow and point-like. Harpsichord depiction taken from VictorianLady (2016)

in or modeled for the near field and the far field case. An example was given in Eq. 4.8 and Fig. 4.21 in Sect. 4.4.2. What remains ambiguous is the transition from one to the other. The situation is similar in sound field synthesis applications. In a free field, no reflections occur. Still, musical instruments may sound considerably wider than a point. The main reason for this is the radiation characteristic which creates incoherent ear signals. Only at a distance of several meters, sources tend to sound more point-like. In typical, echoic rooms, the situation is different. Here, the apparent source width is affected strongly by the early reflections. There is wide consensus in the literature that the first 50–80 ms after the direct sound are integrated to build the foundation of auditory width perception. At a large distance, the ratio of reflected to direct sound is rather high. Reflections may then have an even larger effect on ASW than the radiation characteristics. It has been reported that the auditory source width may be even wider than the visual source width.[21]

In the examples given above—e.g. in Figs. 9.1, 9.2 and 9.4—the RC of a source is measured with 128 microphones at a distance of 1 m, so one microphone is placed every 0.05 m. That means when extrapolating a sound field from a virtual source, listeners can have a distance of up to 3 m before both ears lie between the same pair of complex radiation factors and are thus affected mostly by interpolation. So

[21]Sometimes, ASW is even defined as the auditory perception of a source which is wider than its optical size, see e.g. Blau (2004), p. 720.

complicated radiation patterns can be reconstructed for sources with a distance of up to three meters before they start to become monopole-like. This is a reasonable distance. In natural sound fields, this is the order of magnitude in which reflections start to dominate over radiation characteristics in terms of width perception. If available, a higher precision is certainly welcome.

9.3.3 Implementation of the Precedence Effect

In the wave front synthesis approach, the sound propagation direction is considered. Loudspeakers whose normal direction varies from the desired propagation direction by more than 90° are muted. This suppresses echoes and helps listeners to localize the sound source properly. But due to this spatial windowing technique it may happen that much less than half of the available loudspeakers are actually active to synthesize a virtual source. An example had been given already in Fig. 8.10 in Sect. 8.3.2. Compared to that, the psychoacoustic sound field synthesis approach has the advantage that all loudspeakers are active for all virtual source angles. The same is true for many ambisonics approaches. However, it is criticized that source localization does not work for listeners who are situated even slightly beyond the optimal listening position, a sweet area which often has the size of no more than a listener's head. For off-centered listeners the precedence effect becomes active. They perceive the source at the origin of the first wave front reaching them, i.e. the location of the nearest loudspeaker. Likewise, it could be demonstrated that localization in the psychoacoustic sound field synthesis does not work until the precedence effect is leveraged by technical means. This was deliberately demonstrated in Ziemer (2011b) and can be seen in Fig. 9.14 showing the same setup in the background as the simulation illustrated in Fig. 9.5. Listening test results of an early stage of during the development of the psychoacoustic sound field synthesis are plotted in the foreground. This polar plot shows how many out of 10 subjects marked the corresponding angle as source location during the performance of a violin piece. This polar plot looks rather random. Listeners localized the source at one or several positions or regions somewhere within 230°. Due to the extent and position of the listening area, no subject was standing near to the loudspeaker, whose position almost coincided with the virtual source position. Consequently, the first wave front arrived from one of the other loudspeakers. Still, only some point at the discrete loudspeaker angles. Obviously, the subjects did not necessarily localize the virtual source position as coinciding with a discrete loudspeaker position. The reason for that certainly is that localization is not only a matter of the first wavefront. The psychoacoustic sound field synthesis approach also creates interaural level and phase differences which resemble the natural binaural information of a real source. But in this case, the cues from the first wavefront and the spectral cues during the stationary part of the sound were contradicting. This may have confused the listeners. When implementing the precedence effect, the location of the source became distinctly clearer for the subjects, as demonstrated in Fig. 9.15. At this early stage of development systematic localization

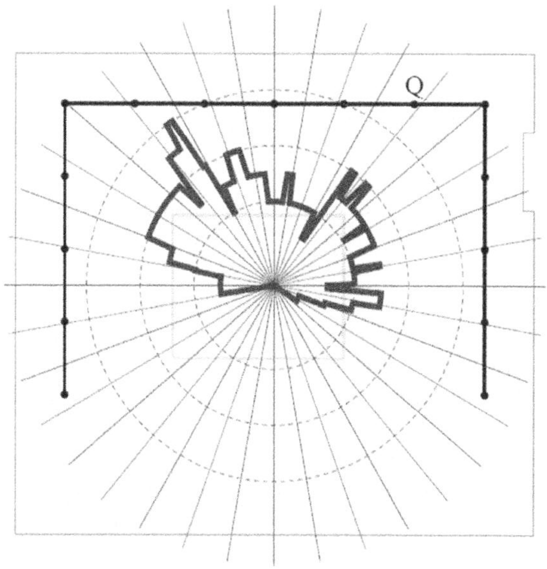

Fig. 9.14 Virtual (Q) and perceived source location (polar plot) for a spectral sound field synthesis without the implementation of the precedence effect. From Ziemer (2011b), p. 194, with the permission of the Audio Engineering Society

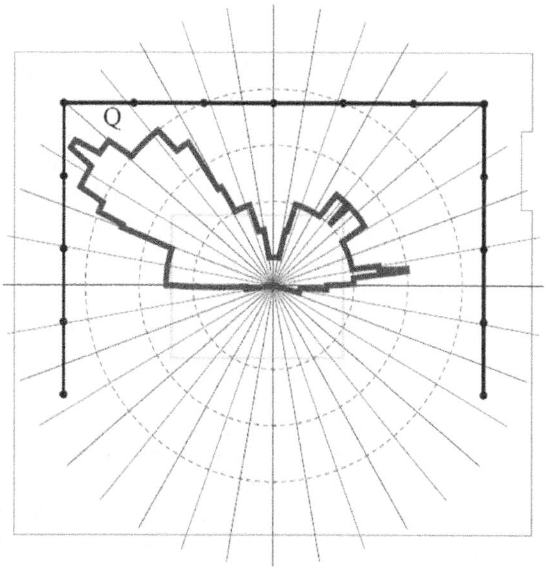

Fig. 9.15 Virtual (Q) and perceived source location (plot) for a spectral sound field synthesis when implementing the precedence effect. From Ziemer (2011b), p. 194, with the permission of the Audio Engineering Society

errors occurred because not all psychoacoustic principles had been implemented at that time.

Localization suffers under the precedence effect. So it is a logical consequence to control the first wave front in the sound field synthesis approach to gain control over the virtual source location. Especially at note onsets the direction of the first wave front is important. The law of the first wavefront determines the source localization in a natural, reverberant sound field. But also under unnatural listening conditions an effect related to the precedence effect has been observed: the interaural onset delay determined the source lateralization in experiments with headphones, even when it was followed by contradicting interaural time delays.[22]

Control over the first wave front at note onsets is achieved by the *precedence fade*. The loudspeaker which is situated most proximate to the virtual source position always plays the complete notes. This loudspeaker is referred to as *precedence speaker*. The signals of all other loudspeakers are gradually faded in. This ensures that all listeners inside the listening area will perceive the same source location due to the precedence effect. Here, the fading duration is critical. It has to be at least long enough so that the wave front from the precedence speaker reaches all listening points before the other loudspeaker signals are faded in completely. And it has to be short enough so that neither the faded signals appear to be echoes nor the fading as such becomes audible as a softened attack. This is a matter of integration times of the auditory system and of masking phenomenons, as will be discussed subsequently.

To achieve a rather linear increase of loudness, an amplitude power function of form

$$\hat{A}(t) = \frac{t^3}{T_{\text{fade}}^3} \tag{9.10}$$

is suitable. Here, T_{fade} is the chosen fading duration. An example of the precedence fade is illustrated for a 5.1 setup in Fig. 9.16. The precedence fade was also included in Fig. 9.11 and eliminated the impulsive attack of a bass drum.

The downside of the precedence fade is that the played signal needs clear onsets. A crescendo at note onsets, as often played by bowed instruments like the violin, makes the precedence effect less robust if the the gradual amplitude increase creates a transition from silence to a quasi-stationary sound which lacks an obvious attack. On the other hand, such types of signal are difficult to localize in a natural listening situation as well. Furthermore, the note onsets either need to be encoded or detected in the source signal. Luckily, the precedence effect can last for seconds. So the location of the source does may remain even if one onset is not detected or encoded.

It is interesting to see that applying the precedence fade only— i.e., routing the unfiltered source signal to all loudspeakers and applying the precedence fade—is also insufficient for a proper localization. The reason for that may be that the precedence speaker signal is not necessarily the first wavefront to arrive at all listening points. It is just the first signal to arrive with full amplitude. Fig. 9.17 shows again the same listening test setup as before. Here, the precedence fade is applied on 15 exact copies of the source signal, a harpsichord piece. It is presented to 30 subject. Many subjects localized the virtual source is the correct region. But this region is rather wide and many "false" localizations can be observed as well. Compared to that, the combination of spectral sound field synthesis and the precedence fade creates a robust source location for all listeners, as shown in Fig. 9.18. It is the combination

[22]See Freyman et al. (1997).

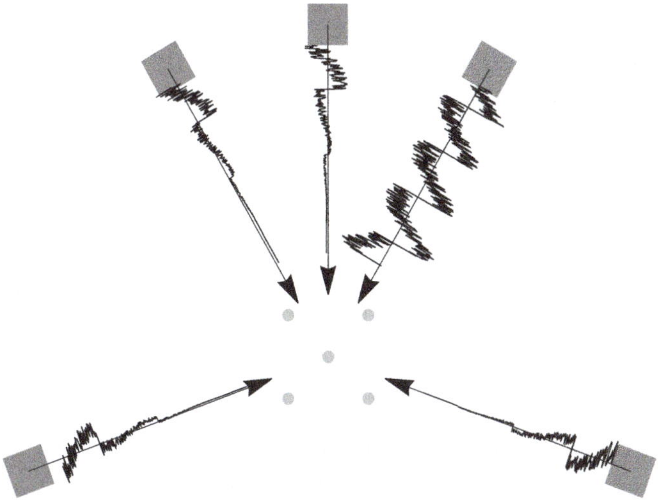

Fig. 9.16 Demonstration of the precedence fade in a 5.1 loudspeaker setup. The virtual source is situated at the front right. From Ziemer and Bader (2017), p. 489, with the permission of the Audio Engineering Society

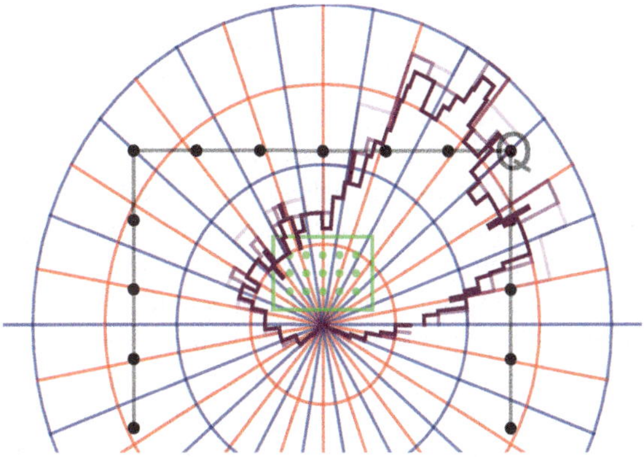

Fig. 9.17 Perceived source locations when applying the precedence fade on the unfiltered source signal. From Ziemer and Bader (2017), p. 492, with the permission of the Audio Engineering Society

of the spectral sound field synthesis and the precedence fade which make source localization possible for listeners. The sound field synthesis delivers more static cues, like interaural level and phase differences, whereas the precedence fade delivers the the temporal cue. Due to the fading, the interaural time differences at the note onsets may deviate from the intended source location. The arrival of multiple wave fronts is similar to the aliasing effect in wave front synthesis systems, so they may appear like very early reflections. After the arrival of the first wave fronts the synthesized sound field delivers additional cues that affect localization. It has been found that ILD cues are more robust in echoic environments than ITD cues (Klockgether and van de Par 2016). These ILD-cues are delivered right after the arrival of the first wave front by

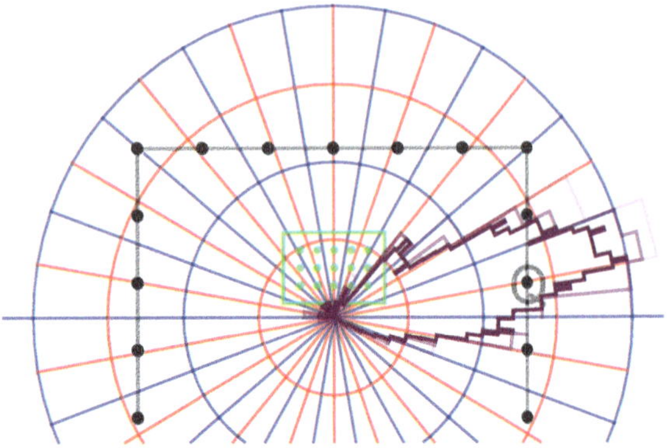

Fig. 9.18 Perceived source locations in the psychoacoustic sound field synthesis system. From Ziemer and Bader (2017), p. 492, with the permission of the Audio Engineering Society

the synthesis of the sound field. Together, they deliver robust and stable localization cues that are consistent in the complete listening area.

9.3.4 Implementation of Integration Times

In the presented sound field synthesis approach a multitude of wave fronts arrive at the listeners from different locations and at different points in time. The auditory system has several integration times. Perceptions of loudness, timbre, pitch and source extent emerge after an integration process of up several milliseconds, say 50–200. All loudspeaker signals should at least arrive at each listening position within that time window. In this case, they can be integrated and create a common loudness, timbre and pitch perception. For source localization, however, the auditory system evaluates interaural time differences in the order of microseconds. Reflections that arrive within about 4 ms after the direct sound shift the perceived source location towards the origin of the reflection. At larger delay times, the law of the first wave front takes effect and the reflections contribute to the perception of loudness and source extent rather than to localization. When the delay time surpasses the echo threshold t_e of about 50 ms, it can clearly be distinguished from the direct sound and is heard as echo. This had been discussed in Sect. 7.2.2 in the context of stereophonic audio systems and illustrated in Fig. 7.7 and in Chap. 6 in the context of room acoustics. It is the reason why the precedence fade is applied. This way, the first wavefront approximately delivers the interaural time differences that the auditory system takes into consideration for source localization. The echo threshold dictates an upper limit to the psychoacoustic sound field synthesis approach. First of all, the duration of the precedence fade must be shorter than that, so the faded signals don't appear as echoes at any listening point. And second of all, since all loudspeaker signals start at the same time, there must not be any listening point at which the time delay between the wave front arriving from the nearest and the farthest loudspeaker is larger than t_e. That is if a listening point exists, from which the distance to the nearest loudspeaker deviates more than $t_e/c = 17$ m from the distance to the farthest loudspeaker, only a subset of all loudspeakers can be used. This is necessary to make sure that all loudspeaker signals arrive within this temporal upper bound.

9.3.5 Implementation of Masking

One reason why only the loudest frequency of each critical band is considered is because these 25 frequencies tend to mask a large portion of the audible spectrum in terms of simultaneous masking. Other frequencies may at least be partially masked by them. So they may be heard and contribute to the perceived sound but they are heard with little precision. When introducing the precedence fade to the psychoacoustic sound field synthesis system, temporal masking becomes effective as well. As mentioned in Sect. 4.3.1.4 backwards masking can be effective for a duration of 10–100 ms, depending on signal features, like the frequency content, the level and duration of the masker. Due to the overshoot phenomenon, discussed in Sect. 4.3.1.3, the masking threshold rises by 3–26 dB, again depending on signal properties. An exemplary envelope of temporal masking including pre-masking and the overshoot phenomenon was plotted in Fig. 4.11 in Sect. 4.3.1.5. It is due to backward masking and the overshoot phenomenon at note onsets that the precedence fade is inaudible even if the faded sound arrives shortly before the signal of the precedence speaker. In Fig. 9.19 the time signal from the precedence speaker and from another loudspeaker are plotted for one point in the listening area. The figure includes the temporal development of the masked threshold. The soft sound right before the arrival of the precedence speaker signal is completely masked due to backwards masking. The weak attack of the faded signal is mostly masked due to the overshoot phenomenon right at the onset of the precedence speaker signal. During the quasi-stationary phase the precedence speaker signal barely masks the other signal.

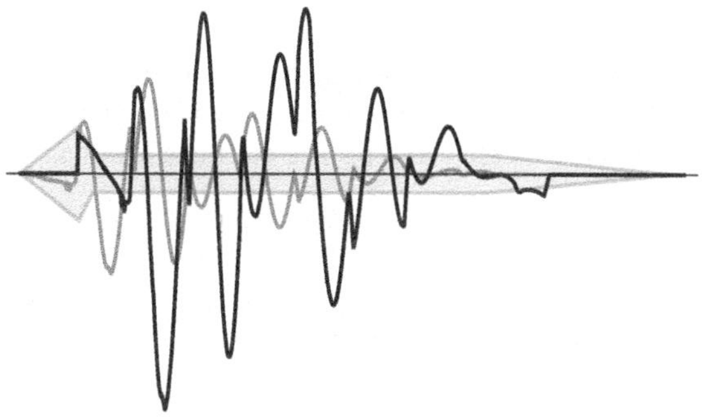

Fig. 9.19 Masked threshold (light gray area) of the precedence speaker signal (black) that partly masks another loudspeaker signal (gray) in the psychoacoustic sound field synthesis

9.3.6 Implementation of Timbre Perception

Timbre is mostly independent of pitch and duration, loudness, source location and width. As already discussed in Sect. 2.1, timbre perception is multi-dimensional. One dimension is of rather stationary nature, related to auditory brightness, bandwidth and balance of the spectral components. This aspect of timbre is barely impaired by the psychoacoustic sound field synthesis approach. Indeed, up to 25 frequency components are manipulated in amplitude and phase. The manipulation may be even rather drastic, so single loudspeaker signals can show massive deviations from the original source spectrum. But loudspeaker signals superimpose in the listening area to recreate the original spectrum. As the sound field is calculated in frequency domain, however, it is only valid for stationary signals. Synthesis errors may occur at signal transients. When dealing with musical instruments, the source sounds tend to consist of a transient note onset which is followed by a quasi-stationary phase. This is obvious for sustained sounds like a bowed cello note. But it is also valid for decaying sounds. Although a guitar sound is only sustained shortly, spectral changes are gradual and relatively slow compared to the contained frequencies. So a large portion of a note can be considered as quasi-stationary. The only exception is the attack transient.

One temporal timbre feature is onset-synchrony of partials. Note that this feature is also largely spared by the manipulations inherent in the psychoacoustic sound field synthesis approach. The attack of many instrumental sounds tends to have a dense, broad spectrum. Many spectral components dye out quickly. This is true for percussive sounds, like hit drums, plucked-string instruments and stroked idiophones. But it can also be observed for instrumental sounds which have a rather sustained, stationary regime, like bowed strings or blown wind instruments. An impulsive strike or a noisy rubbing or blowing contains way more than 25 frequency components. The psychoacoustic sound field synthesis approach only affects the onset-synchrony of a small subset of partials.

Informal listening to an electronic drumkit—including bassdrum, hihat and snaredrum—revealed that even these rather percussive sounds can by synthesized well. Due to their steep attacks, the precedence effect works very robust and ensured localization. Formal listening test, ideally including a comparison between original source sound and psychoacoustic sound field synthesis sound, will reveal how natural the timbre is perceived. The lack of a quasi-stationary phase may affect the perceived tone color.

9.3.7 Implementation of Auditory Scene Analysis

Just like most other audio systems, the psychoacoustic sound field synthesis only works due to auditory scene analysis. Stereo panning works because two or more loudspeaker signals are not perceived individually. Instead, they are integrated into one common auditory stream, having one perceived source location and timbre. This location can be manipulated by changing their signal amplitude or their arrival time ratio. This is referred to as amplitude based and time based panning. The perceived source extent is mainly affected by the degree of coherence that the loudspeaker signals exhibit. Incoherent loudspeaker signals tend to create rather incoherent ear signals. As long as the signals share enough features, they are perceived as one source with a certain extent. Features could be pitch, synchrony, continuity and common fate etc. Only if their signals are too incoherent, i.e., if they do not share enough sound aspects, they split up into separate auditory streams, being localized individually, each source having its own timbre. The same is true for wave field synthesis systems. Due to spatial aliasing, high frequency echoes arrive at the listeners right after the synthesized broadband wavefront. In the case of focused sources these high-frequency echoes even arrive before the synthesized wavefront and are referred to as pre-echoes. They largely arrive within the temporal echo threshold, so they are not heard as echoes. They contain only partials of the synthesized wavefront, so they certainly provoke the same pitch perception and the harmonicity principle of auditory scene analysis. It is likely, however, that they do affect the perception of loudness, brightness and roughness, i.e., that they affect the perceived timbre. Even though the synthesized wavefront and the high frequency echoes do not arrive synchronously,.[23]

In the psychoacoustic sound field synthesis approach, a number of manipulated source signals arrive at the listeners from different directions and at slightly different points in time. To be integrated into one common auditory stream, these signals must share enough features other than spatial location. It certainly helps that during large portions of the quasi-stationary part of the instrumental sound the sound field synthesis creates plausible interaural level- and phase differences that indicate the intended source location and support the localization process that is initialized by the precedence effect during the note onset. But more importantly, the loudspeaker spectra are only altered at 25 frequencies. So they still share a common fundamental and have a similar timbre and temporal envelope. Even in an extreme case, where one loudspeaker signal strongly emphasizes these 25 frequencies and another one almost mutes them, the signals would still likely provoke the same pitch perception, have a similar spectral and temporal contour and have a common fate, i.e., play the same notes. Manipulating more frequencies could diminish their community and provoke stream segregation. It is the integration that makes the signals appear as one auditory stream having one common perceived sourcelocation. So choosing 25 frequencies is not only meaningful in terms of the critical band width and simultaneous masking but in terms of auditory scene analysis. Features of an individual stream can be heard more easily compared to details within one stream, as this detail is just a small

[23]In Cusack and Carlyon (2004), p. 23 it is described that music ensembles exhibit asynchronies between 30 and 50 ms. Still, this is not perceived by the audience, probably because sequential grouping—as discussed in Sect. 4.5.1—captures more attention than simultaneous grouping, so details on timing are lost.

element of a larger stream. This lack of detail is of course helpful to achieve a natural sound impression even if the synthesized sound field differs from a natural sound field during a short period between successive time windows.

9.4 Summary

The presented psychoacoustic sound field synthesis approach samples a listening area into discrete listening points. The desired sound field at these points is calculated by simplifying musical instruments as complex point sources. The temporal source signals are transferred into frequency domain via discrete Fourier transform. Then, each considered frequency is propagated to each listening point. The result is a number of complex amplitudes which represent the sound field for that frequency. Loudspeakers are supposed to create this desired sound field by superposition of their radiated driving signals. Just like the musical instruments, the loudspeakers are considered as complex point sources. A transfer matrix describes the propagation of each complex loudspeaker amplitude to each listening point. The complex loudspeaker amplitudes are necessary to create the desired sound field are calculated by solving a linear equation system. The solutions are the complex amplitudes for each loudspeaker for the considered frequency. Considering loudspeakers as complex point sources is referred to as radiation-method, which is a powerful regularization method dur to the complicated sound radiation behavior of loudspeakers. An inverse Fourier transform yields the temporal loudspeaker driving signal for the corresponding frequency. This approach neglects time. It is assumed that the source signal is quasi-stationary. Only under this assumption a Fourier transform delivers the correct spectrum. And only under this assumption, it is valid to create a transfer matrix that neglects propagation time. Many musical instrumental sounds have sustained quasi-stationary regimes. Phase manipulations, as resulting from the solution of the linear equation system, are practically inaudible in stationary signals. They become audible during transients. Especially at note onsets, phase manipulations audibly affect the timbre; percussive attacks become blurred without phase synchrony. Therefore, only 25 frequencies are manipulated in terms of amplitude and phase. Attack transients of most musical instruments have a relatively high spectral bandwidth. Many frequencies decay fast, while mostly a harmonic spectral series sustains over longer quasi-stationary phases. So manipulating the phase of 25 frequencies typically leaves the majority of the spectral content unchanged. This way the timbre remains natural but the dominant frequency in each critical band create interaural level and phase differences that indicate the sound radiation characteristics as well as the location of the virtual source. Since time is neglected, interaural time differences are missing, especially at the note onsets. The precedence fade is utilized to add interaural time difference cues and stabilize localization. Solutions for successive time windows may exhibit a phase jumps, which could be audible as clicks. Therefore, overlapping time windows are chosen and cross-faded. This way, a phase jump is smoothed and sounds like a subtle beating instead of a click. The approach has been simulated and validated in several listening experiments.

In many contexts, what really matters to a listener is the perceived sound, not its physical features. The auditory system has certain thresholds and a limited resolution, which may be considered as a spatial and/or temporal integration and measured in terms of just noticeable differences. A limited portion of the original sound field is sufficient to replicate one or more of its perceptual qualities.

9.5 The Future of Psychoacoustic Sound Field Synthesis

The presented psychoacoustic sound field synthesis approach creates a robust source localization and the impression of a wide source for listeners within an extended listening area and with little computational effort and a comparably small number of loudspeakers. It leverages numerous psychoacoustic effects so that the sound field is controlled with a precision that is sufficient for the most important aspects of human sound perception. Thresholds and integration times, as well as auditory scene analysis principles are considered. Although only implemented in the horizontal plane so far, the method could be extended to three-dimensional loudspeaker setups. In contrast to time and amplitude based panning, the precedence effect is very effective and robust in the median plane as well.[24] Psychoacoustic sound field synthesis is certainly an approach which can be developed further. Many researchers emphasize the potential of psychoacoustics in auralization and sound field synthesis applications.[25] One could combine it with conventional wave front synthesis. Above the aliasing frequency of a wave front synthesis system the psychoacoustic sound field synthesis approach could deliver cues for localization and source width perception. This is similar to the OPSI approach discussed in Sect. 8.3.2.2. The grid that samples the listening area could be chosen frequency-dependent. This way, just as in ambisonics systems, a sweet spot for high frequencies would increase towards lower frequencies because wider sample point distances are sufficient for large wavelengths. This idea is depicted in Fig. 9.20. In this figure sample points for four different octaves are coded by the gray level. In an actual implementation one would chose one grid of listening points per critical frequency band so that each linear equation system to be solved contains an individual set of listening points.

Furthermore, the number of listening points could be increased. This makes the linear equation system, Eq. 9.8, an over-determined equation system. A small number of loudspeaker signals is searched which recreate the desired sound field at a higher number of listening points. Here, as in many ambisonics approaches, approximate solutions can be found. This way, one could increase the extent of the listening area or reduce the number of loudspeakers even further. However, finding an approximate solution which minimizes the squared synthesis error may not be the best choice.

[24]See e.g. Litovsky et al. (1997) and Tregonning and Martin (2015).

[25]See e.g. Fastl (2010), Blauert (2008), Spors et al. (2013), Ziemer (2018).

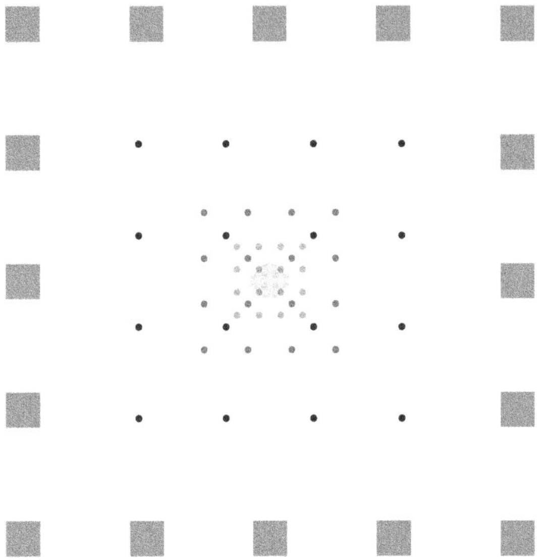

Fig. 9.20 Example of a frequency-dependent listening area extent. The gray level of the listening points that sample the listening area denotes the frequency

Instead, one should implement psychoacoustic considerations, again. For example, the synthesized phase of high frequencies may have large errors without perceivable effects, because the auditory system is insensitive for phase and interaural phase differences of frequencies above 1.2 kHz.[26] Here, minimizing amplitude errors should have the higher priority. Similarly, very low frequencies cannot be localized well. So instead of concentrating on creating plausible ILDs and IPDs in the listening area, one should concentrate on finding a solution which minimizes amplitude- and phase manipulations so that the filtered source signal preserves its transients as good as possible. In the field of subjective room acoustics, LEF in the frequency region from 125Hz–1 kHz and the $\mathrm{BQI}_{\mathrm{early},500-2000\mathrm{Hz}}$ are considered when it comes to the apparent source width, as discussed in Sect. 6.2.2. So to control the perceived source extent, one might need to concentrate on this frequency region. Unfortunately, three to four octave bands cover a large portion of the audible frequency region. Further studies concerning the perception of source extent is necessary to reveal which of the two proposed frequency regions affects ASW more.

Until now, most sound field synthesis approaches apply the Rayleigh integral of either form, as discussed in Sect. 8.3.2. This is true for ambisonics, wave field synthesis and the presented psychoacoustic sound field synthesis approach. The Rayleigh integrals allow for the choice between secondary monopole or dipole sources. Both types of radiation can be approximated fairly well by different types of loudspeakers. It is, however, also possible to approximate a cardioid by means of one loudspeaker or a combination of one monopole and one dipole speaker which are closely spaced.

[26]Explained in Sect. 4.4.2, mainly based on Hall (2008), p. 343, Davis (2007), p. 750 and Ross et al. (2007).

With an array of such loudspeakers, a discretized version of the K-H integral could be implemented. Surrounding a listening area with inwards-facing cardioid loudspeakers has the advantage that only a small portion of sound will radiate outside the listening area. This would make listening room compensation much easier. If the loudspeakers are rotatable, both sound pressure and gradient could be controlled. This approach is worth thinking about, as it actually comes up to the K-H integral in contrast to the Rayleigh integral based approaches. Of course, many issues need to be solved here as well. For example a high rotation velocity may be necessary when several sources or moving sources are synthesized. Psychoacoustic considerations will reveal the necessary rotation speed. Furthermore, cardioids are usually not energy efficient radiators. And often only a narrow frequency band can be radiated with a cardioid characteristic by a loudspeaker. Other frequencies tend to be radiated in a more complicated fashion. Loudspeakers with cardioidic sound radiation reduce the number and amplitudes of early reflections. A low number of first-order reflections can be implemented in the sound field synthesis calculation. This approach is comparable to adaptive wave field synthesis. But as the current psychoacoustic sound field synthesis approach allows for wavefronts to arrive within a delay of several milliseconds before or after the direct sound of the precedence speaker. This is a much higher order of magnitude compared to conventional wave front synthesis. For psychoacoustic sound field synthesis controlled early reflections are not that disturbing, because localization is mainly a matter of the precedence speaker.

There are still so many aspects of sound perception that have not been implemented systematically in audio systems and psychoacoustics offer countless starting points for alternative approaches. Focusing on control over the three dimensions of timbre space could be a starting point for an audio system from a generally new perspective. In discotheques and live concerts, sound pressure levels are often way beyond the 85 dB_{SPL} that are assumed as the audio engineer's mixing level and the common listening level in MP3. Consequently, masking thresholds are even higher and one could save computation time when reconsidering masking effects under these circumstances. Until now, many auditory scene analysis principles are only known from a phenomenological point of view. They have been studied by means of test signals which tend to differ from the sound of musical instruments. Deeper insight into the underlying mechanisms will certainly reveal new ways of recording and reproducing sound. And vice versa, building an audio system based on ASA principles may generate new knowledge in auditory perception.

Rather than copying a complete wave field, the vision of psychoacoustic sound field synthesis is to deliver the necessary cues for the perception of a desired source location, source extent, loudness, timbre, pitch, melody, degree of consonance, roughness etc. The necessary precision is defined by spatial, spectral and temporal resolution and thresholds of the auditory system and by organization principles in terms of auditory scene analysis.

References

Ahrens J (2015) Perceptual evaluation of the diffuseness of synthetic late reverberation created by wave field synthesis at different listening positions. In: Fortschritte der Akustik—DAGA '15, Berlin

Ahrens J, Spors S (2009) Sound field reproduction employing non-omnidirectional loudspeakers. In: Audio engineering society convention 126

Ahrens J, Geier M, Spors S (2010) Perceptual assessment of delay accuracy and loudspeaker misplacement in wave field synthesis. In: Audio engineering society convention 128

Ajdler T, Faller C, Sbaiz L, Vetterli M (2008) Sound field analysis along a circle and its application to hrtf interpolation. J Audio Eng Soc 56(3):156–175. http://www.aes.org/e-lib/browse.cfm?elib=14380

Aures W (1985) Berechnungsverfahren für den sensorischen Wohlklang beliebiger Schallsignale (a model for calculating the sensory euphony of various sounds). Acustica 59(2):130–141. https://www.ingentaconnect.com/content/dav/aaua/1985/00000059/00000002/art00008

Bader R (2013) Nonlinearities and synchronization in musical acoustics and music psychology. Springer, Berlin Heidelberg. https://doi.org/10.1007/978-3-642-36098-5

Bader R (2014) Microphone array. In: Rossing TD (ed), Springer handbook of acoustics. Springer, Berlin Heidelberg, pp 1179–1207. https://doi.org/10.1007/978-1-4939-0755-7_29

Bai MR, Chung C, Wu P-C, Chiang Y-H, Yang C-M (2017) Solution strategies for linear inverse problems in spatial audio signal processing. Appl Sci 7(6):paper number 582. https://doi.org/10.3390/app7060582

Berkhout AJ (1988) A holographic approach to acoustic control. J Audio Eng Soc 36(12):977–995. http://www.aes.org/e-lib/browse.cfm?elib=5117

Böhlke L, Ziemer T (2017a) Perception of a virtual violin radiation in a wave field synthesis system. J Acoust Soc Am 141(5):3875. https://doi.org/10.1121/1.4988669

Böhlke L, Ziemer T (2017b) Perceptual evaluation of violin radiation characteristics in a wave field synthesis system. Proc Mtgs Acoust 30(1):035001. https://doi.org/10.1121/2.0000524

Blau M (2004) Correlation of apparent source width with objective measures in synthetic sound fields. Acta Acust united Ac 90(4):720–730. https://www.ingentaconnect.com/content/dav/aaua/2004/00000090/00000004/art00015

Blauert J (2008) 3-d-Lautsprecher-Wiedergabemethoden. In: Fortschritte der Akustik—DAGA '08. Dresden, pp 25–26

Brandenburg K, Stoll G (1994) ISO/MPEG-1 audio: a generic standard for coding of high-quality digital audio. J Audio Eng Soc 42(10):780–792. http://www.aes.org/e-lib/browse.cfm?elib=6925

Cusack R, Carlyon RP (2004) Auditory perceptual organization indside and outside the laboratory. In: Neuhoff JG (ed), Ecological psychoacoustics, chapter 2. Elsevier, Oxford, pp 15–48. https://doi.org/10.1016/b978-012515851-0/50003-5

Daniel P, Weber, R (1997) Psychoacoustical roughness: implementation of an optimized model. Acta Acust united Ac 83(1):113–123. https://www.ingentaconnect.com/contentone/dav/aaua/1997/00000083/00000001/art00020

Davis MF (2007) Audio and electroacoustics. In: Rossing TD (ed), Springer handbook of acoustics, chapter 18. Springer, New York, pp 743–781. https://doi.org/10.1007/978-0-387-30425-0_18

de Vries, D (1996) Sound reinforcement by wavefield synthesis: adaption of the synthesis operator to the loudspeaker directivity characteristics. J Audio Eng Soc 44(12):1120–1131. http://www.aes.org/e-lib/browse.cfm?elib=7872

Dubnov S, Rodet X (2003) Investigation of phase coupling phenomena in sustained portion of musical instruments sound. J Acoust Soc Am 113(1):348–359. https://doi.org/10.1121/1.1518981

Fastl H (2010) Praktische Anwendungen der Psychoakustik. In: Fortschritte der Akustik—DAGA '10. Berlin, pp 5–10

Fischer JL, Bader R, Abel M (2016) Aeroacoustical coupling and synchronization of organ pipes. J Acoust Soc Am 140(4):2344–2351. https://doi.org/10.1121/1.4964135

Fletcher NH (1978) Mode locking in nonlinearly excited inharmonic musical oscillators. J Acoust Soc Am 6(64):1566–1569. https://doi.org/10.1121/1.382139

Fletcher NH (2008) Fletcher and thomas D rossing, 2nd edn. The physics of musical intstruments. Springer, New York

Freyman RL, Zurek PM, Balakrishnan U, Chiang Y-C (1997) Onset dominance in lateralization. J Acoust Soc Am 101(3):1649–1659. https://doi.org/10.1121/1.418149

Geier M, Wierstorf H, Ahrens J, Wechsung I, Raake A, Spors S (2010) Perceptual evaluation of focused sources in wave field synthesis. Audio Eng Soc Conv 128

Grantham DW (1986) Detection and discrimination of simulated motion of auditory targets in the horizontal plane. J Acoust Soc Am 79(6):1939–1949. https://doi.org/10.1121/1.393201

Hall DE (2008) Musikalische Akustik. Ein Handbuch. Schott, Mainz

International Organization for Standardization/International Electrotechnical Commission. ISO/IEC JTC1/SC29/WG11 MPEG, IS11172-3, information technology—coding of moving pictures and associated audio for digital storage media at up to about 1.5 Mbit/s – part 3: Audio (1993)

Jensenius AR, Wanderley MM, Godøy RI, Leman M (2010) Musical gestures. concepts and methods in research. In: Godøy RI, Leman M (eds) Musical gestures. Sound, movement, and meaning. Routledge, New York, NY, pp 12–35. https://www.taylorfrancis.com/books/e/9781135183639/chapters/10.4324%2F9780203863411-8

Klockgether S, van de Par S (2016) Just noticeable differences of spatial cues in echoic and anechoic acoustical environments. J Acoust Soc Am 140(4):EL352–EL357. https://doi.org/10.1121/1.4964844

Lindau A, Erbes V, Lepa S, Maempel H-J, Brinkman F, Weinzierl S (2014) A spatial audio quality inventory (SAQI). Acta Acust united Ac 100(5):984–994. https://doi.org/10.3813/AAA.918778

Litovsky RY, Rakerd B, Yin TCT, Hartmann WM (1997) Psychophysical and physiological evidence for a precedence effect in the median sagittal plane. J Neurophysiol 77(4):2223–2226. https://doi.org/10.1152/jn.1997.77.4.2223

Meyer J (2009) Acoustics and the performance of music. manual for acousticians, audio engineers, musicians, architects and musical instrument makers, 5th edn. Springer, Bergkirchen. https://doi.org/10.1007/978-0-387-09517-2

Nuttal AH (1981) Some windows with very good sidelobe behavior. 29:84–91. https://doi.org/10.1109/tassp.1981.1163506

Patel AD (2002) Music, language, and the brain. Oxford University Press, Oxford

Ross B, Tremblay KL, Picton TW (2007) Physiological detection of interaural phase differences. J Acoust Soc Am 121(2):1017–1027. https://doi.org/10.1121/1.2404915

Spors S, Wierstorf H, Raake A, Melchior F, Frank M, Zotter F (2013) Spatial sound with loudspeakers and its perception: a review of the current state. Proc IEEE 101(9):1920–1938. https://doi.org/10.1109/JPROC.2013.2264784

Tregonning A, Martin B (2015) The vertical precedence effect: utilizing delay panning for height channel mixing in 3d audio. In: Audio engineering society convention 139, New York, NY

VictorianLady (2016) cembalo-geige-notenständer-möbel. https://pixabay.com/de/cembalo-geige-notenst%C3%A4nder-m%C3%B6bel-1412019/

von Bismarck G (1974) Sharpness as an attribute of the timbre of steady sounds. Acustica 30:159–172. https://www.ingentaconnect.com/contentone/dav/aaua/1974/00000030/00000003/art00006

Wierstorf H (2014) Perceptual assessment of sound field synthesis. PhD thesis, University of Technology Berlin, Berlin

Wierstorf H, Raake A, Geier M, Spors S (2013) Perception of focused sources in wave field synthesis. J Audio Eng Soc 61(1/2):5–16. http://www.aes.org/e-lib/browse.cfm?elib=16663

Wierstorf H, Raake A, Spors S (2017) Assessing localization accuracy in sound field synthesis. J Acoust Soc Am 141(2):1111–1119. https://doi.org/10.1121/1.4976061

Wittek H (2007) Perceptual differences between wavefield synthesis and stereophony. PhD thesis, University of Surrey, Guilford

Wittek H, Kerber S, Rumsey F, Theile G (2004) Spatial perception in wave field synthesis rendered sound fields: distance of real and virtual nearby sources. In: Audio engineering society convention 116, Berlin. http://www.aes.org/e-lib/browse.cfm?elib=12711

Ziemer T (2009) Wave field synthesis by an octupole speaker system. In: Naveda L, (ed) Proceedings of the second international conference of students of systematic musicology (SysMus09), pp 89–93. http://biblio.ugent.be/publication/823807/file/6824513.pdf#page=90

Ziemer T (2001a) Wave field synthesis. theory and application. (magister thesis), University of Hamburg

Ziemer T (2011b) A psychoacoustic approach to wave field synthesis. In: Audio Engineering society conference: 42nd international conference: semantic audio. Ilmenau, pp 191–197. http://www.aes.org/e-lib/browse.cfm?elib=15942

Ziemer T (2011c) Psychoacoustic effects in wave field synthesis applications. In: Schneider A, von Ruschkowski A (eds), Systematic musicology. Empirical and theoretical studies. Peter Lang, Frankfurt am Main, pp 153–162. https://doi.org/10.3726/978-3-653-01290-3

Ziemer T (2011d) A psychoacoustic approach to wave field synthesis. J Audio Eng Soc 59(5):356. http://www.aes.org./conferences/42/abstracts.cfm#Tim_Ziemer

Ziemer T (2014) Sound radiation characteristics of a shakuhachi with different playing techniques. In: Proceedings of the international symposium on musical acoustics (ISMA-14). Le Mans, pp 549–555. http://www.conforg.fr/isma2014/cdrom/data/articles/000121.pdf

Ziemer T (2015a) Exploring physical parameters explaining the apparent source width of direct sound of musical instruments. In: Jahrestagung der Deutschen Gesellschaft für Musikpsychologie. Oldenburg, pp 40–41. http://www.researchgate.net/publication/304496623_Exploring_Physical_Parameters_Explaining_the_Apparent_Source_Width_of_Direct_Sound_of_Musical_Instruments

Ziemer T (2015b) Spatial sound impression and precise localization by psychoacoustic sound field synthesis. In: Deutsche Gesellschaft für Akustik EV, Mores R (eds), Seminar des Fachausschusses Musikalische Akustik (FAMA): "Musikalische Akustik zwischen Empirie und Theorie". Hamburg, pp 17–22. Deutsche Gesellsch. f. Akustik. https://www.dega-akustik.de/fachausschuesse/ma/dokumente/tagungsband-seminar-fama-2015/

Ziemer T (2016) Implementation of the radiation characteristics of musical instruments in wave field synthesis application. PhD thesis, University of Hamburg, Hamburg

Ziemer T (2017a) Source width in music production. Methods in stereo, ambisonics, and wave field synthesis. In: Schneider A (ed) Studies in musical acoustics and psychoacoustics, vol 4 of Current research in systematic musicoogy, chapter 10, pages 299–340. Springer, Cham. https://doi.org/10.1007/978-3-319-47292-8_10

Ziemer T (2017b) Perceptually motivated sound field synthesis for music presentation. J Acoust Soc Am 141(5):3997. https://doi.org/10.1121/1.4989162

Ziemer T (2017c) Perceptual sound field synthesis concept for music presentation. In: Proceedings of meetings on acoustics, page paper number 015016, Boston, MA. https://doi.org/10.1121/2.0000661

Ziemer T (2018) Wave field synthesis. In: Bader R (ed) Springer handbook of systematic musicology, chapter 18. Berlin Heidelberg, pp 175–193. https://doi.org/10.1007/978-3-662-55004-5_18

Ziemer T, Bader R (2015a) Complex point source model to calculate the sound field radiated from musical instruments. In: Proceedings of meetings on acoustics, volume 25. https://doi.org/10.1121/2.0000122

Ziemer T, Bader R (2015b) Implementing the radiation characteristics of musical instruments in a psychoacoustic sound field synthesis system. J Audio Eng Soc 63(12):1094. http://www.aes.org/journal/online/JAES_V63/12/

Ziemer T, Bader R (2015c) Implementing the radiation characteristics of musical instruments in a psychoacoustic sound field synthesis system. In: Audio engineering society convention 139, page paper number 9466, New York. http://www.aes.org/e-lib/browse.cfm?elib=18022

Ziemer T, Bader R (2015d) Complex point source model to calculate the sound field radiated from musical instruments. J Acoust Soc Am 138(3):1936. https://doi.org/10.1121/1.4934107

Ziemer T, Bader R (2017) Psychoacoustic sound field synthesis for musical instrument radiation characteristics. J Audio Eng Soc 65(6):482–496. https://doi.org/10.17743/jaes.2017.0014
Zwicker E, Fastl H (1999) Psychoacoustics. Facts and models. Springer, Berlin, Heidelberg, second updated edition. https://doi.org/10.1007/978-3-662-09562-1

Index

© Springer Nature Switzerland AG 2020

T. Ziemer, *Psychoacoustic Music Sound Field Synthesis*, Current Research in Systematic Musicology 7, https://doi.org/10.1007/978-3-030-23033-3

J

L

M

N

O

P

R

S

The manufacturer's authorised representative in the EU is Springer Nature Customer Service Centre GmbH, Europaplatz 3, 69115 Heidelberg, Germany. If you have any concerns regarding our products, please contact ProductSafety@springernature.com

Printed and bound by CPI Group (UK) Ltd, Croydon, CR0 4YY
30/09/2026
02225495-0010